THE HANDBOOK OF ENVIRONMENTAL VOLUNTARY AGREEMENTS
Design, Implementation and Evaluation Issues

ENVIRONMENT & POLICY

VOLUME 43

The titles published in this series are listed at the end of this volume.

The Handbook of Environmental Voluntary Agreements

Design, Implementation and Evaluation Issues

Edited by

Edoardo Croci

Vice Director of IEFE
Università Bocconi, Milan,
Italy

Springer

A C.I.P. Catalogue record for this book is available from the Library of Congress.

ISBN-10 1-4020-3355-9 (HB) Springer Dordrecht, Berlin, Heidelberg, New York
ISBN-10 1-4020-3356-7 (e-book) Springer Dordrecht, Berlin, Heidelberg, New York
ISBN-13 978-1-4020-3355-1 (HB) Springer Dordrecht, Berlin, Heidelberg, New York
ISBN-13 978-1-4020-3356-8 (e-book) Springer Dordrecht, Berlin, Heidelberg, New York

---

Published by Springer,
P.O. Box 17, 3300 AA Dordrecht, The Netherlands.

Printed on acid-free paper

Printed in the Netherlands.

To my wife Marina
and to my daughters
Lavinia and Eugenia

TABLE OF CONTENTS

PREFACE

EDOARDO CROCI

IEFE - Università Bocconi, Milano, Italy

Voluntary approaches in environmental policy represent a "third wave" of regulation in the environmental field.

"Command and control" was the first wave. Its core is based on uniform emission standards, the respect of which needs to be enforced through extensive monitoring and severe sanctions. The expected cost of sanction for non-compliance, calculated as its amount multiplied for the probability to be caught, must be superior to the benefits of non-compliance, in order to let the sanction be effective. As the benefits of non-compliance can vary among firms, sanctions need to be very high in order to be effective. In fact sanctions are ordinary correlated to environmental damage and not to the benefits of non-compliance. But very high sanctions can be difficult to enforce as they appear unfair and can lead to dramatic consequences on firms and workers, up to shut-downs of plants. Ambient standards reduce these problems, but oblige the regulator to know a huge amount of information, regarding the specific contribution of each polluter to the polluted body. Information is difficult to obtain because of asymmetric information and costly to produce because it requires large and skilled regulating and enforcing organizations.

Nevertheless complex regulation is the base of any environmental policy framework, as it allows the policy maker to fully exercise its power of composition of various interests in a relatively transparent way.

Economic instruments were the second wave. They are based on altering "natural" market mechanisms in order to include externalities in the decisions of production and consumption of economic players. Taxes, subsidies and "artificial" markets where environmental goods or bads are negotiated build incentives and disincentives to modify the behaviour of economic players. Economic instruments show a higher flexibility than command and control. Unfortunately the level to which taxes and subsidies should be set in order to be effective is often so high to be unpracticable. Industrial opposition in the first case and public expenses constraints in the second case cannot be easily ignored. Equity concerns also arise, for example in the definition of criteria of allocation of marketable emission allowances. Moreover the administration of economic instruments ordinary implies heavy bureaucratic structures.

Economic instruments are in most cases limited to specific and limited areas of application.

Voluntary approaches were the third wave. They require the creation of specific benefits (or the avoiding of specific costs) to firms who decide to commit in voluntary pollution reduction or other forms of environmental improvements. The benefits (or the costs) have not the form of clear-cut public monetary incentives (or disincentives) but assume the form of competitive advantages acquired by the committed firms with respect to the uncommitted ones. These advantages are ordinary based on signals to relevant stakeholders (citizens, consumers, the P.A., employees, the financial system, etc.) which modify their behaviour in more favourable ways for the committed firms. Voluntary approaches are highly flexible and can reduce regulatory costs, but there is no guarantee that all relevant parts commit as participation is not mandatory. Moreover non-compliance is in most cases punished only through moral sanctions and the loss of the advantages linked to the signals. So relevant free riding problems arise in both the participation phase and the compliance phase.

Environmental voluntary agreements (VAs) are voluntary approaches where a public and a private counterparts are identified. The public side uses the agreement as a policy tool to reach a goal of environmental improvement.

The specific framework and rules of VAs strongly influence their effectiveness. So the analysis of their characteristics and ways of application can help in determining the best conditions and designing the best rules for their use.

This book provides a comprehensive analysis of VAs, using standard concepts of the economics of the environment. A theory of VAs is built, taking into consideration the still young, but promising, literature on the topic. Both a positive and normative perspectives are included. Case analysis complements the theoretical analysis. A European and an American approach to VAs are distinguished. National and sector experiences are investigated in order to consider the full range of applications which the flexibility of VAs allows. Opportunities and risks in the use of VAs are examined. Their evaluation, also in comparison and in conjunction with other policy tools, is performed. VAs are still an instrument in evolution, so the trends in their design and enforcement rules are considered.

Authors are economists from Universities, research centers, environmental agencies and international institutions. VAs experiences considered mainly regard industrialized countries: the US, Europe and Australia. A Chinese case is also presented.

The book is destined to researchers, scholars and graduate and post-graduate students. Most contributions can be of great interest also for environmental officers in various P.A. administrative and technical bodies and for environmental managers and consultants.

Chapter 1 "Voluntary agreements in environmental policy" presents basic concepts of VAs and provides a general analysis of their use in environmental policy. Both a Government and a firm perspective are considered.

Croci provides taxonomies and models for VAs, in order to contribute in including economics of voluntary environmental agreements into the standard framework of economics of the environment. VAs is a general category which includes public voluntary schemes, negotiated agreements, and industry's unilateral commitments. The motivations of both the P.A. side and the industrial side to enter a VA can vary depending on specific circumstances. Typical circumstances are categorized. Efficiency and effectiveness of VAs also depends on various conditions limiting the regulator capture and free riding.

Mazurkiewicz provides an overview of various forms of self-regulatory practices the private sector has been applying in recent decades. He also presents the main drivers stimulating business to take up self-regulatory measures, and vehicles for establishing better corporate environmental behaviour, with special attention to multi-stakeholder dialogue.

Chapter 2 "The political economy of voluntary agreements" presents models of use of VAs as regulatory instruments. Depending on model hypothesis VAs can result more or less efficient and effective than alternative policy tools.

Glachant analyses whether VAs are able to achieve an efficient level of environmental protection when they are obtained under the legislative threat of an alternative stricter policy option. The threat is the outcome of a rent-seeking contest between a green and a polluter lobby group influencing the legislature. He shows that a VA systematically emerges in equilibrium and that it leads to a more efficient level of pollution abatement than the legislative pollution quota. However this level is lower than the first best level of environmental protection.

Hansen interprets empirical evidence from case studies of environmental voluntary agreements using a political economy model. Data suggests that VAs may often be chosen in order to shift the responsibility for implementation to industrial organizations that are less sensitive to criticism from powerful environmental interest groups. When this explanation of an environmental voluntary agreement applies, the model predicts that the agreement will be less cost effective and achieve lower environmental performance than the traditional regulatory alternative which would otherwise have been adopted.

Chapter 3 "The European and American approach to environmental voluntary agreements" analyses the differences in the two approaches.

Schnabl analyses the role of the European Community in structuring its approach to VAs since the mid 80s regarding both VAs in Member States and VAs at Community level. Recent cases of VAs at the Community level are taken into consideration. Specific legal and institutional constraints at EU level are examined. The concepts of self-regulation and co-regulation are detailed.

Brouhle, Griffiths and Wolverton perform a comprehensive analysis of the use of voluntary approaches for environmental policymaking in the U.S. from the origin to nowadays. The use of voluntary approaches to achieve environmental improvements has grown dramatically in the U.S. since they were first introduced. As of 2004, there are over 50 voluntary programs in the U.S. at the federal level alone. These programs take a variety of forms. Despite the diversity of voluntary approaches in the U.S., they often pursue common, and sometimes overlapping environmental objectives and use similar methodologies to achieve such goals. While most voluntary initiatives in the U.S. state an explicit environmental goal, they may also have less direct policy objectives such as enhancing innovation or increasing awareness of environmental issues.

Ashford and Caldart distinguish among negotiated regulation, implementation and compliance in the U.S.. Their analysis accords to the following taxonomy: (a) negotiated regulation (either preceding formal regulation or as a substitute for formal regulation); (b) negotiated implementation (negotiations with an individual firm to establish the timetable and/or the means for meeting a particular regulatory standard); and (c) negotiated compliance (negotiation in the context of an enforcement action in which the firm is out of compliance with an applicable standard and there is an opportunity for extra-statutory environmental gains.

Chapter 4 "Design, negotiation and implementation of environmental voluntary agreements: national and sector approaches" considers experiences of VAs which can be potentially replied in other contexts.

Menanteau compares efficiency standards and energy labelling versus negotiated agreements in the European electrical appliance sector. Standards are widely adopted, but suffer from long and often difficult implementation periods because of the resistance of the industrial sector. VAs offer flexibility margins in the achievement of commitments. They can be an effective instrument if certain conditions are respected. In particular, the alternative of regulatory measures must remain a credible, realistic threat if voluntary agreements are to have a really significant impact on performance improvement.

De Clercq and Bracke focus on the implementation process of environmental voluntary agreements based on a Flemish case study concerning the introduction of the duty of acceptance in Flemish waste policy. The duty applies for paper, batteries, vehicles, tyres and electrical and electronic equipment. For the practical execution of the basic rules laid down in the legal framework, VAs are concluded with sector associations. The relationship between the underlying environmental legislation and the voluntary character of the agreements are particularly relevant. It appears that the policy process entails much more than just fixing overall collection and recycling targets, but should be regarded as an on-going process with evaluations and consultations with all stakeholders. In this regard, the importance of alternative instruments in case of non-compliance to the agreement and back-up policies to create market opportunities for recycled materials is stressed.

Cabugueira considers the Portuguese approach to environmental voluntary agreements. Since the end of the 80s this approach has evolved through various phases. The Portuguese experience represents a reference to make a theoretical review of the VAs characterization and to emphasise the factors that motivate public and private participation on this "voluntary" environmental regulation process.

Price, Worell and Sinton analyse an experience of VA in China. Fostering innovative approaches that are tailored to China's emerging market-based political economy to reduce the use of polluting energy resources and to diminish pollution from industrial production is one of the most important challenges facing the nation today. The use of VAs as a policy for increasing energy-efficiency in industry, which has been a popular approach in many industrialized countries since the early 1990s, is being tested for use in China through a pilot project with two steel mills in Shandong Province. The pilot project was developed through international collaboration with experts in China, the Netherlands, and the U.S.. Designing the pilot project involved development of approaches for energy-efficiency potential assessments for the steel mills, target-setting to establish the VA energy-efficiency goals, preparing energy-efficiency plans for implementation of energy-saving technologies and measures, and monitoring and evaluating the project's energy savings.

Chapter 5 "Evaluation of environmental voluntary agreements" describes methodologies for the assessment of efficiency and effectiveness of VAs. Results deriving for the use of these methodologies in national cases are also provided.

De Clercq and Bracke perform a comparative case study analysis covering twelve voluntary agreements from six different European countries. First, a general evaluation framework for assessing the performance of environmental voluntary agreements is presented. This framework takes into account three different evaluation dimensions: application, impact and resource development. Second, the factors explaining the level of performance are examined. The comparative case study shows that taken individually each of the factors is not as such a necessary condition for the success of an environmental voluntary agreement. Rather it is the combination of these success factors that is ultimately decisive for the performance of an agreement.

Bressers and De Bruijn analyses the use of environmental voluntary agreements, or covenants, in Dutch environmental policy. Covenants have become a widely used policy instrument in the Netherlands. This trend reinforces the strong neo-corporatist traits of Dutch society with its tendency towards bargaining and cooperation with interest groups. Through negotiations between sectors of industry, the Ministry of the Environment, and regional governments, agreements are sought concerning the contribution of specific industrial sectors to the goals of the National Environmental Policy Plan. These goals aim for 50-90 percent emission reductions for specified pollutants. Since 1989 many such agreements have been reached. The effectiveness of covenants is evaluated from the authors and success and fail factors are identified. The central conclusion on the use and effects of the covenants is quite positive, although also several constraints are identified. The implementation context results highly relevant for covenant success.

Burritt, Lewis and James analyse the effectiveness of the Australian national packaging covenant (NPC) signed on 27 August 1999. Based on the notion of "effectiveness analysis" they examine whether the NPC can be seen to be effective in relation to a set of criteria that address the following aspects: the confining of corporate action, institutional structuring of agreements and checking of performance outcomes.

Couder and Verbruggen build an integrated performance indicator for energy benchmarking covenants. Voluntary approaches play an important role in reducing industrial energy use and CO_2-emissions. Benchmarking can provide a starting point for negotiating targets, and are an added value to a monitoring program. Indicators identify performance gaps and track performance over time. However, indicators at the firm level are still characterized by a low degree of standardization. Lack of comparability makes benchmarking very difficult. Indicators measure changes in one aspect (e.g. energy use) as if they were completely independent of changes in other aspects (e.g. waste generation). Integrated indicators, based on micro-economic productivity theory, may assume the role of certified tools.

Chapter 6 "Environmental voluntary agreements in policy mixes" evaluates the use of VAs I conjunction with other environmental policy tools, like standards and economic instruments.

Braathen discusses the impacts on environmental effectiveness and economic efficiency of applying environmental agreements in combination with other instruments in environmental policy. The findings are relatively negative: while the administrative costs of such combinations can be high, there is a distinct possibility that the environmental effectiveness will be lower than if the voluntary approach had not been part of the policy mix.

Van Dril examines the possibility to use the benchmarking covenant for allocating emission allowances for the European CO2 emission trading systems in the Netherlands. The regulation of the eighties and VAs of the nineties concerning energy efficiency meet most of the conditions regarding policy effectiveness. The current Benchmarking Covenant suffers from lack of transparency and the absence of an emission reduction target. The EU emissions trading scheme that is currently developed, at least has a well defined cap, but whether it will enforce real emission reduction in the future remains to be seen. The conversion of the Benchmarking Covenant into a cap for emissions trading currently does not reduce CO_2 emissions but increases emissions of the participants involved. The cap that is derived exposes the lack of stringency of this policy.

CHAPTER 1

VOLUNTARY AGREEMENTS IN ENVIRONMENTAL POLICY

THE ECONOMICS OF ENVIRONMENTAL VOLUNTARY AGREEMENTS

EDOARDO CROCI

IEFE - Università Bocconi, Milano, Italy

Abstract. This paper provides taxonomies and models, in order to contribute in including the economics of environmental voluntary agreements (VAs) into the standard framework of the economics of the environment.
Environmental voluntary agreements try to remedy market failures differently from traditional regulatory and economic instruments. In fact, they are based on the exchange between the P.A. and firms and on the design of a framework of incentives to parties in a context of negotiation and cooperation. Efficiency and effectiveness of VAs depend on specific features, which can be evaluated only by a case by case analysis. VAs can effectively be included in the tool-kit of the environmental policy-maker if some conditions regarding their design and implementation are respected, in order to limit the risks of regulator capture and free riding.

1. CONDITIONS FOR EFFICIENCY AND EFFECTIVENESS OF GOVERNMENT ACTION IN POLLUTION CONTROL

Economists consider pollution as a negative externality, deriving from the activity of an economic agent causing an uncompensated loss of welfare to another agent. The absence of markets for externalities is the cause of an inefficient allocation of assets and resources, as agents cannot improve their welfare through market exchanges.

In presence of externalities generated by a polluting firm, the price of goods produced by the firm reflects only the marginal costs of production and does not include marginal external costs, violating the condition for Pareto efficiency (price equals marginal social cost in perfect competition).

So Government intervention will be necessary to correct externalities[1]. The "ideal" goal of the Government should be to "impose", through some instruments, to attain the level of pollution which would be reached if a market for externalities existed, i.e. an efficient level of pollution.

An efficient level of pollution, ordinary different from zero, can be defined in correspondence to a situation where benefits of pollution abatement (equal to the reduction of external costs) minus costs of pollution abatement are maximized. This implies the equalization of marginal pollution abatement costs with marginal external costs for each pollution source.

E. Croci (ed.), The Handbook of Environmental Voluntary Agreements, 3 - 30

The aim of an efficient level of pollution is very difficult to attain, as it requires Government to know the marginal cost of pollution abatement and the marginal external cost for all pollution sources.

A less ambitious goal is cost-effectiveness, which implies to reach a certain level of pollution abatement at the least cost (alternatively to reach the maximum level of abatement at a given cost). This corresponds to the minimization of total abatement costs through the equalization of the marginal abatement cost for each polluter[2]. This is not an easy task anyway, as to attain effectiveness Government needs to know the marginal abatement cost for each polluter.

Severe hypothesis about the structure of markets and the behaviour of agents also need to be respected in order to reach efficiency and/or effectiveness.

Overall, instruments for pollution control can be efficient and effective only if all the following hypothesis are respected:

1. Government has a perfect knowledge of all relevant variables - like, identity of pollutees, marginal abatement costs of individual polluters and (for efficiency) effects and damage costs of pollution,
2. Government acts in the "general interest", which means aims at maximizing social welfare,
3. all economic agents are perfectly informed of Government actions and of the consequences on their utility and react in a rational manner,
4. there are no transaction costs.

On this basis, the Government will make the best possible choices in applying appropriate policy instruments and will be able to forecast and induce the desired changes in the behaviour of all affected agents.

So, market failures due to the presence of externalities can be corrected by an omniscient and benevolent regulator[3] (Pigou, 1932).

The comparison among command and control, economic and voluntary instruments is ordinary performed under these hypothesis. This normally implies the superiority of command and control and economic instruments over voluntary instruments, as voluntary instruments, by definition, don't involve all relevant polluters and being not based on mandatory participation require more sophisticated forms of incentive to participate. Moreover enforcement is ordinary less strictly performed under voluntary instruments as sanctions for non compliance are rarely envisioned.

Unfortunately the satisfaction of all these hypothesis looks quite unrealistic. Situations of optimal efficiency or maximum effectiveness can be only abstractly reached. So it is not possible anymore to provide an absolute judgement on the superiority of a certain environmental policy instrument over another one. Different instruments can show a superior efficiency and effectiveness depending on specific factors. The appropriateness of an instrument depends on the specificities of the situation considered.

2. COOPERATIVE ENVIRONMENTAL POLICY

Cooperative policies are founded on the idea that Government has a limited capability to orientate the development of society and is just one of the parts in the social arena where decisions of public interest are assumed.

Social change in cooperative models occurs through the interaction of the Public Administration and institutionalized private parts which express their direct interests in social disputes. These interactions between various bodies of the P.A. and different parts of the civil society generate a mix of conflict and consensus which, if well structured (Enevoldsen, 1998), can lead to negotiated settlements[4].

Cooperative policies offer specific advantages compared to traditional policy making models.

The main advantages are:

- the institution of structured frameworks to show and compose different interests,
- the promotion of social learning,
- the extension of the policy alternatives being considered, taking into account data, information and scenarios provided by interest groups,
- more consensus on adopted solutions,
- the reduction of risks deriving from the adoption of solutions in which excessive costs are imposed on one party,
- the reduction of public administrative costs in the implementation phase as private "intermediate subjects", representatives of group interests, play regulatory functions.

Cooperation also implies a few critical aspects:

- the opportunity of considering all relevant parts and interests,
- the risk that a few big stakeholders might dominate the process,
- the elusion of traditional democratic decision processes,
- the Government's ability to represent collective interests,
- the ability of "intermediate subjects" to represent the interests of their members and to guarantee the implementation of the assumed decisions.

Cooperative solutions ordinarily show their superiority, when compared to traditional policy tools, with respect to equity. Their evaluation is more problematic in terms of efficiency and effectiveness. On one side the process of cooperation reveals information and reduces uncertainty, so that solutions can be improved. Consensus increases enforceability and reduces non-compliance and, consequently, improves effectiveness. On the other side, asymmetric information is a relevant obstacle to reach efficiency: the regulator attempts to set the abatement level in order to equalize marginal social costs and marginal social benefits, but firms have an incentive to submit overestimated costs of abatement, since the regulator cannot ascertain them accurately (Segerson, 1999). Transaction costs can vary enormously.

Overall the specific design of the framework and rules for cooperation strongly affect the outcome of the process[5]. Moreover, institutional and social country-specific factors can strongly affect the effectiveness of cooperative environmental policies (Delmas and Terlaak, 2001).

2.1 Models of cooperation

Among the various models of public-private cooperation, we can distinguish three main categories depending on the degree of Government influence on final decisions:

1. consultation, where the P.A. consults private parts before making a decision,
2. negotiation, where the P.A. negotiates decisions with private parts,
3. framework setting, where the P.A. defines the rules that the private parts must follow in order to make a decision, with the guarantee that whatever the decision, the Government will respect it.

The three forms find growing application opportunities in real contexts.

In fact, consultation is the basic element of any public participation process. Public participation is a principle affirmed in international declarations and protocols like Agenda 21 (UN, 1992) and the Aarhus Convention (UNECE, 1998). Many kinds of environmental choices implying a high impact on the environment require, by national legislations, forms of consultation (like in the Environmental Impact Assessment). In many cases the P.A. has to keep public comments and observations into account when making final decisions

Public-private negotiation is a very frequent way of keeping into account stakeholder interests in public decision making. Negotiation is a general category and can be applied to various phases of environmental policy[6]. Negotiation can either be an institutionalised process required by law and rules, or an informal process involving the main stakeholders. Negotiation can start either by the P.A. initiative or by private parties request.

Framework setting is a typical coasian approach, where the role of the Government is limited to the definition of some of the elements of the decision framework (for example, actors admitted to take part in the process, basic rules to assume the decision, time limits). The Government is not even a party in the negotiations and leaves it up to the private parts to find a solution. The solution is assumed and enforced by the Government. Framework setting is applied to various policy areas, especially water management (Massarutto et al., 2004).

3.VOLUNTARY APPROACHES

Voluntary approaches in environmental policies are based on the idea that, under certain conditions, firms can decide to commit themselves to go beyond regulation. Chapter 6 analyses the reasons which induce firms to assume such a commitment. In general this commitment is taken on the basis of a cost and benefit analysis, where relevant information to perform the analysis is acquired by firms through negotiation with the P.A.. In many cases the commitment is assumed as the consequence of an explicit negotiation process. In other cases a spontaneous commitment is a way to hide negotiations where the P.A. threatens to adopt regulatory alternatives. Finally, there are cases where the P.A. designs standard agreements on the basis of an estimation of costs and benefits to firms.

A comprehensive categorization of voluntary approaches includes[7]:

1) voluntary public schemes,
2) negotiated agreements,
3) unilateral commitments recognized by the P.A.,
4) unilateral commitments,
5) third party initiatives,
6) private agreements[8].

Voluntary public schemes are standardized schemes, designed by regulators. Individual firms are given the choice of whether or not to participate in such schemes. Participation can be restricted to firms with certain characteristics (like operating in a specific sector, using a certain technology or certain substances, being located in a certain area, etc.). If a firm decides to participate, it must respect the program's prescriptions like, for example, targets, timing, monitoring, data disclosure. The participation to voluntary public schemes can be defined as an "optional regulation" (OECD, 1999). Administrative (easier permitting procedures, less frequent inspections), economic (tax exemptions, subsidies, access to credit) or information access (technical assistance, training) benefits can be reserved to participants. In most cases, participating firms can also use a specific logo, indicating the participation to the program. Monitoring can be performed by the regulator, third independent parties or by the firm itself. Sanctions for not respecting the program are ordinarily limited to the expulsion from the program or moral sanctions. Voluntary public schemes have been widely adopted in the US (see Brouhle, Griffiths and Wolverton in this book). Also the European EMAS and ecolabel schemes fall into this category.

Negotiated agreements are agreements struck as the result of a negotiation process between the Public Administration on one side and one or more firms or an association of firms on the other side. Environmental goals and means to achieve them are set during the negotiation. Bargaining can be started by each of both sides. Third parties (like NGOs) can participate to negotiations. The agreement can be open to participation by other parties in the future. The agreement involves obligations for both parties. Firms should either reach specified targets or adopt specified measures. The P.A. should provide administrative, economic or information access benefits. If participants represent a large share of potential participants, the P.A. often commits to not regulate the matter. Other terms of the agreement are also fixed. Negotiated agreements can be binding (like Dutch covenants) or non binding. In most cases they are not, so they cannot be enforced by Courts of Justice.

Unilateral commitments are set by the industry (either individual firms or associations of firms) without a public counterpart. They can assume the form of either structured programs or codes of conduct aimed at improving environmental performances. Unilateral commitments are a form of self-regulation. Targets and measures to be adopted are defined by the industry. Monitoring can be performed by the regulated firms or by third parties identified by the firms. Firms can adopt a logo to identify the program. The "Responsible Care" by the chemical industry association and the "Global Environmental Management Initiative" (GEMI) by the World Business Council on Sustainable Development are examples of unilateral commitments.

To give more credibility to their commitment[9], firms can ask the Public Administration to recognize such commitment. In this case the P.A. can perform monitoring or define guide-lines regarding the implementation of the commitment, like in the case of European Community "self-regulation" and "co-regulation" introduced with a Communication on "environmental agreements at the Community level within the framework of the action plan on the simplification of the regulatory environment" (CEC, 2002; Croci, 2003).

Third party initiatives are programs designed by third parties open to the participation of individual firms. They are similar to public environmental schemes, but here the scheme and/or the obligations are not designed by a public body, but by private organizations, like the ISO or NGOs, or even international organizations (not possessing any authority to regulate the matter). These private organizations cannot give anything in exchange for participation except image benefits and access to management improvement procedures. ISO 14000 and the UN "Global compact" are examples of third parties initiatives.

Private agreements are reached through direct bargaining between polluters and pollutees. They identify common solutions and provide direct compensation without any public intervention. The Coase theorem affirms that if property rights on the environment are clearly assigned, negotiation between polluters and pollutees will lead to a condition of social optimum (Pareto efficiency) independently from the initial allocation of property rights (Coase, 1960). Coase argues that negotiations take place only if the expected benefits are higher than the transaction costs. A relevant area of application of private agreements is between firms and trade unions regarding the health of workers.

"The rich diversity of voluntary initiatives is essential to meet the different needs of an industry or country, which may vary according to their socio-economic context and stage of responsible entrepreneurship" (UNEP, 2000).

4. ENVIRONMENTAL VOLUNTARY AGREEMENTS

In the following we will restrict our attention to voluntary approaches where a private and a public counterpart are clearly identified. Three categories respect these criteria: voluntary public schemes, negotiated agreements and – under certain conditions[10] – unilateral commitments recognized by the P.A (Croci, 1995; Croci and Pesaro 1998; Amadei, Croci and Pesaro, 1998). We will use the term "environmental voluntary agreements" (VAs) when the discussion applies to all three categories. We will distinguish among them when we examine their distinctive features.

In general, VAs can be considered as contracts "between individual companies and/or association of companies, on the one hand, and public authorities, on the other hand, concluded with the aim of protecting or restoring the environment" (CEC, 1996a).

The EU Commission has defined the typical elements of these contracts (CEC, 1996b) – see Table 1.

Table 1. *Elements of a voluntary agreement*

1.	Parties to the agreement (associations and/or individual firms)
2.	Subject
3.	Definition of terms
4.	Quantified objectives
5.	Staged approach
6.	Specification of obligations
7.	Monitoring of results
8.	Periodic reporting
9.	Access to information
10.	Arrangements for collection/evaluation/verification of results
11.	Sanctions
12.	Accession of third parties
13.	Duration
14.	Revision
15.	Termination
16.	Legal nature of the agreement
17.	Jurisdiction

Source: (Commission of European Communities, 1996)

For their characteristic of flexibility, VAs can be applied to a variety of fields, like sectors, specific territorial areas, activities and events generating environmental impacts, scarce resources, use of toxic substances, products and services, technologies, adoption of environmental management tools and international issues.

4.1. VAs as processes

In general the definition of VAs requires the succession of the following phases:

1) design. The VA can be designed either by the public or by the private part, or even jointly. In this phase the main elements of the agreement are set (parts, rules, etc.). In the case of public voluntary schemes, precise commitments are also determined, as phase 3 is not included.

2) identification of relevant parts. Relevant parts are all parts which are potentially affected, either directly or indirectly, by the agreement. Some parts (environmental associations, affected firms operating in sectors which are not directly involved, etc.) could participate to the negotiations as "third parties" without signing the agreement Relevant parts could change through the process.
3) negotiation. In this phase all the elements of the agreement are determined through negotiation. Parts can be assisted by negotiation experts. Timing, monitoring and reporting are also established. New parts can be identified and admitted to negotiations.
4) signature. Formal commitments and rules are accepted with the signature of the participants, which can partly differ from the negotiating parts.
5) compliance. Parts implement actions in order to comply with their commitments. For example, they adopt specific technologies, or procedures, or measures.
6) monitoring. Monitoring can be performed either by the private part, following certain rules, or by a public agency, or by a third independent part.
7) evaluation. Committees formed by representatives of parts and third parties can evaluate the performances and the respect of time schedules.
8) sanctioning. Sanctions could be applied in case of non-compliance. Depending on the institutional context they can be of a legal, economic or moral nature. Administrative sanctions can also be taken by the P.A., for example through stricter regulations.
9) revision. The agreement may be revised in case of emergence of relevant external factors (like the availability of a new technology or "force majeure"),
10) data disclosure. This phase is transversal to all others, as disclosure regards the structure and the rules of the agreement, the negotiating parts, the signing parts, the commitments, the results, etc.. All data are ordinarily disclosed to the public, with the exception of specific information subject to industrial secret.

In environmental policy the following moments are ordinarily separated: a) definition of the policy goal, b) choice of the best instrument to reach that goal under given conditions.

This is not true for VAs. In negotiated agreements the policy goal is not precisely determined (both with regard to its quantitative and qualitative aspects) until the end of the negotiation, as it is an outcome of the negotiation process. Even in voluntary public schemes, where the commitments of individual participants are set in advance, the global policy goal depends on the number and size of participants, which is not predetermined.

4.2 Individual and collective agreements

We can distinguish two main typologies of negotiated agreements:

a) individual agreements,

b) collective agreements (or branch agreements).

Individual agreements are agreements signed by one or more firms. Firms commit themselves individually.

Collective agreements are agreements signed by one or more branch organizations representing their associated firms. First the branch organization negotiates the agreement. Then it allocates targets among members. In this case, we can say that Government delegates quasi-regulatory powers to the branch organization. This implies some advantages and some disadvantages.

Among the former:

- a better allocation efficiency, as the branch organization knows the marginal abatement costs of individual firms better than the Government does[11];
- the activation of a collective learning process, as the branch organization can promote exchange of information and access to the best practices and technologies.

Among the latter:

- the branch organization cannot use means of compulsory enforcement like the Government can. Its action is based more on "moral suasion";
- the possibility of "free riding", as some firms can choose not to implement their commitments and to try to leave all the efforts to others. This is a different form of free riding from the one affecting individual agreements, where it refers to either firms which simply do not sign the agreement, so as not to commit themselves, or firms not complying with their obligation expecting they will not be sanctioned anyway.

A way to keep the advantages and avoid the disadvantages is represented by "two level" agreements (Croci, 2003). In that case, first a framework agreement with a collective target is negotiated and signed at the branch level, then firms negotiate their individual commitments inside the branch organization and finally individual agreements between firms and the P.A. are signed based on the agreed allocation scheme.

5. WHY DO ENVIRONMENTAL VOLUNTARY AGREEMENTS EXIST AND HOW DO THEY WORK?

Government will rationally enter a VA if evaluates it can be more efficient - or at least effective - than alternative instruments (unless a situation of "capture of the regulator" exists). But why firms should enter a VA and commit to go beyond regulation?

The commitment of firms in VAs can regard: a) the reduction of polluting emissions, b) the adoption of clean technologies or processes, and c) the introduction of cleaner products.

Commitments can regard quantitative goals or the development of procedures and systems that facilitate the improvement of environmental performance (Khanna, 2001).

The neoclassical economic theory assumes that economic players are rational, so they will behave in such a way as to maximize their utility. Under this assumption, voluntary commitments to reduce pollution or improve the environment shouldn't exist. In fact, firms should spontaneously carry out all actions for which abatement costs are inferior to expected benefits. Any voluntary action - besides those - would be irrational.

As VAs exist we have to individuate the rational mechanisms which they are based on. These are basically of two kinds:

1) VAs imply an exchange, where polluting firms bear costs in exchange for some kind of superior benefit from the P.A.;
2) VAs are learning processes, during which participating firms, under conditions of shared uncertainty (regarding the evolution of standards, technologies or markets), perform a collective search, sharing costs and gains, where expected gains are superior to expected costs.

These two mechanisms operate in various situations, most of which have already been analyzed in economic literature, regarding the motivations of firms for entering VAs (OECD, 1999, Brau and Carraro, 2001). We found seven of them, characterized by different conditions. We analyze them separately, even if a combination of them can operate at the same time in the real world.

5.1 To avoid (or procrastinate) stricter regulation

Suppose the P.A. intends to adopt a new or stricter standard (or a tax). Firms will sign a VA if the voluntary standard is inferior to the expected standards (at least for a certain period). In this way, firms try to pre-empt public regulation.

Let's assume that marginal abatement costs (MaC) are increasing[12] (which means decreasing in pollution level). Assume Q* is the standard before the agreement. If the target to which the firm commits itself under the agreement Qva is inferior to the standard the Government would set under an alternative regulation Q_r, then the gain of the VA for the firm is represented by the area ABQvaQr (Figure 1).

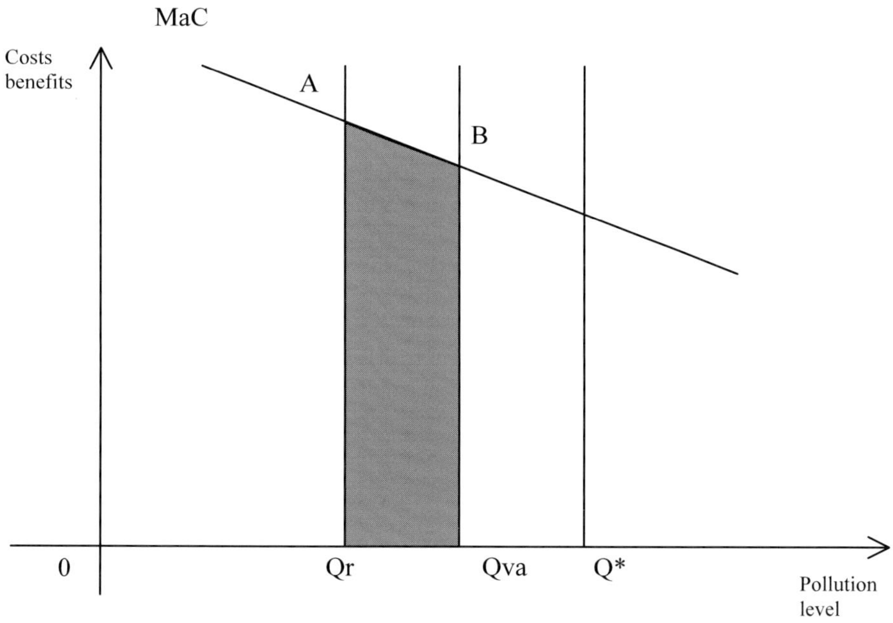

Figure 1. Gain of the VA for the firm avoiding stricter regulation

5.2 To obtain flexibility by complying with the regulation

Suppose the P.A. intends to adopt a new or stricter standard. Another reason which can induce firms to sign a VA is if through it they can reach the standard at a minor cost having the freedom to choose specific technologies or measures.

Let's assume MaC be the curve of marginal abatement costs and Q* is the standard before the agreement. The target to which the firm commits itself under the agreement Qva is the same as the standard the Government would set under an alternative regulation Q_r, but under the agreement the firm obtains more flexibility in the way to implement the pollution abatement. This means there is a downward shift of the marginal abatement costs under the VA (MaCva). The gain of the VA for the firm is represented by the area ABCD (Figure 2).

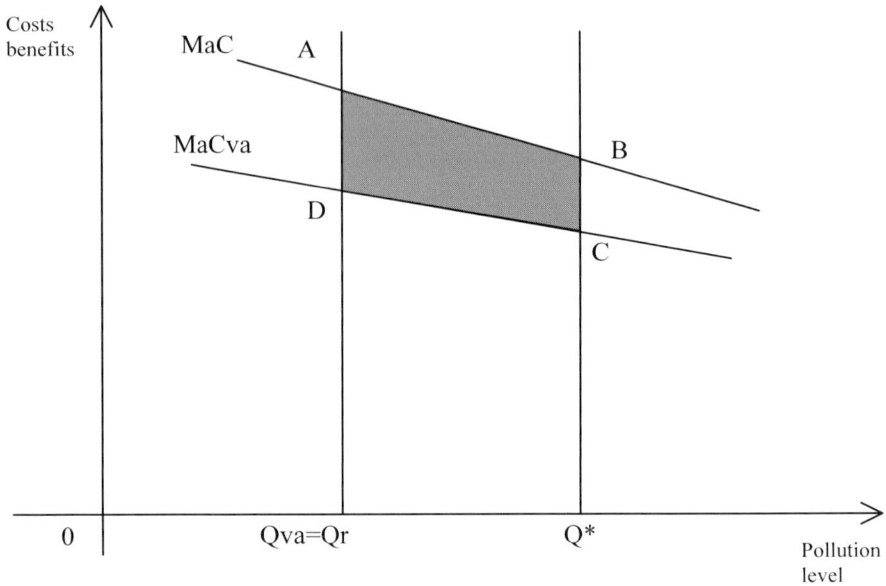

Figure 2. Gain of the VA for the firm getting flexibility by complying with the regulation

5.3 To induce the P.A. to adopt a stricter regulation

Suppose a firm exclusively owns a new technology (or has exclusive access to a cleaner production input) thanks to which it can adopt a cleaner production process or produce a more "environmental friendly" product at no extra cost (the results can hold, under certain conditions, even if the firm incurs in positive costs, but inferior to the costs of competitors). Competitors would have to bear relevant costs to reach the same standard. The firm can subscribe a VA in order to demonstrate the feasibility to respect the new standard and generate public pressure to adopt it (and possibly to prescribe that specific technology or the use of that specific production input). If the new standard will be adopted, the firm will gain a competitive advantage, because its competitors will incur in extra costs. Eventually competition in the industry will be restricted and the firm can get a monopolistic rent (Barrett, 1991; Arora and Cason, 1995).

Let's assume MaC be the curve of marginal abatement costs for all firms except the one entering into the VA. Let's also assume Q* is the standard before the agreement. The firm commits to the target Qva under the agreement. Let's assume the marginal abatement costs are 0 in the trait Q*Qva for the firm only. The firm will reduce pollution from Q* to Qva. If the Government subsequently sets a new standard at level Qva, all competitors will suffer a loss represented by the area ABQ*Qva (Figure 3). Depending on the market structure, this could alter competition.

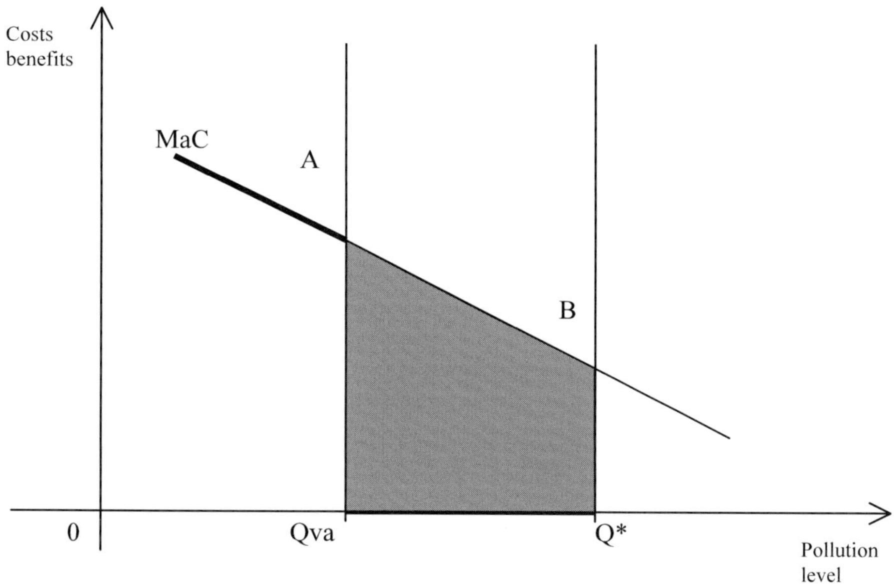

Figure 3. Gain of the VA for the firm inducing the P.A. to adopt a stricter regulation

5.4 To cut costs through pollution prevention

The neoclassical condition of perfect information is unrealistic. To collect information is a money and time consuming activity. So firms base their action on incomplete information. As Simon (1955; 1979) argues we live in a world of "bounded rationality"[13]. In other words we drop the initial hypothesis 3, following a behavioral approach. Under this assumption firms maximize their utility basing their choices on a subset of all relevant information, as information searching and gathering is a costly activity and satisfactory solutions, instead of optimal, are pursued in reality[14]. By entering into a VA, a firm can gain, in exchange, access to information about technological options, either through direct access provided by the P.A. or by sharing resources with other firms in a collective search. This extra information can lead to "more rational" choices (regarding, for example, saving of inputs or recycling, or other eco-efficiency improvements), so the firm can abate pollution and get higher profits at the same time.

Let's assume MaC be the curve of marginal abatement costs and Q* is the standard before the agreement. The firm gets information about more technological options through a VA, so discovering it can reduce pollution improving profits, thanks to pollution prevention measures. The gain is represented by the area below the curve of marginal private benefits (MaB). This is due to the exploitation of opportunities unknown to the firm before the agreement and made available to the firm because of it. Under these conditions an optimal VA exists. In fact, the firm will maximize its utility (benefits minus costs) committing to a target Qva, where MaC equals MaB. The conditions for the firm to get a net gain is that MaC crosses MaB (i.e. at some point marginal benefits of pollution abatement are over marginal costs) and that the area BCQ* (loss) is inferior to the area between MaB and MaC in the trait AB (gain)[15] (Figure 4).

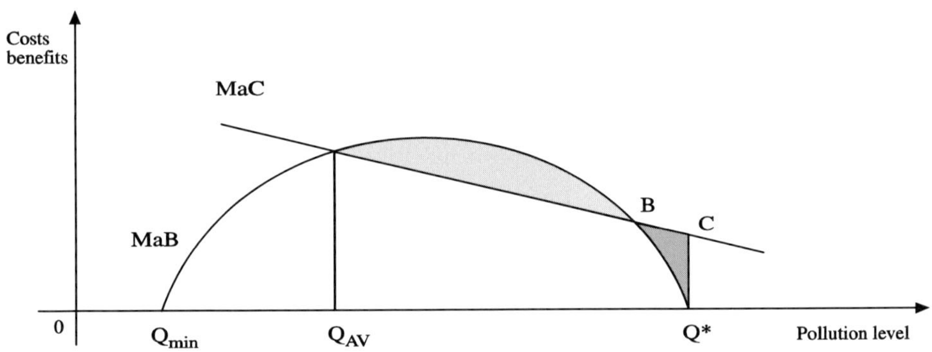

Figure 4. Gain of the V.A. for the firm cutting costs through pollution prevention

5.5 To get access to credit for profitable investments

A situation similar to the previous one regards access to credit. Some investments which are profitable and, at the same time, reduce pollution are not made because of the relevant up-front expenditures implied, while benefits come over a long period. Firms may not have access to credit for such investments. The P.A. can grant access to credit to those firms participating to the VA, thanks to specific agreements between the P.A. and financial institutions. In this way, the firm can pursue profitable business opportunities thanks to the VA.

This situation can be represented with the previous figure, where the curve of marginal benefits represents the actualization of future profits.

5.6 To obtain tax exemptions or incentives

The P.A. can provide: a) tax exemptions, or b) economic incentives to firms subscribing a VA. If the VA provides superior net gains to the firm than the tax exemption (or the incentive), then the firm will enter the agreement.

a) Let's assume MaC be the curve of marginal abatement costs and Q* is the standard before the agreement. Let's also assume the Government applies a new tax on pollution $T = tQ$. Then the firm will reduce pollution up to the level Q_t at which MaC equals t. The cost to the firm is represented by the area Q^*Q_tBA (corresponding to the cost of abatement) plus the area $0tBQ_t$ (corresponding to the tax). Suppose the firm is granted a tax exemption if it subscribes a VA so as to commit itself to reduce pollution to the level Qva. The cost of the VA to the firm is represented by the area Q^*Q_tBA plus the area Q_tQvaCB (corresponding to the total cost of abatement passing from a pollution level Q* to a pollution level Qva). Then the firm will enter the agreement if its cost is inferior to the cost of the tax, i.e. if Q_tQvaCB is inferior to $0tBQ_t$. The gain for the firm is represented by the area Qva0tS minus the area BSC Note that Qva is always inferior to Q_t, which means that the firm will always prefer to attain a given target of pollution abatement through a VA than through a tax or, in other words, that the firm will be eager to commit to a more severe target of pollution abatement under a VA, in order to get a tax exemption, than the target correspondent to the avoided tax (Figure 5.1).

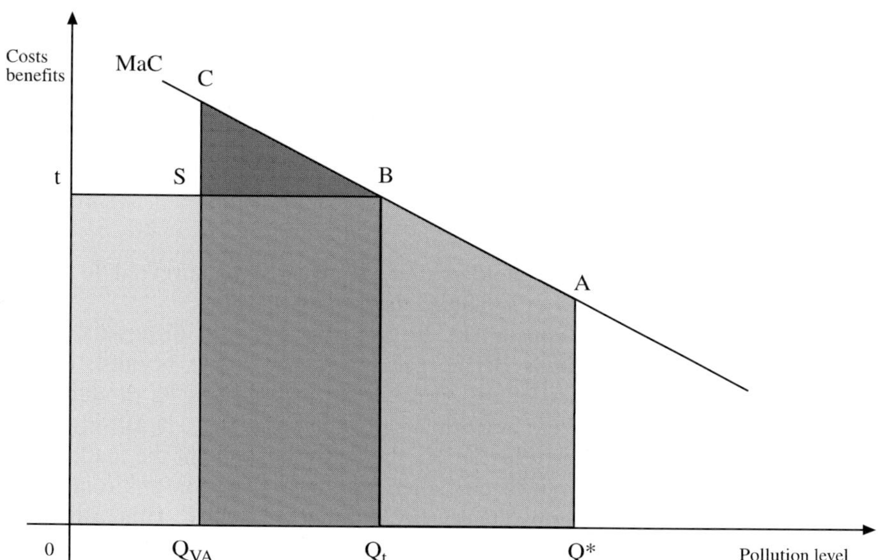

Figure 5.1. Gain of the V.A. for the firm obtaining a tax exemption

b) Let's assume MaC be the curve of marginal abatement costs and Q* is the standard before the agreement. Let's also assume the Government offers a firm to enter a VA to reduce pollution to level Qav. The Government offers the firm an incentive S=sQ to accept the agreement. The cost of the VA to the firm is represented by the area Q*QvaBA, while the benefit is represented by the area Qva0sB. Then the firm will enter the agreement if its cost is inferior to the benefit (Figure 5.2).

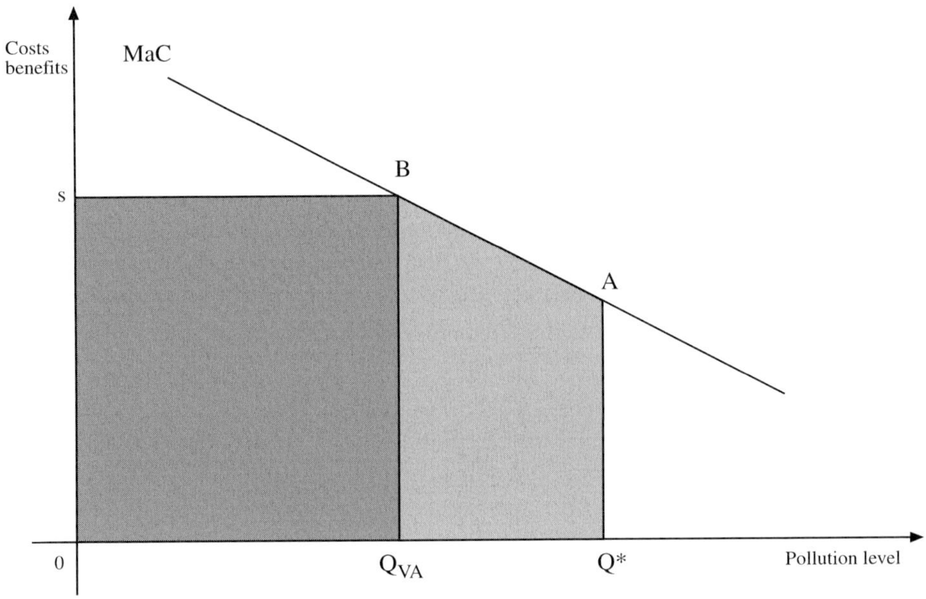

Figure 5.2. Gain of the V.A. for the firm obtaining an incentive

5.7 To gain reputation

Entering a VA is a signal to stakeholders. The signal can be directed to either consumers or other stakeholders. Let's analyze the two possibilities:

a) the signal is directed to consumers. In this case the firm differentiates its product from other products of the same category either because of the environmental performance of the firm (which also affects its products) or because of the environmental performance of the product itself. Basing itself on the theory of product differentiation, the firm can segment the market and raise the price of its product characterized by a superior environmental quality (Arora and Gangopadhyay, 1995, Stoeckl, 2004). This aspect is particularly relevant when firms participating in the agreement can use a specific green label or obtain other forms of public recognition (prizes, quotations, etc.).

b) the signal is directed to other stakeholders, for example local communities, financial and insurance institutions, permitting or controlling authorities, employees. In this case the firm can improve its relationships with these stakeholders and gain competitive advantages regarding, for example, less local opposition, better access to credit, lower insurance rates, easier permitting or control procedures, easier recruiting or qualified employees, better employee satisfaction.

Let's examine situation a). Let's assume MaC be the curve of marginal abatement costs and Q* is the standard before the agreement. The VA provides a signal to consumers. Consumers value the environmental benefit attributed to the product of the firm. This causes an increase in their "willingness to pay", reflecting in a change of the slope (price sensitivity) of the demand curve, which becomes steeper: at any production level corresponds a higher market price. So the firm will raise prices and earn higher profits. A curve of marginal benefits (MaB) can consequently be drown. Suppose that after the agreement the curve MaB crosses the curve MaC at points A and B. Then the firm will get a net gain entering an agreement to reduce pollution up to Qa, as in the trait between Q* and Qa the curve MaB lies over the curve MaC. The VA corresponding to a pollution abatement up to Qa represents an optimal agreement for the firm as it maximizes the net gain for the firm, represented by the difference between the area below MaB and the area below MaC (Figure 6).

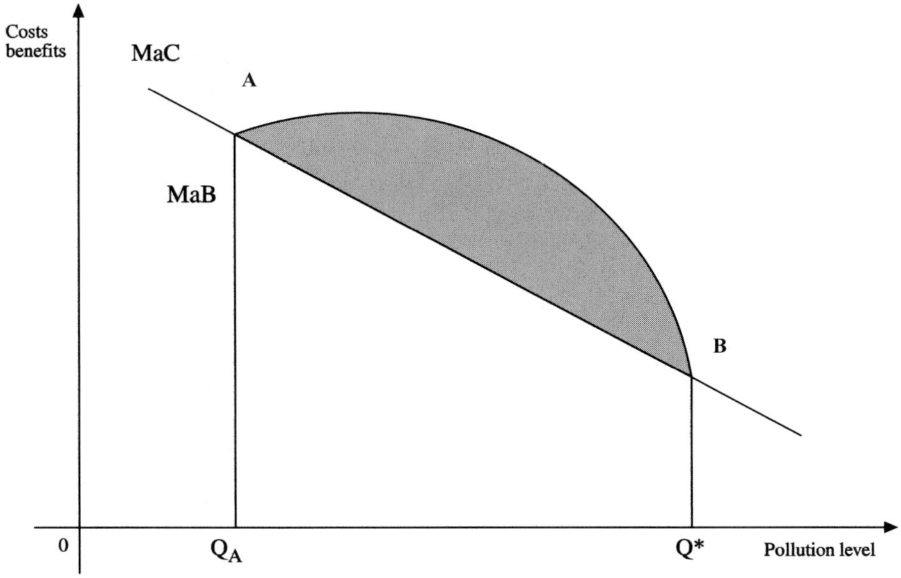

Figure 6. Gain of the V.A. for the firm gaining reputation

Situation b) can be similarly modeled to the case of pollution prevention.

Similar considerations to the ones made for gaining reputation can be made considering avoiding adverse publicity.

5.8 A few general considerations

Depending on the specific circumstances, a VA can improve or lower efficiency, in comparison with alternative instruments, or even in comparison with the original situation.

When a VA is applied to reduce market imperfections, like in the fourth case, where it reduces uncertainty, its effect will be to improve efficiency. The same is true when a VA improves flexibility, so minimizing abatement costs, like in the second case.

When the pollution level that would be reached under alternative instruments (or the original level) is efficient and the VA is struck thanks to market imperfections or regulatory capture[16], the VA will reduce efficiency. This can happen, for example, in the first and the third cases, if the standard to be adopted (in the first case) or the present standard (in the third case) are efficient. If a tax or incentive is set at an efficient level, the VA will normally be comparatively inefficient, like in the sixth case (unless it exactly replicates the pollution abatement obtained through the tax or the incentive). Efficiency in the seventh case depends on the reliability of the signal, i.e. if a higher price corresponds to real superior environmental attributes.

Even when a VA is not an efficient solution, it can be the only feasible one due to political constraints or insufficient consensus on alternative solutions.

5.9 The relationship between VAs and regulation

In economic literature, VAs are often considered alternative to regulation. This is a narrow view. VAs have been developing mostly in countries where a well-structured regulatory framework already exists. In fact VAs are an advanced policy tool which can assume various relationships with regulation.

With respect to regulation, the aims of VAs can be:

a) substitutive, in that they are adopted as an alternative to regulation,
b) integrative, in that participating firms commit to respect stricter standards than regulation,
c) anticipatory, in order to provide information to the P.A. about the costs of future regulations and gradually predispose the industry to stricter standards,
d) applicative, as means of implementing a law (for example a European directive at the domestic level).

The policy-maker stimulates different motivations on the side of the industry, depending on the aims of the VAs with respect to regulation. So, substitutive agreements are normally associated with the motive of avoiding stricter regulation; integrative agreements with the motive of inducing pollution prevention, or gaining access to credit, incentives, or reputation; anticipatory agreements with the motive of fostering stricter regulation; and applicative agreements with the motive of obtaining flexibility. In Figure 7 the correspondence between industry motives and the aims of VAs with respect to regulation are illustrated. Industry motives will be analyzed in the next chapter.

***Table 2**. Correspondence between industry motives and the policy-maker aims with respect to regulation for VAs*

	no regulation	flexibility	stricter regulation	pollution prevention	access to credit	incentives	reputation
substitutive	x						
integrative				x	x	x	x
anticipatory			x				
applicative		X					

Only through a case by case analysis is it possible to classify VAs with respect to regulation.

6. VAS UNDER THE THREAT OF REGULATION: A SIMPLE MODEL

6.1 Non participation

In the first case examined in the previous chapter, a VA is reached under the threat of regulation. This is a common situation. Firms must bear a cost, but this cost is estimated to be inferior to the alternative under regulation. So they choose to sign the agreement in order to avoid the regulation.

More realistically, firms will make their decision regarding the participation to the VA by assigning a weight to the threat of regulation. If the threat is judged unrealistic, because of lobbying ability or other opposition instruments, firms could prefer not to enter the agreement and risk a stricter regulation.

A very simple model can be provided for this case.

We consider just two actors, the Government and a firm, where the Government offers the firm the opportunity of entering a VA. We assume the VA involves costly obligations for the firm. In the model, by "costs" we mean the firm's "net costs" (costs minus benefits) involved by the VA.

We also assume the VA has already been designed and the commitment of the firm (in terms of a pollution abatement target) has already been set; in other words there is no possibility of further negotiation. The firm can only accept or refuse to enter the agreement.

Then, the firm will enter the agreement if:

$$\exp \mathrm{Cv} < w\, \mathrm{Cr} \qquad 0 < w < 1$$

where exp Cv is the expected cost of the VA to the firm, Cr is the cost of alternative regulation to the firm and w is the probability assigned by the firm that a regulation will be issued in case of refusal to sign the agreement.

6.2 Non compliance

Let's now admit a new option: the firm can enter the agreement, but decide not to comply with it. Let's assume that the only sanction could be that of adopting a stricter regulation[17]. Rational firms will compare the benefits deriving from the non-compliance with the expected loss.

Formally, after the agreement has been signed, the firm will respect its commitments if:

$$\text{Cv} < p\,\text{Cr} \qquad 0 < p < 1$$

where Cv is the cost of the VA to the firm (we suppose that the firm is able to calculate exactly this cost after entering the agreement), Cr is the cost of the regulation to the firm and p is the probability assigned by the firm that a regulation will be issued in case of non compliance.

Non compliance can be difficult to detect because of "imperfect information" (Akerlof, 1970). This is particularly relevant in the case of agreements requiring a superior environmental performance of products. In fact the environmental quality of a product can be difficult and costly to assess. These products can be classified as "experience goods", as their quality can be discovered only after purchase, or even "credence goods", as their quality remains uncertain even after purchase (Cavaliere, 2000).

So the firm will assume that the Government will decide to apply the regulation threat on the basis of an "observed" compliance and not of a "real" compliance. Assuming the Government identifies compliance with the observation that the firm sustains a cost of at least Cv, the firm will respect its commitments if:

$$\text{obs Cv} < p\,\text{Cr}$$

where obs Cv is the cost to respect the VA "observed" by the Government.

If the firm is able to "manipulate" the Government's observation, because of uncertainty, the observed costs to respect the VA will be a function of costs "revealed" by the firm. So:

$$\text{obs Cv} = g\ \text{rev Cv} \qquad 0 < g < 1$$

where rev Cv is the cost revealed by the firm to respect its commitments and g is a coefficient representing the ability of the firm to make the Government believe that the information it reveals about the cost to respect its commitments. If $g = 0$, the firm is unable to manipulate information. If $g = 1$, the firm is able to make the Government believe that the information it reveals is true. In that case the firm has a strong incentive to do nothing, sustaining 0 costs, and making the Government believe it has fully respected its commitment. So the firm can "free ride" without any sanction[18].

This brief analysis puts in evidence the relevance of monitoring performed by the P.A.. In fact, improvements in monitoring lower g. The P.A. should improve monitoring up to the point where the marginal cost of monitoring equals the marginal benefits provided by non manipulation of information. These benefits refer to the prevention of free riding and the ability to sanction it.

Public transparency also reduces "strategic (or opportunistic) behaviors" on behalf of firms. Thanks to asymmetric information, firms can induce the Government and citizens to think that they are implementing measures to abate pollution, while they are not. To this end, firms can also advertise investments as "green" even if they are made for other reasons. Information disclosure and public scrutiny can lower this risk.

7. EVALUATION OF ENVIRONMENTAL VOLUNTARY AGREEMENTS

VAs can be tailored to specific conditions and aims. Flexibility is the main advantage of VAs with respect to command and control and also to economic instruments. Flexibility contributes to efficiency and effectiveness, even if VAs cannot guarantee a full internalization of externalities and the respect of the principle of equimarginality of abatement costs among polluters.

Flexibility is particularly relevant in keeping into account local and branch peculiarities.

The process in which stakeholders are involved to reach a VA can contribute to build trust among actors and consensus on targets, which can positively affect the implementation phase and reduce monitoring and enforcement costs.

Interaction among actors is also a means to incentivize revelation of information regarding policy goals on the public side and costs of compliance on the private side.

Finally, if the targets go beyond "business as usual", VAs can stimulate technological innovation and diffusion. Diffusion can particularly be stimulated if the P.A. or the branch organization (in collective agreements) provides assistance and access to information.

Both because of cost and market competitive advantages, VAs can generate a virtuous circle of imitation among firms, which progressively improve environmental performances.

Other specific features of VAs generate risks. As VAs do not pass through traditional legislative, regulative or administrative procedures, they show a lower "social legitimacy". In fact this can induce the definition of modest targets, even below "business as usual", often through the "capture of the regulator" (Stigler, 1971). The process itself leading to a VA offers opportunities of collusion between the P.A. and firms, unless the following conditions are granted: a) public transparency and third party participation, b) availability of shared information about potential environmental performances and abatement costs of firms, c) the Government is structured in a plurality of organizational levels and of administrative and technical bodies, which make collusion difficult (OECD, 1999).

7.1 VAs performances against "business as usual"

ten Brink (2002) has elaborated a scheme to assess the performance of a VA (Figure. 7): a target T for a pollution reduction is set at time Ts to be reached at time Tt (an intermediate target can be set at time Ti). The target can be expressed in absolute terms or as a reduction percentage with respect to a reference year (which does not necessarily coincide with the start date). The scheme shows the reference level, the "business as usual" trend of pollution, the performance under the voluntary agreement and the performance under alternative policy instruments. The relative positions of these curves can vary. Except for the reference level (which is fixed), other curves can be decreasing, stable or increasing. An evaluation can be performed ex ante, based on projections, in itinere, or ex post. If we perform an intermediate evaluation at time Ti, the environmental effectiveness of the agreement is measured by D-C with respect to the "business as usual" scenario, or by E-C with respect to a counterfactual scenario. In many cases neither a "business as usual", nor a "counterfactual" trend is calculated, so the only quantitative evaluation is performed as A-C with respect to the reference level of pollution. This can cause an "over-evaluation" of the impact of the VA.

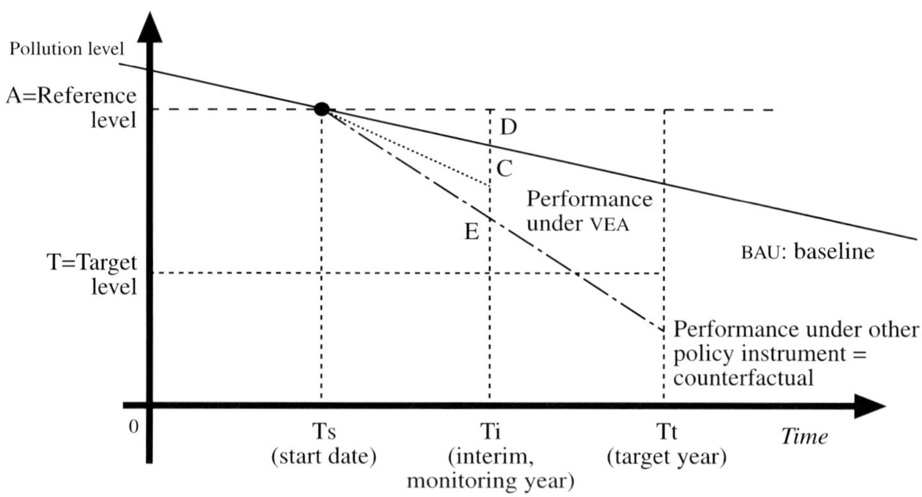

Figure 7. VA performance assessment

Asymmetric information plays a central role in this context. Firms tend to over-estimate their abatement costs. Public agencies, even when owning strong technical expertise, don't have reliable data about this. This aspect makes it very difficult to provide an ex ante evaluation of VAs, especially if, at the time they are signed, the "business as usual" and alternative scenarios are not available.

Most studies providing an ex-post evaluation of VAs show that set targets have been reached, but that these targets were set at low levels, in many cases close or under "business as usual" (OECD, 2003; Croci, 2003).

Under this condition, most of the beneficial mechanisms activated through VAs are compromised, in particular environmental effectiveness, economic efficiency and technological innovation and diffusion.

Recently "benchmarking agreements" have been proposed and signed in order to overcome this problem (see Bressers and De Bruijn in this book). In benchmarking agreements, firms do not make a commitment to reach targets quantified ex ante, but their commitment is to classify among the top ranking after a certain period of time. Targets become stricter over time, so increasing pressure on firms and fostering innovation and competition[19]. Environmental effectiveness and economic efficiency increase.

Alternatively, the possibility to update targets in response to the availability of new technical solutions can offer some guarantees with respect to this problem.

7.2 VAs and competition distortion

Environmental aims can also mask other issues at stake when the P.A. negotiates directly with industry. VAs can distort fair competition through oligopolistic agreements among a set of producers in order to restrict competition, State aids, barriers to free trade. VAs connected to the adoption of a stricter regulation or public incentives can involve these risks. Even reputation effects gained through the use of signaling power by the P.A. can operate in this direction. A way to lower these risks is through public participation and information disclosure from the beginning of the process up to monitoring and reporting.

These risks regard the possibility that participants to VAs gain unfair advantages. On the contrary, another risk category is the possibility that participants to VAs are damaged in comparison to their competitors.

This is the case of "free riding", which can arise in two ways (Gunningham and Rees, 1997). First, especially when VAs are substitutive of a stricter regulation (which means they cause a cost to participating firms) non participating firms get an unfair competitive advantage. In order to avoid this, VAs should involve a number of firms representing a large majority of the sector and should include all the most relevant competitors. Second, free riding can regard non compliance. This case can be particularly relevant when VAs are struck between the Government and branch organizations, if single firms decide not to respect their individual targets (Carraro and Leveque, 1999). Besides altering competition and raising equity concerns, free riding can compromise efficiency and also lead to the failure in meeting the agreed targets.

Again only a case by case analysis can lead to determining the possibility of altering competition plays in favor or against firms participating into a VA.

7.3 Transaction and administrative costs

Transaction costs regard the costs that private and public parts must bear in order to stipulate and manage an agreement. They include administrative costs which the P.A. bears in order to enforce the agreement[20].

VAs can involve high transaction costs. They involve: the drafting of the agreement, the individuation of parts, negotiations, data gathering, data checking, evaluation and diffusion, monitoring procedures, revisions and sanctioning (Krarup and Rameshol, 2000, Paton, 2000). It is not easy to quantify these costs ex ante. It is possible to say that they grow with the number of participants on both the public and private side.

On the other side, VAs can reduce administrative costs for regulatory agencies, in particular for enforcement, as forms of self enforcement performed by committed firms or branch organizations are envisioned.

The consideration of transaction costs can lead to a trade-off between effectiveness and efficiency. Effectiveness requires complex and prolonged data gathering, negotiation and monitoring in order to set credible targets. This causes an increase in transaction costs, thus reducing the efficiency of VAs compared to other environmental policy tools (Croci, 2003).

Social and institutional factors can condition the effectiveness of VAs. Among them, the allocation of institutional competencies between administrative and technical bodies and among various government levels, the existence of a tradition of public-private cooperation, business culture and public environmental awareness.

8. CONCLUSIONS

Environmental voluntary agreements try to remedy market failures differently from traditional regulatory and economic instruments. In fact, they are based on the exchange between the P.A. and firms and on the design of a framework of incentives to parties in a context of negotiation and cooperation.

VAs can also be analyzed as learning processes, where the P.A. and the firms widen the set of information on which they base their choices and reduce information asymmetries.

Flexibility is the main feature of VAs. Flexibility facilitates the application of VAs to a wide variety of circumstances, but flexibility also makes it difficult to determine univocal motives and mechanisms playing in the use of VAs. In fact only a case by case analysis can reveal the peculiarities of each agreement, even if some general rules can be applied.

The evaluation of VAs is also a difficult task, because their efficiency, effectiveness, equity and other features can be determined only if based on a set of data which is ordinarily lacking.

Nevertheless VAs can effectively be included in the tool-kit of the environmental policy-maker if some conditions regarding their design and implementation are respected.

9. ACKNOWLEDGEMENTS

While I keep full responsibility for this paper, I thank Prof. Luigi De Paoli for his helpful suggestions.

10. NOTES

[1] Coase considers government intervention unnecessary if property rights are clearly assigned either to the polluter or to the pollutee. If transaction costs are sufficiently low to be irrelevant, social optimum will be reached through directly negotiated compensations (Coase, 1960).
[2] Government can also pursue other goals in pollution control. An important goal can be equity. At this purpose who bears the costs and who enjoys the benefits of pollution control becomes of great relevance.
[3] We consider Government and the regulator as coincident here.
[4] Lafferty and Meadowcroft call these frameworks "co-operative management regimes" (Meadowcroft, 1998). Liefferink talks about "joint environmental policy making" (Liefferink, 1997) and considers two policy style elements, characterizing most industrialized countries, encouraging it: the tendency to deregulate and the growth of corporative models in public decision making. Glasbergen sees the emergence of a new "architecture of decision making", where civil organizations and the business community "share the task of public service" (Glasbergen, 2000).
[5] For this reason game theory provides an effective methodology to analyze cooperation processes.
[6] Public-private negotiation in environmental policy finds three kinds of applications:
1) negotiated rule-making,
2) negotiated implementation,
3) negotiated compliance.
The first one refers to negotiation in the setting of laws and regulations, where the Government anyway has the last word. The second one refers to a firm's voluntary commitments in exchange for something else by the Government. The third one refers to the consensual definition by enforcers or Courts and violators of an alternative penalty with respect to criminal or civil sanctions in case of violations of environmental rules (for an exhaustive analysis see Ashford and Caldart, this book).
[7] This categorization is a development of a basic taxonomy, including the first four categories, elaborated by Croci (Croci, 1995; Croci and Pesaro 1998; Amadei, Croci and Pesaro, 1998), which has been widely adopted in economic literature.
[8] We don't include negotiated compliance, which is a form of sanctioning for violators of environmental regulations, as we focus on implementation of public policies.
[9] Unilateral commitments promoted by branch associations are often subject to adverse selection. In absence of serious mechanisms to monitor and enforce compliance and in the presence of reluctance to expel free riders, poor performing firms seek to join in order to gain the signalling benefits deriving from this form of differentiation, without costs (Lenox and Nash, 2003).
[10] The main condition is that recognition by the P.A. is not just a moral encouragement, but implies a form of public "guarantee" for the credibility of the commitment, so that the P.A. acts as a counterpart and reacts to violations of the commitment.
[11] Large firms can be more effective in defending their interests inside the organization, but this happens also in individual agreements through lobbying.
[12] This is a common hypothesis: in general even if marginal abatement costs may be flat over a range of pollution abatement, they increase over a certain level.
[13] Behavioural economics removes the hypothesis of perfect rationality of economic agents. Choices are supposed to be assumed following a bounded (or limited) rationality, as information gathering is a costly activity and agents accept satisfactory solutions, even if they are not optimal in absolute (Simon, 1955; Stigler, 1961; Arrow, 1974).
[14] Sub-optimal solutions can be pursued also because of principal – agents problems between manager and employees or between shareholders and managers generating organizational inefficiency (Howarth, Haddad and Paton, 2000).
[15] Access to information can also lead to a new curve of marginal abatement costs, lower than the initial, as the firm can discover methods to reduce pollution at lower costs.

[16] Other possible explanations regard the drop of initial hypothesis 2. They refer to the Government pursuing its "private agenda" and principal-agent problems if the Government is a structured organization.
[17] This is quite a realistic assumption. Except for the few cases where sanctions can be enforced in Courts (like Dutch covenants) the main sanctions regard the adoption of a stricter regulation, the loss of agreed benefits and reputation.
[18] A formal model of the conditions of participation and compliance to a VA is provided by Xepapadeas and Passa (2004).
[19] At the same time benchmarking agreements allow to keep the international framework into account and to make comparisons with competitors' performance.
[20] Williamson distinguishes between "ex ante" and "ex post" transaction costs. The first one refer to the contracting phase, the second ones to the post-contracting phase (Williamson, 1985).

11. REFERENCES

Akerlof G. (1970), The market for lemons; quality uncertainty and the market mechanism, *Quarterly journal of economics*, n.84.

Amadei P., Croci E. and Pesaro G. (1998), *Nuovi strumenti di politica ambientale, gli accordi volontari*, Franco Angeli, Milano.

Arora S. and Cason T.N. (1995), An experiment in voluntary environmental regulation: participation in EPA's 33/50 program, Journal *of environmental economics and management*, n.28.

Arora S. and Gangopadhyay S. (1995), Toward a theoretical model of voluntary overcompliance, *Journal of economic behavior & organization*, vol.28.

Arrow K.J. (1974), Limited knowledge and economic analysis, *American Economic Review*, n.64.

Ashford N.A. and Caldart C.C. (2004), Negotiated regulation, implementation and compliance in the United States, in *The handbook of environmental voluntar agreements*, (edited by Croci E.), Kluwer Academic Publishers, Dordrecht.

Barrett S. (1991), Environmental regulation for competitive advantage, *Business Strategy Review*, spring.

Brouhle K. Griffiths C. and Wolverton A. (2004), The use of voluntary approaches for environmental policymaking in the U.S., in The *handbook of environmental voluntary agreements*, (edited by Croci E.), Kluwer Academic Publishers, Dordrecht.

Bressers H. and de Bruijn T (2004), Environmental voluntary agreements in the Dutch context, in *The handbook of environmental voluntary agreements*, (edited by Croci E.), Kluwer Academic Publishers, Dordrecht.

Brau R. and Carraro C. (2001), Are VAs a threat to competition?, in (ed.) Highley C.J and Leveque F., *Environmental voluntary approaches: research insights for policy-makers*, FEEM report, Milano.

Carraro C. and Leveque F. (1999), The rationale and potential of voluntary approaches, in (ed.) Carraro C. and Leveque F., Voluntary *approaches in environmental policy*, Kluwer, Dordrecht.

Cavaliere A. (2000), Overcompliance and voluntary agreements, *Environmental and resource economics*, n.17.

CEC (1996), *Communication on environmental agreements*, COM(96) 561, Brussels (a)

CEC (1996), *Reccomendation of the Commission concerning environmental agreements implementing Community directives*, 96/773/EC, (b).

CEC (2002), *Communication on environmental agreements at the Community level within the action plan on the simplification of the regulatory environment*, Brussels.

Coase R.H. (1960), The problem of social cost, *Journal of law and economics*, n.3.

Croci E. (2003),Voluntary agreements for CO2 emissions reduction: evaluation and perspectives, *Energy & Environment*, vol.14, n.5.

Croci E. (1995), Managerial and competitive implications of voluntary and negotiated agreements, *Economia delle fonti di energia e dell'ambiente,* n.3.

Croci E. and Pesaro G. (1998), Voluntary agreements in the environmental sector – the Italian experience, *CAVA working paper*, CERNA, Paris.

Delmas M. and Terlaak A. (2001), The institutional context of voluntary environmental agreements, in Hoffman A. and Ventresca M. (editors), *Organizations, policy and the natural environment: institutional and strategic perspectives*, Stanford CA, Stanford University Press.

Enevoldsen M. (1998), Democracy and environmental agreements, in Glasbergen P. (edited by), *Co-operative environmental governance: public-private agreements as a policy strategy*, Kluwer Academic Publishers, Dordrecht.

Glasbergen P. (2000), Voluntary environmental agreements as institutional change, *CAVA working paper*, n. 2000/2/3, CERNA, Paris.

Gunningham N. and Rees R. (1997), Industry self-regulation: an institutional perspective, *Law and policy*, n.19(4).

Howarth R.B., Brent M.A. and Paton B. (2000), The economics of energy efficiency: insights from voluntary participation programs, *Energy policy*, n.28.

Khanna M. (2001), Non-mandatory approaches to environmental protection, *Journal of economic surveys*, vol.15, n.3.

Krarup S. and Rameshol S. (2001), *Voluntary agreements in energy policy – implementation and efficiency*. Final report from the VAIE project, AKF Forlaget, Copenhagen.

Lenox M.J. and Nash J (2003), Industry self-regulation and adverse selection: a comparison across four trade association programs, *Business strategy and the environment*, n.12.

Liefferink D. (1997), Joint environmental policy making: the emergence of new interactive approaches in environmental policy, in AA.VV., *The innovation of environmental policy, summer symposium acts, Bologna University*, Bologna, July 21st-25th.

Massarutto A. et alii (2004), *Public participation in river basin management planning in Italy*, working paper, www.harmonicop.org.

Meadowcroft J. (1998), Co-operative management regimes: a way forward?, in Glasbergen P. (edited by), *Co-operative environmental governance: public-private agreements as a policy strategy*, Kluwer Academic Publishers, Dordrecht.

OECD (2003), *Voluntary approaches in environmental policy: environmental effectiveness, economic efficiency and usage in policy mixes*, OECD publications, Paris.

OECD (1999), *Voluntary approaches for environmental policy*, Paris.

Paton B. (2000), Voluntary environmental initiatives and sustainable industry, *Business strategy and the environment*, n.9.

Pigou A.C. (1932), *The economics of welfare*, MacMillan, London.

Segerson K. (1999), Do voluntary approaches lead to efficient environmental protection, *CAVA working paper*, n. 99/10/10, CERNA, Paris.

Simon H.A. (1955), A behavioral model of rational choice, *Quarterly Journal of Economics*, vol. LXIX, February.

Simon H.A. (1979), Rational decision making in business organization, *American Economic Review*, vol.69, issue 4, September.

Stigler G.J., The economics of information, Journal of political economy, n.69, 1961.

Stigler G.J. (1971), The theory of economic regulation, *Bell Journal of economic and management science*, n.2 (1).

Stoeckl N. (2004), The private costs and benefits of environmental self-regulation: which firms have most to gain?, *Business strategy and the environment*, n. 13.

ten Brink P. (ed.) (2002), Voluntary *environmental agreements: process, practice and future use*, Greenleaf, Sheffield, UK.

United Nations (1992), *The Rio Declaration, Agenda 21*, Earth summit on sustainable development, Rio de Janeiro.

United Nations Economic Commission for Europe (1998), *Convention on access to information, public participation and access to justice in environmental matters*, Aarhus.

UNEP (2000), *Voluntary initiatives: current status, lessons learnt and next steps*, UNEP discussion paper, Paris.

Xepapadeas A. and Passa C (2004), Participation in and compliance with public voluntary environmental programs: an evolutionary approach, *FEEM working paper*, n. 67,Milan.

Williamson E.O. (1985), *The economic institutions of capitalism*, The Free Press, London.

CORPORATE SELF-REGULATION AND MULTI-STAKEHOLDER DIALOGUE

P. MAZURKIEWICZ*

World Bank, Washington D.C., USA

Abstract. This paper provides an overview of various forms of self-regulatory practices the private sector has been applying in recent decades. It also presents the main drivers stimulating business to take up self-regulatory measures, and vehicles for establishing better corporate environmental behavior, with special attention to multi-stakeholder dialogue.

1.INTRODUCTION

The decade since the Earth Summit in Rio de Janeiro has been characterized by overarching global economic, political and social change. This period has seen a major shift in the role of the state and increasing reliance on market forces with an emphasis on self-regulation (Jenkins, 2001).

Environmental protection, traditionally, has been considered to be "in the public interest" and generally external to private sector functions. As a result, governments have assumed principal responsibility for assuring sound environmental management, and have focused on creating and preserving a safe environment as a public good. They have attempted to direct the private sector and civil society to adopt environmentally sound behavior through regulations, sanctions, and, occasionally, through offering incentives. When environmental problems have arisen, the public sector has most frequently been responsible for mitigation. In this approach, unrestricted private sector behavior has typically been considered as presenting the "environmental problem". It has also been considered that, by respecting environmentally sustainable practices, the private sector incurred an uncompensated financial cost (Mazurkiewicz, 2003).

However, the roles of the public and private sectors have been changing, with the private sector and civil society becoming active partners in environmental protection projects that were previously the exclusive domain of the public sector. Many governments and businesses are now realizing that environmental protection and economic growth are not always in conflict. Moreover, the "regulate, enforce and

E. Croci (ed.), The Handbook of Environmental Voluntary Agreements, 31 - 45

mitigate" approach, which often involves long legal processes, is becoming increasingly costly to both the public and the private sector, and may eventually be ineffective. Often the private sector prefers voluntary initiatives to taxation owing to lower enforcement costs while the authorities do the same because of lower transaction and abatement costs.

Within this changing context one of the key institutional arrangements to emerge on the corporate agenda for promoting business responsibility and self-regulation has been voluntary activities (Utting, 2000). Evidence of this trend is seen in the increasing variety of corporate codes of conduct, certification schemes, auditing and reporting initiatives, voluntary environmental agreements, and social investment initiatives.

Voluntary environmental initiatives are increasingly being looked to as a possibly appropriate instrument to help address environmental problems covering a broad range of pollutants and natural resources. ten Brink (2002) observes that the motivations for voluntary environmental initiatives were not only to avoid arguably cumbersome and slow-to-develop legislation or costly taxes, but also reflected a number of other issues such as: a) the concept of shared responsibility, b) the concept of stakeholder involvement and c) the principle that a problem should be solved at the level that can most effectively address the problem.

Effecting changes in environmental practices through voluntary initiatives is of growing importance in the developed world. However, middle-income countries, where investment and private consumption represent significant shares of GDP, also have significant potential. An additional opportunity for such an approach to take place is the political shift happening in many middle-income countries, which are democratizing and fostering much more active engagement of civil society actors. These factors have provided channels and mechanisms whereby environmental issues can more easily reach decision-makers and influence economic and sectoral policies. However, a more sustainable situation could be created if business were to realize that it, too, could benefit from employing environmentally sustainable practices, as has been the case in many developed countries.

The challenge is: a) to make voluntary initiatives a regular feature of the business environment, and b) to improve public sector capacity.

Making the transformation from a strict "regulate and enforce" approach to a "facilitate and verify" situation would require clarification of policy and expectations of behavior, and the creation of supporting institutional arrangements. The proponents of voluntary initiatives generally consider them as a pragmatic and innovative way of enhancing the contribution of the private sector to sustainable development. Many, as mentioned in the section on drivers, also see voluntary initiatives as an alternative to government regulation. Others consider them rather as complements to regulations.

Moreover, while companies and their stakeholders are often attracted to the concept of corporate self-regulation there is uncertainty about what it means and how to initiate it. They might agree that building strong, mutually beneficial stakeholder relationships is a crucial point of departure, but few understand how to establish and maintain win-win associations and sustainable multi-stakeholder dialogue.

Such a dialogue between a multiplicity of interests and perspectives would appear to be crucial to lead to sustainable change in current practices. It allows participants to (OECD, 2001):

1. Tap the collective intelligence of all participants,
2. Evolve a new set of values and perspectives,
3. Understand each other's different experiences and backgrounds,
4. Generate innovative ideas and solutions,
5. Sort through ethical issues and areas of potential conflict,
6. Create a common language, set of assumptions and a collaborative process that works, and
7. Develop stronger, trusting relationships.

For various reasons - such as differences in the socio-political situation, level of economic development, existing legal framework, presence of multinational companies, and strength of civil society sector - the evolution of the dialogue and its participants have specific country characteristics. However, there is a list of common steps that might be used in establishing a dialogue on voluntary initiatives between all interested stakeholders.

2. MULTI-STAKEHOLDER DIALOGUE IN VOLUNTARY INITIATIVES

2.1. Voluntary Initiatives

A review of the literature indicates that voluntary initiatives are now considered one of three general categories of policy instruments, along with traditional regulatory and market-based instruments (OECD, 1997; OECD, 1999). The most commonly cited typology used to distinguish types of voluntary approaches is based on the parties involved in determining the commitments (Acutt, 2002). This common typology usually distinguishes three categories of approaches: unilateral initiatives made by business community (self-regulatory initiatives/codes of conduct), multi-stakeholder agreements (local/national-scale agreements, sectoral agreements, multi-stakeholder agreements), and programs initiated by the authorities or other stakeholders (third-party initiatives) and supported on voluntary basis by the private sectors.

1) Unilateral initiatives

Since the 1980s, there has been a considerable shift in thinking with regard to how to improve the social and environmental performance of companies (UNRISD, 2002). An earlier emphasis on strict governmental regulations has ceded ground to corporate codes of conduct and other voluntary initiatives. Faced with the apparent impossibility of generating a single set of commonly accepted standards, backed by legal sanctions, governments, companies and civil society organizations (CSOs) initiated discussions aimed at developing and supporting non-binding codes of corporate conduct. Basically, there are two main types of codes of conduct, namely collective ones, and those developed and endorsed by a single company.

Additionally, one might also recognize those of international character, and those specific for a particular country. For example, broadly inclusive business organizations such as the International Chamber of Commerce and the International Organization of Employers have been able to define joint positions on social and environmental responsibility and introduce at the international level a code of conduct for their members. Sectoral organizations representing certain industries are also active in this field, for instance, Responsible Care, the code of conduct originated by the Canadian Chemical Producers Association in the aftermath of the disastrous gas leak in Bhopal, India, and adopted by over 40 chemical associations worldwide. Many individual companies adopt their own codes of conduct that address environmental responsibility, sometimes drawing on an industry code or international business principles (UNCTAD, 1999).

The content of existing codes varies widely in purpose, coverage, specificity, implementation and monitoring mechanisms. Companies adopt them for a variety of reasons, ranging from personal interest and beliefs of a CEO to explicit expectations voiced by important governmental or other public interest groups. Many stakeholders, however, consider codes of conduct as another public relations product without transparent monitoring and evaluation mechanisms, that companies use only to improve their image and to avoid potential public criticism. As a response, companies and their stakeholders have started developing new forms of voluntary initiatives in recent years, such as negotiated voluntary initiatives, which attempt to overcome some of the limitations of simple unilateral initiatives.

2) Multi-stakeholder agreements

Multi-stakeholder agreements can be developed between various sectors, such as corporate and government sectors, corporate and non-profit sectors, government and non-profit sectors or even between all three sectors.

Voluntary environmental agreements (VAs) represent a specific kind of multi-stakeholder agreements. VAs are a type of contract between an individual company and/or association of companies and the public authorities, concluded with the aim to protect or restore the environment (CEC, 1996). It is a peculiar kind of contract. In fact, with a few exceptions, voluntary agreements are not legally binding, since they cannot be enforced by the courts of justice (Croci, 2003). Similar to other voluntary initiatives, VAs are being looked to as a possibly appropriate instrument to help address environmental problems. Advocates of this approach claim that VAs are very flexible in nature, can be tailored to the particular circumstances for each stakeholder, can build on business's particular knowledge of its capacity to address environmental concerns, have a participatory character, are potentially cost-effective, and can compensate for limited existing regulatory capacity. Opponents see VAs as a potentially exclusionary instrument to the extent that not all affected parties may be involved in negotiations; it may lead to static efficiency gains if targets are not set appropriately and are not revisable; it needs monitoring of compliance, which can be resource-intensive; it can lead to reduced quality of regulatory control; and it does not raise revenue that could be valuable for other initiatives (ten Brink, 2003).

In 1995 in Germany, the Bavarian State Government and the Bavarian business community entered into the Environmental Pact of Bavaria as a voluntary agreement aimed at greater protection of the environment. Both parties set the deadline of the year 2000 for additional environmental achievements in various sectors of industry, crafts and trade through enforcement of 180 specific commitments. The Pact was the country's first reciprocal cross-sector and multi-discipline agreement between a state government and a business in the environmental sector.[1] In 1994, the New Zealand Government announced that in addition to other policies relating to energy efficiency and energy sector reform, it would seek voluntary agreements with industry to reduce CO_2 emissions. In 1996, 17 voluntary agreements had been signed. The agreements included undertakings with the New Zealand steel and aluminium industries. The agreements were signed by the companies and the Minister of Energy representing the New Zealand Government. Targets were specified as savings achieved or planned over the period 1990-2000 to coincide with the Government set national stabilization objective. Within this framework the actual texts of the agreements vary widely reflecting different company and sector processes and technologies, opportunities for achieving CO_2 savings, views about the agreements themselves and the relationship of CO_2 savings activities to the company objectives. The agreements were specifically not legally binding, and avoid penalty for under achievement (OECD, 1997b).

Public-Private Partnerships (PPPs) for sustainable development can be also considered as another example of multi-stakeholder agreement. PPPs are a form of co-operation in which government and private companies assume co–responsibility and co–ownership for the delivery of negotiated services. Through these partnerships, the advantages of the private sector – dynamism, access to finance, knowledge of technologies, managerial efficiency, entrepreneurial spirit – are combined with social responsibility, environmental awareness, local knowledge and job generation concerns of the public sector. PPPs for sustainable development between companies and the government are a relatively new phenomena, but are a growing feature of both developed and emerging economies. Working together, partners seek to meet the objectives of each, while performing better than if acting alone. In this case, partnering institutionalizes collaborative arrangements in which the differences between the sectors become blurred. In this approach – which is slightly different than in VAs where business is mainly responsible for implementation of agreements -- partners seek not to shift responsibility and risk from one party to another, but to share risks, pool resources and skills, and deliver mutual benefits for each party. For instance, the Indian city of Tiruppur in the State of Tamil Nadu provides an interesting approach to public - private partnership. Tiruppur is India's largest exporter of cotton knitwear. From 1985 to 1996 the value of those exports grew by 200 percent. But since 1990 it was clear to the business and city leaders alike that water supply and sewerage was becoming a constraint on both manufacturing growth and the ability of the city to adequately serve its citizens. As a result of talks between the city leaders and the Tiruppur Exporters Association, a water and sewerage development plan was drawn up with the city and the exporters. The State of Tamil Nadu did not have adequate resources to finance and implement the $200 million plan, nor was it possible for the Municipality, with its small annual budget and limited management capacity. So, the city and its corporate partners decided not to rely on government funding to implement their plan. Instead, in 1995 they created the New Tiruppur Area Development Corporation Limited as a public limited company owned by government, the exporters, and a private financial services company. This innovative venture of both public and private institutions have joined together to finance and implement an urban environmental infrastructure project[2].

3) Public schemes and third-party initiatives
In this category governments, NGOs and multilateral and other organizations encourage companies to participate in schemes that set social and environmental standards, monitor compliance, promote social and environmental reporting and auditing, certify good practices, and encourage multi-stakeholder dialogue. Usually this approach includes a monitoring and reporting scheme that assures information dissemination internally and externally to compensate for the lack of legal obligation and maintain a sufficient level of peer pressure.

The United States employs a broad portfolio of public schemes/programs in the industrial sector. These schemes encourage a specific action by industry based on some agreed upon criteria. Examples include Green Lights, Climate Wise, Motor Challenge, and the Voluntary Aluminum Industry Partnerships (VAIP). The actions are typically to implement cost-effective technologies from some well defined set of feasible technologies. These programs offer a variety of support mechanisms such as public recognition incentives, coupled with education and training, information systems and database support. Climate Wise-Energy Star is a joint program run by the Environmental Protection Agency (EPA) and the Department of Energy (DOE). The program is based on the principle that energy efficiency and pollution prevention are the cornerstones of any strategy to reduce greenhouse gas emissions worldwide. The program is founded on the notion that government can spur innovation by establishing performance-based goals and letting individual organizations determine how best to contribute to meeting those goals. Participants in Climate Wise are given recognition, technical assistance, and financial assistance. Climate Wise companies undertake specific actions that they identify (such as process changes, fuel switching, and new product designs). Participants are also encouraged to participate in various end-use specific greenhouse gas related programs, e.g., Green Lights or Motor Challenge. The EPA Voluntary Aluminum Industry Partnership Program was introduced to engage the aluminium industry in voluntary reductions of PFCs. The program has two elements. The first is the voluntary commitments, the second is an information collection and measurement program. The program aims to accelerate some of the replacement of equipment and practices that directly impact anode effects (OECD, 1997b).

The Better Banana Project represents an interesting case of engagement in a third–party initiative. For instance, Chiquita relies on the strict standards of Rainforest Alliance's Better Banana Project (BBP)[3] to gauge its environmental performance and identify priority areas for improvement. In 2001, it earned re-certification of all owned farms and improved its average BBP audit scores in five of seven divisions, with significant gains in Peru, Guatemala and Honduras. The new Rainforest Alliance scoring system goes beyond basic threshold requirements and challenges farms to focus attention on particular problems and continuously improve their performance. The Rainforest Alliance certifies farms that follow environmentally and socially responsible agricultural practices. Through audits conducted by conservation groups in the Sustainable Agriculture Network, the Rainforest Alliance measures performance against the following nine BBP standards: ecosystem conservation, wildlife conservation, fair treatment and good working conditions, community relations, integrated crop management, integrated waste management, conservation of water resources, soil conservation, and environmental planning and monitoring. Additionally, farms must demonstrate: no evidence of "fatal flaws" or flagrant environmental or social problems; an active program of improvement in all areas; a plan that schedules needed improvements; and record keeping and monitoring systems that can document management practices, changes and impacts. Achieving certification is a real accomplishment, but not an end in itself. Certified farms must commit to continuous improvement. The Rainforest Alliance continually revises the standards to include new technologies and methods, and all certified farms undergo surprise audits annually.

2.2. Promoting dialogue

More and more companies and their stakeholders are attracted to voluntary environmental initiatives, but are often uncertain as to what steps may create an adequate environment for putting the concept into operation. (Mazurkiewicz, 2003). Three such steps could facilitate this process: (i) promote dialogue among stakeholders; (ii) create the actual partnerships necessary for bringing voluntary initiatives to fruition; and (iii) agree on a systematic and monitorable program for establishing and financing voluntary initiative.

It is stakeholders who know best how they affect or are affected by the environmental issues. Moreover, as affected parties stakeholders should have some say on matters concerning their welfare.

Different players have different perceptions, interests and goals for voluntary initiatives. A multi-stakeholder dialogue has proven useful in beginning to establish confidence that a win-win situation can be established and sustained. In practical terms, the first step would be to determine the stakeholders. Bendell observes that this process involves a pictorial representation of primary and secondary stakeholders that depicts which groups influence or are influenced and what their respective interests are. (Bendell, 2000). Further dialogue can assist in a) identifying areas of priority for various stakeholders, b) revealing areas requiring further attention, c) enabling new models of partnership and decision-making that encompass various priorities, and d) highlighting issues and problems.

Ideally, all relevant stakeholders should participate in the development of voluntary environmental initiatives, and during their life-cycle. Their participation could lead to more effective implementation, since multi-stakeholder-led consultations have the potential to ensure realistic targets and the setting up of a reporting and monitoring scheme (ten Brink, 2002). Lack of involvement of various stakeholders can be the result of their not being invited by the primary drivers to the initial conceptualizing phase, or sometimes also the result of the reluctance of stakeholders to become involved, given participation costs and skepticism regarding the benefits. Ideally, development of voluntary environmental initiatives should have clearly defined consultation/participation benchmarks that incorporate stakeholders' inputs into the building phase of the initiative. This should also enhance involvement of stakeholders in implementation part, and push towards transparent monitoring.

Grayson and Hodges (Grayson and Hodges, 2002) stress that building sustainable relationships among these stakeholders requires investment of time and resources. It also requires a genuine willingness on the part of government, business and civil society to listen and learn from their contacts with each other. Dialogue should be sustained in order to facilitate the growth of discovery and understanding, and eventually establish commitment among stakeholders to common goals.

Dialogue would hopefully develop collective intelligence, and a shared set of ideas. Finally, a creative dialogue would encourage participants to suspend their attachment to particular viewpoints so that deeper levels of listening, synthesis and meaning can evolve (Svensen, 1998). Multi-stakeholder dialogue, therefore, can constitute a base for trust-building and further joint initiatives.

As a constructive dialogue matures, it should pass through stages reflecting increased engagement. Bendell specifies eight such levels moving from a state of "manipulation", in which a dominant stakeholder uses its position to direct particular outcomes, to democratization, in which stakeholders share the decision-making (Bendell, 2000). The OECD (OECD, 2001b) cites three key concepts that span the same range:

- *Information giving* as a one-way relationship in which stakeholders produce and deliver information for use by other stakeholders. It covers both passive access to information upon demand, and active measures to disseminate information.
- *Consultation* as a "limited" two-way relationship in which stakeholders provide their feedback. This type of dialogue involves the accessing of stakeholders' opinions through a variety of techniques, such as attitude surveys or meetings.
- *Active participation/partnership* as a relationship based on partnership, in which stakeholders actively engage in defining the process and content of policy-making. It also means that all stakeholders share planning and decision-making responsibilities.

The multi-stakeholder approach is a relatively new phenomena both in OECD countries and middle-income countries. At policy-making level the European Multi-Stakeholder Forum on Corporate Social Responsibility, chaired by the European Commission, brings together European representative organizations of employers, business networks, trade unions and NGOs, and fosters a dialogue between them, to promote innovation, convergence and transparency in existing CSR practices and tools. Examples of the multi-stakeholder dialogue approach in Japan, where local communities and authorities directly negotiated with local polluting industries can be found (ten Brink, 2002); while Acutt (2002) presents examples from South Africa such as Saldanha Steel Environment Monitoring Committee or Island View Forum.

3. DRIVERS OF VOLUNTARY INITIATIVES

Since the World Commission on Environment and Development Report of 1997 (the Brundtland Report) was published, corporate managers and management scholars have been grappling with the question of how and why corporations should incorporate environmental concerns into their own strategic decision-making. And they have been casting environmental protection as a positive action, rather than as a problem. Today many companies have accepted their responsibility to do no harm to the environment (Hart, 2000). The Environment Strategy of the World Bank indicates, too, that the private sector is becoming a decisive factor in influencing environmental performance and long-term environmental sustainability (World Bank, 2002).

Nowadays many citizens, environmental organizations and companies define corporate environmental responsibility as the duty to cover the environmental implications of the company's operations, products and facilities; eliminate waste and emissions; maximize the efficiency and productivity of its resources; and minimize practices that might adversely affect the enjoyment of the country's resources by future generations. In the emerging global economy, where the Internet, the news media and the information revolution shine light on business practices around the world, companies are more and more frequently judged on the basis of their environmental stewardship. Partners in business and consumers want to know what is inside a company. They want to do business with companies they can trust and believe in. This transparency of business practices means that for many companies, environmental responsibility is no longer a luxury, but a requirement. However, the challenge ahead is to create a commonly respected system of monitoring and evaluation that would allow a detailed assessment of business practices in the field of environmental responsibility.

Voluntary environmental initiatives are often considered as "tools" for operationalizing the concept of environmental responsibility. A growing number of companies in a wide range of sectors and geographic regions have discovered concrete value and competitive advantages from taking voluntary environmental initiatives, for example, in areas such as pollution prevention, energy efficiency, environmentally oriented design, supply-chain management and industrial ecology. Companies have found that environmental responsibility has often had a positive impact on corporate profits. Of all the topics related to corporate voluntary initiatives, it is environmental initiatives that have produced, so far, the greatest amount of quantifiable data linking proactive companies with positive financial results. Business for Social Responsibility

(BSR), for example, emphasizes that investment in voluntary environmental initiatives has promoted product differentiation at the product and firm levels. Some firms now produce goods and services with attributes or characteristics that signal to the consumer that this particular company is concerned about certain social and environmental issues. The International Financial Corporation, in its report "Developing Value" (IFC, 2002), concludes , based on the experiences of more than 170 companies, that many businesses in emerging markets have been involved in areas such as social development or environmental improvements and have achieved cost savings, revenue growth and other business benefits.

In doing so companies have established an environmentally responsible corporate image which has have facilitated market penetration. Firms have also found savings in input costs, waste disposal costs, labor costs through reduced absenteeism and increased loyalty, reduced costs of compliance with regulations, and other real but more intangible benefits such as attracting quality investors. Firms also benefit from realizing greater cooperation from their communities, and from building political capital that has been useful when community decisions may affect the enterprise.

Reducing the use of energy and raw materials and limiting emissions and waste from production processes are key contributions business can make to tackle the environmental challenges facing the world. The good practices of leading companies build a base for the behavioral change of others. Moreover, ever more frequently many multinationals adopt environmental policies that extend through their supply chains in the form of requirements for suppliers to adhere to sustainability certifications such as ISO 14001, SA 8000 or FSC[4], etc. (IFC, 2002).

However, in many middle-income countries business leaders see risks in expanding their environmental initiatives in the form of higher costs, reduced profitability, and growing regulatory interference. Results of a survey focused on the perception of the private sector towards corporate social and environmental responsibility, conducted by the World Bank in four Central and East European countries, show that the factor most frequently identified as a barrier to expanding the adoption of CSR[5] practices was the lack of regulatory frameworks and the private cost of making changes. Many business leaders also identified the lack of visible results and of linkages between actions and the firm's financial success as constraints to broader adoption of social and environmental initiatives. Related to these, leaders also cited the absence of adequate institutional arrangements to manage CSR expansion and appropriate government leadership as constraints. They did not, surprisingly, consider short-term profit motives or the reluctance of management and labor leaders as being significant barriers. Leaders felt that actions to provide incentives (tax incentives specifically identified), modernized labor laws, and, to some degree, vehicles for government-business-civil society dialogue would be helpful.

Table 1. *Drivers for voluntary environmental initiatives*

Economic drivers	*Social drivers*	*Political drivers*
- company image/reputation - improved risk management - competitive advantage - pressure from business partners - pressure from costumers - pressure from investors	- pressure from NGO/CSOs -pressure from local communities	- improved standing with government - legal, regulatory drivers - political pressure - license to operate

4. VEHICLES FOR INTRODUCING VOLUNTARY INITIATIVES

4.1. Public sector

Governments have a strong interest in promoting voluntary initiatives as a complement to their ongoing environmental programs to serve long-term national interests. They, often with the support of international institutions, or international and local NGOs, are beginning to play a significant role in building a framework for corporate environmental initiatives through a managed, goal-driven approach. For example, the authorities prefer voluntary agreements because they involve both lower transaction and abatement costs (Grepperud, 2003). Moreover, governments may see their interests in achieving improved environmental management in a less conflictive manner, at lower cost and with more impact on job creation, while improving competitive positions in respect to trade, and ultimately making economic and social gains.

Assistance from governments can be planned and programmed as a component in a national environmental program. Usually, governments would plan a three-part approach to the problem: (i) inform, sensitize and engage business in dialogue and negotiations concerning voluntary initiatives, and institutionalize this process; (ii) offer incentives for and assistance to firms seeking to adopt more environmentally responsible business models; and (iii) re-enforce monitoring of environmental conditions and enforce sanctions. More precisely, they can stimulate the private sector by providing funding for research or by leading campaigns, collecting and disseminating information, sponsoring training, and raising awareness. Public bodies can develop or support appropriate management tools and mechanisms, including environmental agreements, voluntary product labeling schemes, benchmarks, and guidelines for company management and reporting systems. They can also create incentives by applying their public procurement and investment leverage. The other crucial role the public sector can play is partnering in environmental initiatives (World Bank, 2002).

4.2. Corporate Internal Approach

The implementation of environmental initiatives usually differs for each company, or even sector, depending on a number of factors such as size and culture. Manufacturing-based companies are confronted by a wide range of environmental challenges, while retail or service-sector companies face these to a lesser extent. Although some companies address environmental issues one facility or department at a time, companies are increasingly integrating the environment into all parts of their operations. Whatever the nature of the commitment, most companies follow a similar series of steps when addressing their impact on the environment:

1. Corporate Environmental Policy: Companies committed to reducing their environmental impact usually create a set of environmental principles and standards, often including formal goals. At a minimum, most such statements express a company's intentions to respect the environment in the design, production and distribution of its products and services; to be in full compliance with all laws and go beyond compliance whenever possible; and to establish an open-book policy whereby employees, community members and others can be informed of any potentially adverse effects the company might have on the environment.
2. Environmental Audit: Before a company attempts to reduce its impact on the environment, it is essential that it first gains a full understanding of it. For most companies, this usually involves some kind of environmental audit. The goal of audits is to understand the type and amount of resources used by a company, product line or facility, and the types of waste and emissions generated. Some companies also try to quantify this data in monetary terms to understand the bottom-line impact. This also helps to set priorities as to how a company can get the greatest return on its efforts.
3. Employee Involvement: Leadership companies recognize that to be effective, an environmental policy needs to be embraced by employees throughout the organization, not just those whose work is related to the environment. To do that, companies engage in a variety of activities, especially education, to help employees understand the environmental impact of their jobs and to support their efforts to make positive changes. Some companies go further by helping employees become more environmentally responsible in all aspects of their lives. Besides education, many companies create incentives, rewards and recognition programs for employees who demonstrate their environmental commitment.
4. Green Procurement: To help ensure that their products and processes are environmentally responsible, many companies seek to buy greener products and materials from their suppliers. Some companies participate in buyers' groups in which they leverage their collective buying clout to push suppliers to consider alternative products or processes.

5. Green Products: Products themselves may be made more environmentally friendly with regard to, for example, the control of emissions, noise, reduced health and safety risks, and reduced energy requirements.

5. CONCLUSIONS

More and more often one can observe a trend in discussion on corporate self-regulation: voluntary measures can help improve private-sector behavior, but voluntary activity is no substitute for regulation.

There are some companies that will only take environmental responsibility on board if they have to. This is why there is still a need for a regulatory framework – a minimum required by law, which cannot be replaced by voluntary initiatives, which should go beyond existing regulations. Voluntary initiatives might constitute a testing field for a new standards that in the future could be included in a legal framework. Such a sequence stimulates innovation and competition.

Multi-stakeholder dialogue seems to be a vital component of self-regulatory process. It helps enhance and sustain operational effectiveness of voluntary environmental initiatives by bringing to light the perspective of stakeholders and providing an enabling social and political environment for exchanging views. It can help in ensuring that the initiative: a) is properly designed, b) its goals are shared by all stakeholders, and c) its implementation is successful. In other words multi-stakeholder dialogue significantly increases the likelihood of initiative's sustainability.

6. NOTES

* This paper is not a publication of the World Bank. The views expressed are solely those of the author and his views and this paper should not be attributed to the World Bank.

[1] See http://www.umweltministerium.bayern.de/agenda/umw_pakt/pakt_en.pdf and http://www.umweltministerium.bayern.de/agenda/umw_pakt/inhalt/engl.htm

[2] See Enterprise for the Environment, Public - Private Partnerships For Environmental Improvement In Asia: The Role Of The U.S. Agency For International Development , http://www.csis.org/e4e/Mayor51Painter.html

[3] The Rainforest Alliance a leading international conservation organization. Its mission is to protect ecosystems and the people and wildlife that live within them by implementing better business practices for biodiversity conservation and sustainability. Companies, cooperatives, and landowners that participate in its programs meet rigorous standards for protecting the environment, wildlife, workers, and local communities.

[4] Standard for Certification of Good Forest Management

[5] Corporate Social Responsibility, or CSR, is the commitment of business to contribute to sustainable economic development, working with employees, their families, the local community and society at large to improve their quality of life in ways that are both good for business and good for sustainable development.

7. REFERENCES

Accut N. (2002), *Voluntary Environmental Initiatives: Case Studies in the South Durban Petrochemical Industry*, Oikos PhD Summer Academy, University of St. Gallen

Bendell J. (2000), *Talking the Change? Reflection on Effective Stakeholder Dialogue*, Academy for Business Information Network

BSR, *Business for Social Responsibility*; www.bsr.org

CEC - Commission of European Communities (1996), *Communication from the Commission to the Council and the European Parliament on Environmental Agreements*, COM(96), 561 def, Brussels

Croci E. (2003) Voluntary Agreements for CO2 Emissions Reduction: Evaluation and Perspectives, *Energy and Environment*, Volume 14, Number 5, Multi-Science Publishing Co. LTD., Brentwood

Grayson D. and A. Hodges (2002); *Everybody's Business*, DK, London

Grepperud S. (2003), and P. Pedersen, Voluntary Environmental Agreements: Taking Up Position and Meeting Pressure, *Economic and Politics*, Volume 15, Issue 3, November

Hart S. (2000), *Beyond Greening in Harvard Business Review on Business and the Environment*, HBD Press, Boston

IFC (2002), Sustainability, Ethose Institute; *Developing Value. The Business Case for Sustainability in Emerging Markets*, Washington D.C.

Jenkins R. (2001), *Corporate Codes of Conduct: Self Regulation in Global Economy*. United Nations Research Institute for Social Development, Paper no. 2, April, Geneva

Kennedy N. (1996), *Voluntary Environmental Initiatives and Environmental Policy: Environmental Management Systems, Auditing, and Enforcement*, in Proceeding Paper from the 4th International Conference on Environmental Enforcement, INECE

Mazurkiewicz P. and L. Grenna (2003), *Corporate Social Responsibility and Multi-stakeholder Dialogue – Towards Environmental Behavioral Change*, World Bank , DevComm-SDO, Discussion Paper, May 2003, Washington D.C.

OECD (1997), *Evaluating Economic Instruments for Environmental Policy*, OECD Publications, Paris

OECD (1997b), *Voluntary Agreements within Industry: Annex I Expert Group on the UN Framework Convention on Climate Change*, Working Paper No. 8, Paris

OECD (1999), *Voluntary approaches for environmental policy: An Assessment*, OECD Publications, Paris

OECD (2001), *Citizens as Partners, Information and Public Participation in Policy Making*, Paris

Svensen A. (1998) , *The Stakeholder Strategy, Profiting form Collaborative Business Relations*, Berrett-Koehler Publishers, INC., San Francisco

Ten Brink P. (2002), *Voluntary Environmental Agreements: Process, Practice and Future Use; Institute for European Environmental Policy*, Greenleaf Publishing

UNCTAD (1999), *World Investment Report - Foreign Direct Investment and the Challenge of Development*, New York and Geneva

UNSRID (2002), *Regulating Business via Multi-stakeholder Initiatives: A Preliminary Assessment in Voluntary Approaches to Corporate Responsibility: Readings and Resource Guide*; NGLS, Geneva

Utting P. (2000), *Business Responsibility for Sustainable Development.* UNRISD Occasional Paper no. 2, January 2000, Geneva

World Bank (2002), *Public Sector Roles in Strengthening Corporate Social Responsibility: Baseline Study*, Washington D.C.

CHAPTER 2

THE POLTICAL ECONOMY OF VOLUNTARY AGREEMENTS

VOLUNTARY AGREEMENTS IN A RENT-SEEKING ENVIRONMENT

M. GLACHANT

CERNA, Ecole des Mines de Paris, France

Abstract. The paper analyses whether voluntary agreements with polluters (VAs) are able to achieve an efficient level of environmental protection when they are obtained under the legislative threat of an alternative stricter policy option. We develop a model in which the threat is a pollution quota. The threat is the outcome of a rent-seeking contest between a green and a polluter lobby group influencing the legislature. We show that a VA systematically emerges in equilibrium and that it leads to a more efficient level of pollution abatement than the legislative pollution quota. However this level is lower than the first best level of environmental protection. The paper also discusses various VA design aspects.

1. INTRODUCTION

In the field of environmental policy, the major policy innovation of the nineties is probably the introduction of voluntary agreements (VA). While they were marginal practices in a limited number of countries beforehand (e.g., in Germany, Japan), they are now used almost everywhere. One illustration of this very fast and widespread development is the first generation of climate change policies adopted in OECD countries around the mid-nineties. They mostly relied on voluntary agreements. Japan set the so-called Keidaren voluntary Action Plan covering 37 industry branches and eighty percent of industrial energy consumption. In the US, the Clinton's Administration 1993 Climate Change Action Program was mainly based on voluntary programs including Green Lights, Climate Wise among many others. In the European Union, almost all Member States launched their own voluntary approaches under various names: branch agreements, covenants, environmental agreements, etc.

Although these approaches differ in certain respects, one common feature is that polluters *voluntarily* commit to undertake pollution abatement activities. The term "voluntary" has long been disputed since many agreements are in fact obtained under the threat of an alternative coercive public intervention. The present chapter deals with these voluntary agreements which are obtained under a legislative threat.

The efficiency of the environmental target embodied in VAs is a major practical concern. Many observers suspect that VAs bring very little environmental improvements beyond the Business-As-Usual trend (e.g., see the recent review of OECD, 2003). The suspicion is due to three features of VAs. First, they are voluntary suggesting that the polluters see them as a cheap solution including little

E. Croci (ed.), The Handbook of Environmental Voluntary Agreements, 49 - 65

abatement efforts. However, we have seen that, for certain VAs, the relevant benchmark is not a 'do-nothing' scenario but a legislative threat. In this context, firms does not enter in a VA because the scheme is cheap but because it is cheaper than the alternative legislative option. The second source of suspicion is that VAs are generally non-binding. Therefore, the regulator lacks enforcement tool in case of non-compliance. Third, many VAs are collective in the sense that they gather a coalition of firms, typically represented by the sector association. This creates free riding concerns which can undermine the cooperation between polluters, and in turn can damage the environmental performance.

The chapter only develops a theoretical analysis of the first argument and assumes that VAs are perfectly enforced and that free riding problems have been solved by the polluters. We develop a model of voluntary agreement in which the threat is a pollution quota and is fully endogenous. More specifically, we model the legislative process whose threat is the outcome. We make the hypothesis that rent seeking affects the making of legislation. Note that this assumption is a necessary condition for the existence of a VA. Otherwise, the regulator would be able to implement the first best legislation and would have absolutely no reason to use a VA. In this politically constrained world, the regulator must choose between two evils: either negotiating a VA with the necessity to reduce the environmental strictness of the VA relative to the first best to obtain the consent of the polluter, or implementing a politically distorted pollution quota.

We investigate whether the VA is likely to lead to a more efficient level of pollution abatement than the level that might have been imposed legislatively. We also investigate various design issues – the opportunity to involve the pollution victims in the negotiation and the efficiency potential of veto rights by the Congress over VAs.

We establish that VAs are always more efficient than the legislation in the politically constrained world depicted by the model. The result is very strong in that it holds true when the world is quasi perfect – when the political constraints are very lax. The underlying intuition is that the polluter is ready to accept a VA that is stricter than the legislative quota. And he does so because the legislative route requires him to make rent seeking efforts. In this context, the model establish that the VA benefit in terms of avoided rent-seeking costs is sufficiently high to make him accept a VA stricter than the legislative quota.

The paper is structured as follows. Section 2 presents the model. In section 3, we establish that the VA is systematically more efficient than the legislative abatement. Section 4 makes two simple extensions of the model to address design issues debated in policy circles: the efficiency potential of involving a green group in VA and the interest of granting a veto right to the Congress over newly adopted VA. Section 5 concludes.

1.1. Economic theoretical literature on VAs

As they have only been developed recently and by practitioners, the theoretical literature on VAs is still limited but it is growing rapidly. Some of these papers deal with the case, similar to ours, where the motivation of the polluters to accept voluntary agreements is the pre-emption of future regulations. Hansen (1999) has developed a political economy model in which polluters negotiate with a regulator under a background legislative threat. The key feature of his approach is that the regulator's objective is biased and differs from that of the threat-making entity, the Congress. A model by Maxwell, Lyon and Hackett (2000) considers firms that voluntarily abate pollution to pre-empt lobbying by consumers in favour of environmental policy. In this case, firms do no pre-empt an explicit threat by a regulator but a risk of new legislation possibly triggered by consumer lobbying. In another paper Segerson and Miceli (1999) develop a normative model in which the polluter undertakes voluntary action under the threat made by a benevolent regulator to implement a pollution quota whose adoption is uncertain. In all these papers, political constraints hinder the alternative, legislative, route to the VA, like in ours. In Hansen's paper, the constraint is modelled as a bias of the policy objective of the regulator. In Segerson and Miceli (1999), the constraint is reflected in an exogenous probability of adoption p of the legislation. In the more recent paper by Lyon & Maxwell (2003) the same probability depends on the cost borne by the polluter to meet the legislative requirement, suggesting that the polluters lobby against the legislation. But they do not make explicit the mechanism through which the polluter's cost affects the probability of adoption. Accordingly, these papers yield the same type of ambiguous results: depending on the stringency of the political constraint, the VA is more or less efficient than the legislation.

In these papers, the political constraint is modelled as an exogenous constraint. In particular rent-seeking or lobbying is not explicitly modelled and the polluter incurs no lobbying cost. By contrast, we model explicitly lobbying or rent seeking, even roughly, and show that the endogenous rent-seeking costs are sufficiently high so that the polluter is willing to accept efficient VAs. However, we do not think that the model as such constitutes the proof of the superiority of VAs on legislation. It rather suggests the necessity to make more complex the analysis. In this regard, the fact VAs are non-binding and free riding issues should be taken into account in future works.

2. THE MODEL

The model depicts a situation in which a benevolent environmental regulator R and a polluter P agree to make a voluntary agreement. The VA specifies a pollution abatement level, denoted B, to be met by the polluter. The model is sufficiently general for the polluter to be either a single firm or an industry. In practice, certain VAs are signed with a coalition of polluters, usually represented by an industrial branch association. In that case, the model implicitly assumes that the members of the coalition have solved the free riding problem associated with collective action.

Before going further in the presentation of the model, it is worth making an important remark about the diversity of VAs encountered in reality. The OECD distinguishes three broad categories (1999). Each type ultimately differs with respect to the degree of involvement of the regulator. Under *public voluntary programs*, the firms agree to make abatement efforts to meet goals which are established by the regulator. In a *negotiated agreement*, the polluter and the regulator jointly devise the commitments through bargaining. Under *self-regulation* or unilateral agreements, the polluter takes the initiative. He freely sets up a program of environmental actions without any formal influence from the regulator.

In our model, the agreement is the outcome of a bargaining process between the polluter and the regulator. So, strictly speaking, it is a negotiated agreement. However, our results apply to public voluntary programs. The reason is that public voluntary programs and negotiated agreements share a key feature which ultimately drives the results: the polluter and the regulator's participation constraints must be satisfied in both cases. On the contrary, self-regulation is not a possible application of the model, the main reason being probably that unilateral commitments are usually not triggered by legislative threats.

Consider now the costs and benefits associated with the polluter's commitment to meeting an abatement level B. The abatement cost born by the polluter is described by an increasing and convex function $C(B)$. It also generates a benefit in terms of avoided environmental damage. For the sake of simplicity, we assume that the benefit equals the abatement level B. The linearity of the benefit function simplifies the analysis without altering any of the results. We further assume that $C'(0) < 1$. and $C(0)= 0$. These hypotheses imply that, for low values of B, gross welfare, denoted function $W(B) = B - C(B)$ is positive. Therefore, in the absence of political constraints, the environmental regulator selects the optimal policy level, B^*, defined by the condition:

$$C'(B^*)\equiv 1$$

The agreement is obtained under the threat of an alternative policy. More specifically, we assume that the regulator is the agenda setter of the Congress. He can thus threaten the polluter with a new legislation. The threat consists in a pollution quota that prescribes a minimal level of abatement L for the polluter. We do not assume any cost advantage for the VA: the polluter has the same cost function under the VA and under the legislative quota. Doing otherwise would make it too easy to reach conclusions about the superiority of voluntary agreements.

The threat is uncertain.[1] This is a crucial feature of this type of models: if the benevolent regulator was able to pass any new legislation with certainty, he would be able to implement the first best policy B^* through the Congress and would have no reason to use a VA instead. Let π be the probability of adoption of the legislative quota. It cannot be an exogenous parameter. It certainly depends on the contents of the Law proposal. For instance, for a given level of environmental benefit, a more costly threat has lower chance of being adopted. One central reason for that is that the potential losers of the policy are trying to influence the legislative process through lobbying, media campaigns, etc.

To account for that, we model the legislative process as follows. We suppose that the proposal of legislation is subjected to a rent-seeking contest involving two lobby groups as popularised by the rent-seeking literature. A first group G (the "greens") is concerned by the policy benefit L associated with the pollution quota whereas the second group is simply the polluter P who bears the policy cost $C(L)$. Group G and the polluter P make rent-seeking expenditures in order to influence the Congress' voting process. Expenditures may be campaign contributions (monetary or in kind), or may correspond to the cost of transmitting strategic information to the "median" legislator on the consequences of the Law proposal. Denote x_G and x_P, the green group's and polluter's rent-seeking expenditures, respectively. These expenditures affect the probability of adoption π via a so-called *contest success function*. Such functions are routinely used in the rent-seeking literature to model lobbying in noisy political environments.[2] As to the functional form, we use a variant of the standard unit logit function pioneered by Tullock (1980):

$$\pi(x_G, x_P) \equiv \begin{cases} \pi^\circ + (1-\pi^\circ)\dfrac{\lambda x_G}{\lambda x_G + x_P}, & \text{if} \quad x_G + x_P > 0 \\ 1, & \text{if} \quad x_G + x_P = 0 \end{cases} \tag{1}$$

where λ is a parameter introducing a heterogeneity in lobby groups' influence technology. It is a routine assumption in the rent seeking literature. When λ lies in between 0 and 1, the green group is less influential than the polluters whereas the contrary holds true beyond 1. π° is a parameter reflecting the responsiveness to lobbying of the Congress. It prevents the probability π to fall below π°. Put differently, whatever the intensity of lobbying, any welfare-improving policy is adopted *at least* with a probability π°. This is a less classical assumption aiming at introducing some concern for the general interest in Congress' behaviour.

To summarize the model, figure 1 describes the VA policy game.

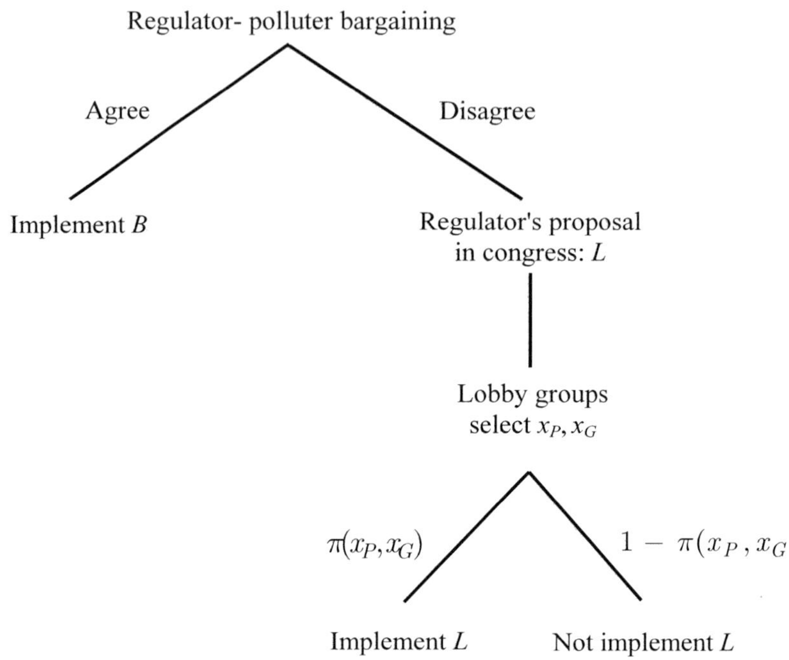

Figure 1. *The decision tree of the VA policy game*

3. RESOLUTION

In this section, we solve the model reasoning backward. We start with the analysis of the rent-seeking sub-game.

3.1. The rent-seeking stage

Consider any Law proposal involving an abatement quota L. What will be its probability of adoption? According to Eq.(1), it is determined by the rent-seeking expenditures of the two groups. Each simultaneously and non co-operatively selects its level of expenditures by maximizing its expected utility, taking the other's level of expenditures as given. The corresponding maximization problem is thus:

$$\max_{x_G} \quad \pi(x_G, x_P) \cdot L - x_G$$

$$\max_{x_P} \quad -\pi(x_G, x_P) \cdot C(L) - x_P$$

The solution to this problem is very classical in the rent-seeking literature (see Nitzan, 1994, for instance) and we will go very fast here. In fact, there are no corner solutions and the equilibrium rent-seeking expenditures is given by the first order conditions:

$$(1-\pi^{\circ}) \cdot L \frac{\lambda x_P}{(\lambda x_G + x_P)^2} = 1$$

$$(1-\pi^{\circ}) \cdot C(L) \frac{\lambda x_G}{(\lambda x_G + x_P)^2} = 1$$

Algebraic manipulations of these two conditions then lead to the following levels of expenditures:

$$x_G(L) = \frac{(1-\pi^{\circ})}{\lambda} \cdot \frac{L^2 \cdot C(L)}{(L + C(L))^2} \tag{2}$$

$$x_P(L) = \frac{(1-\pi^{\circ})}{\lambda} \cdot \frac{L \cdot C(L)^2}{(L + C(L))^2} \tag{3}$$

Finally, plugging these expenditures in Eq.(1) yields the equilibrium probability of adoption of the rent-seeking game:

$$\pi(L) = \pi^{\circ} + (1-\pi^{\circ}) \frac{\lambda L}{\lambda L + C(L)} \tag{4}$$

3.2. The agenda-setting stage

Having characterized the equilibrium probability $\pi(L)$, we identify now the legislative quota that will be proposed to the Congress. The regulator takes into account the fact that adoption is uncertain; he makes a Law proposal that maximizes expected gross welfare:

$$\max_{B} \quad \pi(L) \cdot [L - C(L)]. \tag{5}$$

Note that rent-seeking expenditures are not an argument in the welfare function. The reason is that we consider that such expenditures are transfers between lobby groups and others (legislators, lawyers, experts, etc.). Another possible hypothesis is to consider rent-seeking as a wasteful activity, which leads to include the corresponding expenditures in the social welfare function. We consider this alternative assumption below in section 3.4. We will see that it does not change the results. We now come back to the maximization program (5). Its first order condition implicitly defines the abatement level under legislation:

$$\pi(L)\cdot[1-C'(L)] = -\pi'(L)\cdot[L-C(L)]. \tag{6}$$

We then have a very simple lemma which establishes that this level is lower than the first best abatement level.

Lemma 1. *The equilibrium regulatory policy under legislation is strictly lower than the first best policy:* $L < B^*$.
Proof. First we show that π' is negative for all λ, π° and L. Differentiating (4) yields

$$\pi'(L) = (1-\pi^{\circ})\lambda[C(L) - L.C'(L)]/(\lambda L + C(L))^2.$$

Furthermore $C(L)$-$LC'(L) < 0$ because of the convexity of the cost function. It follows that the right-hand side of Eq.(6) is negative. Hence $C'(L) < 1$, or alternatively $C'(L) < C'(B^*)$. It implies $L < B^*$. □

This lemma states that the first best policy is not attainable under the legislative route. The intuition is simple. The existence of political constraints lowers the probability of adoption. To mitigate the problem, the environmental regulator needs to make a law proposal departing from the first best optimum. This proposal is lower than B* because of the negative sign of the marginal probability. It is ultimately rooted in the fact that increasing L leads to larger losses in marginal terms than benefits due to the convexity of the cost function. It then provides the polluter with more incentives to increase rent-seeking expenditures.

3.3. The bargaining stage

Note that L, the equilibrium policy under legislation, corresponds to the disagreement point of the bargaining game, which we consider now. In this game, polluter and regulator's payoffs are the differences between their expected utility under legislation and their utility under the VA:

$$U_P(B) \equiv \pi(L).C(L) + x_P(L) - C(B) \tag{7}$$

$$U_R(B) \equiv [B - C(B)] - \pi(L).(L - C(L)) \tag{8}$$

Looking at these payoffs, it is immediately obvious that any feasible agreement is more efficient than legislation since it satisfies the participation constraint of the welfare-maximizing regulator. The following result establishes the existence of a unique Nash bargaining solution of the game.

Proposition 1. *Let* $\Omega = \{B : U_P(B) \geq 0$ and $U_R(B) \geq 0\}$. *There exists a unique Nash bargaining solution that solves the following maximization problem*

$$\max_{B\in\Omega} \quad \Pi(B) \equiv (W(B) - \pi(L).W(L))\left(\pi(L).C(L) + x^P(L) - C(B)\right)$$

Proof. First we establish that Ω is not empty. It is convenient to denote B_P the maximal level the polluter is willing to accept, which is defined by $U_P(B_P)= \pi(L)C(L) + x_P(L) - C(B_P) = 0$. Also, denote B_P the minimal level that the regulator is ready to accept. It is implicitly defined by $U_R(B_R) \equiv W(B_R) - \pi(L)\cdot W(L) = 0$. We have $W(B_R) = \pi(L)W(L) < W(\pi(L)L)$ since $W''(L)=-C''(L)< 0$. Then $W(B_R) <W(\pi(L)L$ implies $B_R < \pi(L)L$ since W is strictly increasing below B^*. And so $C(B_R) < C(\pi(L)L) < \pi(L)C(L)$ since C' and C'' are strictly positive. Hence, $C(B_R)+x_P(L) < \pi(L)C(L)+x_P(L)=C(B_P)$. Since $x_P(L)>0$, we finally obtain that $C(B_R)< C(B_P)$ and thus $B_R< B_P$. Hence, $\Omega =[B_R, B_P] \neq \varnothing$.

The second step of the proof is to show that the Nash product is strictly concave. This is straightforward since the second derivative of the Nash product, $\Pi''(B) = W''(B)U_P(B) - 2C'(B)W'(B) - C''(B)\, U_R(B)$, is strictly negative.

Finally, let h be the function describing the utility the regulator obtains for a given utility level of the polluter u_P. The last step of the proof consists in establishing that h is strictly decreasing and concave. From the strict monotonicity of U_P, there exists a unique abatement level $\hat{B} \in \Omega$ such that $U_P(\hat{B}) = u_P$; i.e., $\hat{B} = U_P^{-1}(u_P)$, where U_P^{-1} denotes the inverse utility of the polluter. We can easily get $\hat{B} = U_P^{-1}(u_P) = C^{-1}[\pi(L)C(L)+x_P(L) - u_P]$. Using this expression, the utility the regulator obtains when the polluter obtains u_P is then given by:

$$h(u_P) \equiv U_R(\hat{B}) = U_R\left(U_P^{-1}(u_P)\right) = C^{-1}(K - u_P) + u_P - \pi(L)L - x_P(L),$$

where $K = \pi(L)C(L)+x_P(L)$. Having characterized h, we can now study the sign of its first and second derivatives. We have:

$$\frac{dh(u_P)}{du_P} = 1 - C^{-1}{}'(K - u_P) = 1 - \frac{1}{C'\left(C^{-1}(K - u_P)\right)} = 1 - \frac{1}{C'(\hat{B})}.$$

$$\frac{dh^2(u_P)}{du_P{}^2} = \frac{2}{C'(\hat{B})^2.U_P{}'(\hat{B})} = -\frac{2}{C'(\hat{B})^3}.$$

As $\widehat{B} < L < B^*$, it follows $C'(\widehat{B}) < 1$ and thus $dh/du_P < 0$. Hence h is strictly decreasing. The second derivative is obviously negative and h is therefore strictly concave. □

Proposition 2 is the key result of this paper. It establishes the existence of a VA that is more efficient than the legislative (regulatory) option. The result is very robust in that it does not depend on the stringency of the political constraints, as reflected by the values of λ and π^o. In particular, it still holds true when the Congress is very weakly responsive to lobbies' pressure ($\pi^o \to 1$) or when the polluter is much less efficient than the green group in influencing the Congress ($\lambda \to +\infty$). However, the VA cannot yield the first best optimum as stated by this simple corollary.

Corollary 1. *Let $\hat{B}$ be the abatement level corresponding to the Nash bargaining solution. We have $\hat{B} < B^*$.*

Proof. Obvious since $\hat{B} \leq B_P < L < B^*$.□

The result of proposition 2 is very strong in favor of VAs since it still holds true in a quasi perfect world where $\pi^o \to 1$ and $\lambda \to +\infty$. As stated in the introduction, it is clearly counter-intuitive since the majority of the observers agree that, in practice, VAs are frequently poorly environmentally effective (see for instance OECD, 2003). Therefore, it is essential to cautiously discuss the underlying intuition. The key point in the proof is that the maximal abatement level the polluter is willing to accept, B_P, is lower than the minimal level of abatement for the regulator B_R. The key reason for this is the fact that signing a VA provides a specific benefit for the polluter which lowers his reservation level: avoiding the rent-seeking expenditure $x_P(L)$.

3.4. A variant: rent-seeking is a wasteful activity

So far we have assumed that rent-seeking expenditures do not enter in the welfare function. We now relax that assumption. Reasoning backward, a first remark is that this does not change the equilibrium probability of the rent-seeking sub-game since the game only involves the polluter and the green lobby group. But, at the agenda-setting stage, the regulator now maximizes:

$$\max_{B} \quad \pi(L)\cdot[L - C(L)] - x_P(L) - x_G(L)$$

We are then able to show that the equilibrium abatement quota is lower than the first best quota like in the previous case.

Lemma 2. *When rent seeking is considered as a wasteful activity, the equilibrium legislative quota is still lower than the first best quota: $L < B^*$.*

Proof. The first order condition of the welfare maximization program is $\pi(L)(1 - C'(L)) = -\pi'(L)[L - C(L)] + x_G'(L) + x_P'(L))$. Differentiating (2) and (3) yields:

$$x_G'(L) + x_P'(L) = \frac{1-\pi^\circ}{\lambda} \cdot \left[\frac{L^3.C'(L)+L.C^2(L)+L^2.C(L).C'(L)+C^3(L)}{(L+C(L))^3} \right]$$

which is positive. As $\pi' < 0$, then the right hand side of the FOC is positive. It implies $1 - C'(L) > 0$ and thus $L < B^*$. □

In fact, the legislative abatement quota is even lower that the one proposed to the Congress when excluding rent-seeking expenditures from the welfare function. This is so because the sum of rent-seeking expenditures is increasing with abatement ($x_G'(L) + x_P'(L) > 0$) as established in the proof of lemma 4. It provides the regulator with an additional incentive to lower B relative to the previous case.

Consider now the bargaining stage. In comparison with the previous version, the change in the assumption only affects the regulator's payoff function which is now:

$$U_R(B) \equiv [B - C(B)] - \pi(L) \cdot (L - C(L)) - x_G(L) - x_P(L) \tag{9}$$

Looking at this function immediately suggests that the situation is even more favorable to the emergence of a VA than the previous case since the legislative option now entails an additional cost, $x_G(L) + x_P(L)$, to the regulator. A simple comparison of (9) and (7) establishes that the minimal level the regulator can accept, B_R, is lower than in the previous case. In other respects, the maximal abatement for the polluter B_P remains unchanged. In the end, $B_P > B_R$ still holds meaning that the bargaining set is never restricted to the disagreement point. This is the key element of the proof of a new proposition 5. The rest of the proof is left in appendix.

Proposition 2. *When rent seeking is considered as a wasteful activity, a VA systematically emerges in equilibrium and yields a level of abatement always more efficient than the legislative quota.*
Proof. See in appendix. □

4. DESIGN ISSUES

Having shown that the VA systematically dominates regulation in a second best world where legislative action is constrained by lobby groups' influence, we use the model to analyse two design issues that arise in the policy debate on VAs: the efficiency potential of involving environmental associations in the negotiation, and the interest of an ex post veto right of the Congress over any new VA.[3]

4.1. Associating the green group to the VA

A frequent criticism is that VAs exclude the pollution victims from the negotiation. In this respect, they diverge from classical Coasean bargaining in that not all affected parties are around the table. The Coase theorem then suggests that it would improve welfare to include them in the process. The involvement of green associations in the negotiation of VAs is a recurrent policy recommendation even though it rarely happens in practice (OECD, 2000). Does our model plead for such a recommendation? To answer the question, it is necessary to compare the bargaining outcome of the traditional 2-party VA analysed in the previous section with that of the 3-party VA game involving a green group representing the victims. The payoff to the green group in the bargaining game is

$$U_G(B) \equiv B - \pi(L)\cdot L - x_G(L)$$

No simple equilibrium concept is available for 3-player bargaining games without side payments. To bypass the problem, we assume that bargaining only takes place between the two lobby groups. The environmental regulator only influences the outcome through his participation constraint which still needs to hold. Hence, he has no bargaining power and plays a role of arbitrator (or facilitator) of the negotiation. This hypothesis about the allocation of bargaining power actually corresponds to that of a Coasean negotiation. With this assumption the maximization problem of the three-player game is:

$$\max_{B\in\Gamma} \Pi_{ter}(B) \equiv U_P(B)\cdot U_G(B)$$

where $\Gamma = \{B : U_G(B) \geq 0\} \cap \Omega$.

The following result establishes the existence of the bargaining outcome of the three-player game and states that it is more efficient than the traditional bilateral VA.

Proposition 3. *There exists a unique abatement level, denoted* $\tilde{B}$, *which is the outcome of the 3-party VA. Furthermore, this outcome is always closer to the first level that the bilateral VA, that is* $\hat{B} < \tilde{B} < B^*$ *for any* λ *and* π^o. *Hence associating the green lobby group to the VA negotiation is welfare improving.*

Proof. Establishing the existence and uniqueness of $\tilde{B}$ follows closely the proof of Proposition 1 and is thus left in appendix. As regards the second part of the proposition, consider the Nash product of the 3-party VA game:

$$\Pi_{ter}(B) = \Pi(B) + [\pi(L)\cdot C(L) + x_P(L) - C(B)]^2$$

Its first derivative is

$$\Pi_{ter}'(B) = \Pi'(B) - 2((\pi(L)\cdot C(L) + x_P(L)) - C(B))C'(B).$$

As $\hat{B}$ is the maximum of $\Pi(B)$, then $\Pi'(\hat{B})=0$. Hence we have $\Pi_{ter}'(\hat{B}) = -2((\pi(L)\cdot C(L) + x_P(L)) - C(\hat{B}'))C'(\hat{B}')$, which is strictly negative. It implies $\hat{B} < \tilde{B} < B^*$.□

The involvement of green groups in VAs is a relevant policy recommendation because, in the case of a simple VA excluding the green group, the participants are the polluter – who only cares about abatement costs - and the regulator – who is concerned with both costs and benefits. In this setting, the cost is taken into account twice in bargainers' payoffs while the benefit is only counted once. This is reflected in the bargaining outcome which places more weight on the cost side. Involving the greens - who are only concerned with the benefit - suppresses this distortion since costs and benefits are both taken into account twice in participants' payoffs.

The intuition behind proposition 6 may be used for discussing a related design aspect. In practice, the government often delegates the negotiation of VAs to specialized environmental agencies (e.g., the EPA in the US) or Ministries of the Environment. In comparison with the ideal benevolent regulator of the basic model, it is reasonable to assume that these entities are biased in favour of the environment. Proposition 6 suggests that such a policy delegation and the bias it introduces in the objective of the bargaining regulator in fact promote the efficiency of the VA. Put differently, an inefficient regulator leads to more efficient VA outcomes. The reason is the same as the one justifying the involvement of the greens in VAs. A pro-environment regulator pays more attention to the benefit than to cost, resulting in a more efficient bargaining outcome. There exist instances where VA are in fact delegated to the ministries or agencies in charge of industrial or economic affairs. The Dutch CO2 Long Term Agreements is a possible example. Our model suggests that it is not the best institutional option.

4.2. Granting a veto right to the Congress

A further design question refers to the interest of granting a veto right to the Congress over every new VA. Belgium or the Netherlands are countries which have already adopted this rule. The underlying rationale is to compensate for the lack of democratic legitimacy of the VA process as compared to traditional legislative action. Is it justified on economic efficiency grounds? In our setting, it adds a further (veto) stage to the sequential game. At this final stage, we must assume that, like any proposal made in the Congress, the adoption of the VA is the subject of a further rent-seeking contest between the two lobby groups.

Basically, there is no difference with the rent-seeking sub-game analyzed in section 3.1. The probability that the VA is definitively adopted is therefore equal to $\pi(B)$ given by Eq.(4) and the corresponding rent-seeking expenditures are $x_P(B)$ and $x_G(B)$ for the polluter and the green group, respectively given by Eq.(2) and Eq.(3). Moving on to the bargaining sub-game, the bargainers' payoffs are now:

$$U_P(B) = \pi(L) \cdot C(L) + x^P(L) - \pi(B) \cdot C(B) - x^P(B)$$

$$U_R(B) = \pi(B) \cdot W(B) - \pi(L) \cdot W(L)$$

We then have the following proposition:

Proposition 4. *When the Congress enjoys a veto right, no VA emerges in equilibrium.*

Proof. We keep using B_R to denote the minimal level the regulator is ready to accept with $\pi(L)W(L) = \pi(B_R)W(B_R)$. The left-hand side and the right-hand side of this equation have the same functional form. Furthermore $\pi(.)W(.)$ is strictly monotonic below L. Hence $B_P = L$.

Consider now B_P defined by $\pi(L)C(L) + x_P(L) = \pi(B_P)C(B_P) + x_P(B_P)$. The same argument would apply to establish that $B_P = L$ if $\pi(.)C(.) + x_P(.)$ is monotonic. To show that this is the case, consider the Nash product $\Pi(B) = U_P(B)U_R(B)$. If there exists a VA, the corresponding equilibrium abatement level $\hat{B}$ satisfies $\Pi'(\hat{B}) = U_P'(\hat{B})U_R(\hat{B}) + U_P(\hat{B})U_R'(\hat{B}) = 0$. As $\hat{B}$ is below L, we have $U_R'(\hat{B}) > 0$ and thus $U_P'(\hat{B}) = -(\pi(\hat{B})C(\hat{B}) + x_P(\hat{B}))' < 0$. Therefore, $\pi(.)C(.) + x_P(.)$ is monotonic and $B_P = L$. In the end, $B_P = B_R$ and the bargaining set is restricted to the disagreement point. □

Therefore, introducing a veto right damages social welfare by preventing VAs from emerging in equilibrium. Intuitively, this is so because, in the absence of veto right, the gains for both sides are ultimately rooted in bypassing the legislative route. Offering a veto on the result of the negotiation *de facto* re-introduces the legislative option in the VA route. As a result, the interest for making a VA vanishes for both parties.

5. CONCLUSION

We have developed a model of voluntary agreement under legislative threat wherein the regulator sets the threat while its probability of adoption is the outcome of a rent-seeking contest between the polluter and a green group influencing the legislature. The model establishes that, in this setting, the VA systematically achieves a more efficient level of environmental protection than the pollution quota that might be imposed legislatively. However this level is lower than the first best outcome. This non-ambiguous result is ultimately driven by the fact that the polluter is willing to accept a sufficiently strict VA. The reason is related to the political constraints under the legislative route. Rent seeking in Congress entails rent seeking costs to the polluter that he avoids when making a VA.

We also use the model to analyse a set of design issues that are frequently discussed in the policy arena. First, it is shown that involving a green group in the negotiation of the VA improves welfare. The underlying intuition is that this (partly) compensates for the bias in favour of pollution abatement cost attached to the participation of the polluter in the VA decision process. A second extension of the model assesses the relevancy of granting a veto right to the Congress over each new VA as done in certain countries (e.g., Belgium, the Netherlands). The model demonstrates that this prevents the emergence of any welfare-improving VA.

All in all, these results are quite favourable to voluntary agreements in comparison with the traditional Command and Control approach. However, we do not believe that they establish the superiority of Vas over legislative quota. These results should simply invite us to relax certain oversimplifying assumptions. For instance, the polluters perfectly comply with their commitments in the model whereas, in reality, many commitments are in fact non-binding. Further work should thus include imperfect enforcement of the VA commitments. Furthermore, the model does not address free riding issues which can hinder the emergence of VAs when, as it is frequent in practice, they involve a group of polluters. A further limit providing the opportunity for future work is that we only consider a threat consisting in an abatement quota. It would be interesting to consider more efficient policy options, such as a pollution tax or an emission-trading program.

6. NOTES

[1] The fact that passing a Law is uncertain is definitively supported by evidence. During the last legislative term in France (1997-2002), the Government made 476 Law proposals out of which 351 were finally adopted by the Parliament, corresponding to an average probability of adoption of 0.74. In the US, a paper by Zeckhauser (1981) gives many examples in the field of environmental policy.

[2] Nitzan (1994) is a comprehensive survey of the rent-seeking literature using such contest success functions.

[3] A report by OECD makes a comprehensive review of these policy issues (1999).

7. REFERENCES

Hansen L.G. (1999), "Environmental regulation through voluntary agreements" in *Voluntary Approaches in Environmental Policy*, Carlo Carraro and François Lévêque (eds.), Kluwer Academic Press.

Lyon T., Maxwell J.W. (2003), "Self-regulation, taxation and public voluntary environmental agreements", *Journal of Public Economics*, 87(7-8); pp 1453-86.

Maxwell J.W., T.P. Lyon, S.C. Hackett (2000), "Self regulation and social welfare: the political economy of corporate environmentalism", *Journal of Law and Economics*, 43, pp. 583-618.

Nitzan, S. (1994) "Modeling rent-seeking contests", *European Journal of Political Economy*, 10, pp 41-60

OECD (1999) *Voluntary approaches for Environmental Policy*, OECD, Paris

Segerson, K. and Miceli, T.J. (1999) "Voluntary environmental agreements: good or bad news for environmental protection?", *Journal of Environmental Economics and Management*, 36(2), pp. 109-30

Tullock G. (1980) "Efficient rent seeking", in *Towards and Theory of the Rent Seeking Society*, eds J.M. Buchanan, R. Tollison, and G. Tullock, Texas A&M Press, pp 269-282

Zeckhauser, R. (1981) " Preferred policies when there is a concern for probability of adoption", *Journal of Environmental Economics and Management*, 8(3), Sept., pp. 215-37

8. APPENDIX

Proof of Proposition 1bis

We have already established that the bargaining set is not restricted to the disagreement point. Now we need to show that the Nash product, denoted $\Pi°$, is strictly concave. The second derivative of $\Pi°$ is equal to $\Pi°''(B) = -C''(B)\,[U_P(B) + U_R(B)] - [C'(B)\,(1 - C'(B)]^2$, which is obviously strictly negative. Finally, let f be the function describing the utility the regulator obtains for a given utility level of the polluter u_P. The last step of the proof consists in establishing that f is strictly decreasing and concave. This is immediate since this function is the same as the similar function h of the previous version except a constant term $f(u_P) = h(u_P) - x_P(L) - x_G(L)$. Hence f is strictly decreasing and strictly concave just like h.□

Proof of Proposition 2

First we show that Γ is non-empty. Let B_G denote the abatement level corresponding to the green group's participation constraint, that is $U_G(B_G) \equiv B_G - \pi(L)\cdot L - x_G(L) = 0$. We use the notation as in proposition 1 for the abatement levels corresponding to the polluter's and regulator's participation constraints, B_P and B_R respectively. Proposition 1 has already established that $B_R<B_P$. Therefore, for Γ to be non-empty only requires that $B_G<B_P$, i.e., both polluter's and green group's participation constraints hold. From $U_G(B_G)=0$ follows that $x_P(L) = C(L)(B_G - \pi(L)L)/L$ since $x_G(L) = L\cdot x_G(L)/C(L)$. Plugging $x_G(L)$ in $U_P(B_P) = \pi(L)C(L) + x_P(L) - C(B_P) = 0$, we obtain that $C(B_P) = (C(L)/L)\cdot B_G$ and thus $C(L)/L = (C(B_P)/B_P)\cdot(B_P/B_G)$. From $B_P < L$ and $C''> 0$, it follows that $C(L)/L < C(B_P)/B_P$. Hence $B_G<B_P$. Γ is thus non-empty.

Second it is straightforward to show that the Nash product is strictly concave: $\Pi''(B) = -C''(B)U_G(B) - 2C'(B) <0$. Last, we need to establish the existence and uniqueness of the equilibrium. If g denotes the green group' utility, we have to show that it is a strictly decreasing and concave function of the polluter's utility u_P. We have:

$$g(u_P) \equiv U_G\left(U_P^{-1}(u_P)\right) = C^{-1}(K - u_P) - \pi(L)L - x_G(L),$$

where $K = \pi(L)C(L) + x_P(L)$. The first and second derivatives are respectively:

$$dg/du_P = C^{-1}{}'(K - u_P) = -1/C'(L)$$

and $$\frac{dg^2}{du_P^2} = -2 / C'(L)^3 ,$$

which are both strictly negative. Therefore the Nash bargaining solution exists and is unique.

ASPECTS OF THE POLITICAL ECONOMY OF ENVIRONMENTAL VOLUNTARY AGREEMENTS

A Meta Study

L.G. HANSEN

AKF (Institute of Local Government Studies), Copenhagen, Denmark

Abstract. Environmental voluntary agreements with industries are becoming a popular alternative to traditional regulation. One reason may be that such agreements increase implementation cost efficiency. On the other hand, models of the political economy of environmental voluntary agreements point out that efficiency reducing agreements are also possible under certain conditions.

In this paper we interpret empirical evidence from case studies of environmental voluntary agreements using one such policy formulation and implementation model. When our sample is interpreted in this light the data suggests that environmental voluntary agreements may often be chosen in order to shift the responsibility for implementation to industrial organizations that are less sensitive to criticism from powerful environmental interest groups. When this explanation of an environmental voluntary agreement applies, the model predicts that the agreement will be less cost effective and achieve lower environmental performance than the traditional regulatory alternative which would otherwise have been adopted.

Although our findings are not conclusive nor necessarily representative they do suggest the worrying possibility that many of the environmental voluntary agreements being established today achieve lower environmental performance less cost effectively than the most likely traditional regulatory alternative.

1. INTRODUCTION

Evaluation of policy instrument efficiency is traditionally done by comparing different policy instruments for achieving a given set of policy goals with respect to first and foremost static efficiency, but also dynamic and administrative efficiency. Such evaluations essentially assume a dichotomy between the political process of goal setting and that of policy instrument choice. Given this dichotomy the policy instrument evaluation is robust to imperfections in the political process. Achieving given goals in a least cost way is always welfare improving so that for the problem of instrument choice the instrument evaluation retains its relevance even though policy goals may be set imperfectly.

For voluntary agreements with industrial organizations, on the other hand, macro policy goals are subject to negotiations between authorities and the regulated industry and so the assumption of dichotomy must be rejected. When the assumption of dichotomy is rejected, a meaningful comparison of voluntary agreements with traditional approaches must also include an evaluation of the relative 'efficiency' of the

E. Croci (ed.), The Handbook of Environmental Voluntary Agreements, 67 - 89

underlying political goal setting processes. Policy process efficiency is a new dimension that broadens the usual notion of policy efficiency and should ideally take into account changes in policy process imperfections (e.g. agency capture) as well as transaction costs of the policy formulation process.

In this paper we focus solely on the political economy of voluntary agreements with industrial organizations. The goal of the paper is to confront the policy formulation and implementation model originally presented in Hansen (1999) with empirical evidence collected from a number of case studies of voluntary agreements through a meta study.

In the next section characteristics distinguishing policy formulation through voluntary agreements from the traditional policy formulation process are discussed and in sections 3 to 5 the theoretical model is presented and predictions from the model deduced. This model is a revised version of Hansen (1999) focussing on the particular specification of the model that is relevant for the selection of voluntary agreements studied here. In section 6 the meta study is described and the theoretical predictions confronted with the meta study results. Finally, in section 7 conclusions are drawn.

2. CHARACTERISTICS DISTINGUISHING POLICY FORMULATION THROUGH VOLUNTARY AGREEMENT FROM THE TRADITIONAL POLICY FORMULATION PROCESS

In the following I describe the traditional policy process and the voluntary agreement process based on empirical surveys (e.g. IEA (1995), Glachant (1994)) and point to some key differences.

The traditional process consists of legislation on regulatory instruments where implementation and administration of these instruments are delegated to a regulatory agency. While energy and environmental policy goals may be contested and subject to negotiation, the real battle is over legislation on regulatory instruments. This usually requires a legislative process with direct participation of the executive branch of Government (hereafter just called Government) and the legislative bodies of Government (hereafter called Congress). Affected industrial organizations and other interest groups may indirectly influence the process through quiet lobbying or by participating in the public debate on policy.

Voluntary agreements (VAs) normally only have Government agencies and individual firms or industrial organizations (IOs) as direct participants. Normally, agreements do not result in legislation. The IOs commit to targets and monitoring procedures, but not necessarily to specific instruments or methods of implementation. Normally, no formal sanctions for non-attention of targets are specified.

One apparent difference between the two policy formulation processes is that Congress participates directly in the traditional process, but is excluded from direct participation in the voluntary process. Instead IOs are elevated to a role as direct participants. Clearly, environmental interest groups (EGs) and Congress may still indirectly influence the voluntary policy process.

When considering voluntary agreements with industrial organizations another novelty is that implementation of environmental goals or agreed-on instruments is left to the industrial organizations rather than to public agencies. This can be seen as a necessary consequence of not involving Congress directly since implementation by traditional regulatory instruments through Government agencies would normally require passing of legislation. Though regulatory agencies may still have a monitoring role, Governments must contract with industrial organizations for implementation. Thus the responsibility for and practical implementation of regulatory instruments are shifted to industrial organizations. The reward to IOs for implementing environmental targets is usually implicit in the agreement. One possibility is that Government promises not to push for traditional regulation if targets are met. Though IOs may be able to implement effective regulatory instruments vis-à-vis member firms, the issue of credibility of Government threats/promises and IO - compliance with the negotiated targets is relevant.

Just as implementation through IOs is a consequence of excluding Congress from direct participation so is the voluntary agreement process what makes implementation through IOs possible. If IOs are to take responsibility for implementation, Government must of course negotiate an agreement with them.

In conclusion voluntary agreements can be seen as a policy process with three central characteristics distinguishing it from the traditional policy process:

(i) Statutory sanctions ensuring IO participation and compliance are not possible under voluntary agreements. Instead IOs must be induced to comply through e.g. threats of new regulation in the area covered by the agreement. The question of what government can credibly threaten to do arises.

(ii) Congress is no longer a direct participant in the policy formulation process - instead IOs become direct participants.

(iii) IOs share responsibility for setting goals, and responsibility for implementation of regulation is shifted to industrial organizations.

It seems that a credible threat (i) supporting a voluntary agreement would be the expected result if the traditional policy process was undertaken. We might then expect that VAs must result in situations where both parties of the agreement (Government and the IO) expect to be at least as well off as under the traditional policy formulation process - otherwise they would not have an incentive to enter into the agreement. Since we cannot expect a VA to achieve more than what the regulator otherwise would have been able to achieve through a traditional regulatory process, one might ask why we see so many VAs.

One explanation might be that shifting implementation responsibility to IOs increases efficiency (i.e. reduces the cost of achieving a given environmental goal) thus giving room for both parties to the agreement to become better off. Let us call this *the efficiency explanation*. However, other explanations may also be possible.

The less direct influence of Congress on the VA-process (ii) may in itself be a reason for entering into an agreement. If Government and opposition parties disagree on policy priorities and the traditional policy process necessitates compromises with opposition parties, it may be that compromising with the IO through a VA can get Government a better deal. Let us call this *the policy disagreement explanation*.

The shift of responsibility for goal setting and goal attainment to IOs (iii) may also be a driving force for undertaking VAs. If governments are sensitive to interest group criticism of policy goals and their possible non-attainment, it may be advantageous to shift responsibility to IOs. Let us call this *the responsibility shifting explanation.*

The first explanation of why VAs are made, also implies that they are welfare increasing. The last two explanations, however, open for the possibility that VAs may be entered into for other reasons than increased efficiency and thus might be welfare reducing. This makes it interesting to develop a theory of the VAs and to confront such a theory with empirical evidence.

3. A SIMPLE MODEL

The model has four active agents: the IO representing polluting firms, the environmental interest group, the Government and the Congress. Government may initiate the traditional policy formulation process through Congress or enter a voluntary agreement process with the IO. We assume that a voluntary agreement blocks initiation of the traditional policy formulation process. When negotiating the agreement Government may try to induce the IO to accept terms that are more favourable to Government by threatening to push for traditional regulation in the event that no agreement made. We assume that Government threats of pushing for regulation, which does not maximise the utility of Government, are not credible. In other words the only credible threat that Government can make for the situation where no agreement is made is to maximize Government utility in connection with the traditional policy process.

Initially we give a fairly detailed presentation of a simple model that captures the shift of direct influence away from Congress and towards the IO caused by the VA process (ii). Then we extend the model by introducing interest group criticism thereby making shifting of responsibility (iii) potentially advantageous.

Both policy formulation processes result in the setting of an environmental goal denoted R (indicating amount by which environmental damage is to be reduced) and a tax revenue goal T (indicating the amount of revenue to be collected through regulatory instruments). Implementation of these goals through the available regulatory instruments (taxes, direct regulation or a combination) results in firm compliance costs denoted C in addition to the tax revenue payment.

Let U_f denote the utility effect on firms of regulation and define:

$$U_f = -T - C$$

and assume that the firm IO's utility is equal to the effect of regulation on firms' profits. We assume that the environmental interest groups' utility is equal to regulation effect on environmental damage (i.e. R). Government and Congress are both assumed to take into account the utility effects of regulation on firms, the environmental interest group and the part of the public that might benefit from increased tax revenues. However, they may differ in the relative weights attached to these groups in their respective utility function. Let U_c and U_g denote the utility effects of regulation on Congress and Government respectively and define:

$$U_c = U_f + \lambda_c T + \delta_c R$$

$$U_g = U_f + \lambda_g T + \delta_g R$$

where λ_c and λ_g are the utility weights attached to tax revenue by Congress and Government respectively, δ_c and δ_g are the utility weights attached to environmental damage reduction. The utility function can be interpreted as the first order approximation of the actual utility function and includes all key variables affected by regulation.

The negotiation process between Government and Congress under the traditional policy formulation process is not modelled explicitly. Instead the utility function of Congress should be interpreted as representing the result of this process incorporating the relative power of Government and opposition parties in Congress. If Government's utility function parameters are equal to the parameters of Congress' utility function, this implies agreement between Government and opposition parties or a large relative Governmental negotiation power while unequal parameters indicate disagreement and low Government party negotiation power in Congress.

The traditional policy process sets goals that are implemented through traditional Government policy instruments. Let $C(R, T)$ describe the resulting firm compliance costs when goals are implemented through the available regulatory instruments.

Thus the traditional policy formulation process is assumed to be described by the following maximization problem:

$$\max_{R,T} U_c \quad \text{Under } C = C(R,T)$$

the solution to which is denoted R^* and T^*.

Agent utilities with the traditional regulation process become:

$$U_f^* = -T^* - C(R^*, T^*) \tag{1}$$

$$U_c^* = U_f^* + \lambda_c T^* + \delta_c R^* \tag{2}$$

$$U_g^* = U_f^* + \lambda_g T^* + \delta_g R^* \tag{3}$$

In the voluntary agreement process goals $\tilde{R}, \tilde{T}$ are set through negotiations between Government and the IO and then implemented by the IO. Thus the industrial organization representing firms is assumed to have a regulatory instrument vis-à-vis its members (moral suasion, codes of conduct etc) with which it can ensure attainment of the environmental goals. Clearly, public tax revenues are not generated (i.e. $\tilde{T} = 0$). Further it is assumed that the regulatory costs are described by the functions $C(cR, T)$ where $T = 0$. Thus by assumptions the two regulatory cost functions are identical, save for the cost parameter c and the constraint that $T = 0$ under IO implementation. This simplifies the following derivations while capturing the essential difference in relative efficiency through a single parameter c, indicating the relative cost of IO-implementation.

Agent utilities under the voluntary agreement process become:

$$\tilde{U}_f = -C(c\tilde{R}, 0) \tag{4}$$

$$\tilde{U}_c = \tilde{U}_f + \delta_c \tilde{R} \tag{5}$$

$$\tilde{U}_g = \tilde{U}_f + \delta_g \tilde{R} \tag{6}$$

We assume that the IO as well as government can predict the result of the traditional policy process R^*, T^*. Given this, a necessary condition for a voluntary agreement is that both parties to the agreement experience a non-negative utility gain vis-à-vis the traditional policy process which both parties know is the alternative. In other words, a non-empty set of goals ($\tilde{R}$) must exist for which both the following individual rationality (IR) constraints are satisfied:

IR-firm:

$$\tilde{U}_f = -C(c\tilde{R}, 0) \geq -T^* - C(R^*, T^*) = U_f^* \tag{7}$$

IR-Government:

$$\tilde{U}_g = \tilde{U}_f + \delta_g \tilde{R} \geq U_f^* + \lambda_g T^* + \delta_g R^* = U_g^* \tag{8}$$

The set of parameter combinations in the (δ_g, λ_g) space that for any given combination of parameters (δ_c , λ_c , c) allows a non-empty set of ($\tilde{R}$) satisfying both IR constraints can be found and characterized. When a specific parameter combination (δ_g, λ_g) is a member of this set (hereafter called the VA-set) a voluntary agreement becomes possible. In Hansen (1999) the resulting voluntary agreement for each parameter combination (δ_g, λ_g) is found by assuming that the Nash bargaining solution results whenever a voluntary agreement is possible.

Figure 1 presents a graphical illustration of the VA-set and resulting bargaining solutions (for details of the derivation see Hansen 1999). In figure 1 the VA-set for reduced efficiency (i.e. c>1) is reproduced with its associated interior isoquants indicating the resulting $\tilde{R}$ for the entire VA-set.

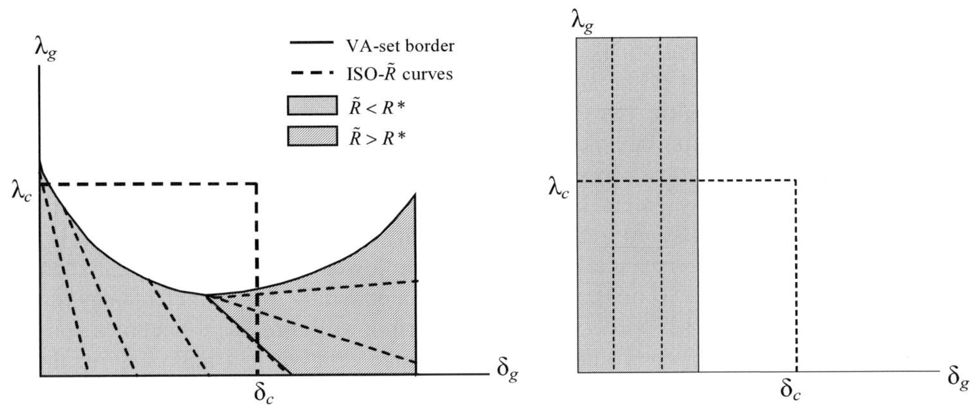

a. Revenue raising alternative

b. Non-revenue raising alternative

Figure 1. *Illustration of the VA-set for a model without interest group signalling power*

The shaded areas of figure 1a is the VA-set, i.e. the (δ_g, λ_g) parameter points where the VA policy process is chosen. In the dark shaded area VAs will result in lower environmental goals than under traditional regulation and in the light shaded area the resulting VAs will have higher environmental goals. As VA implementation costs rise (c increases) the VA-set is shifted down in the figure and when $c>1$ parameter values ($\delta_g = \delta_c$, $\lambda_g = \lambda_c$) are not included in the set.

The intuition is that when the cost of implementing a given goal is higher under a VA than under traditional regulation (i.e. $c>1$) this extra cost reduces both firm and government utility and must therefore be balanced off by some other benefit if the VA is to be chosen. If government agrees with the policy priorities that result from the traditional policy process, a cost increasing VA will not be chosen because government does not gain anything from avoiding the process of compromising with opposition parties. However, VAs that increase implementation costs may be attractive to governments that disagree with the policy priorities that would result under the traditional policy process. If Government is less concerned with the environment and with raising tax revenue than Congress (i.e. $\delta_g < \delta_c$ and $\lambda_g < \lambda_c$) then VAs that lower environmental goals (and yield no tax revenue) may improve both Government and firm utility. Also if Government is more concerned with the environment than Congress ($\delta_g > \delta_c$) VAs are also possible as long as Government is sufficiently less concerned with raising tax revenue than Congress. In this case Government may be willing to accept the reduction in tax revenue collected from firms that result from a VA if firms agree to a higher environmental goal than would result from the traditional policy process.

Model Predictions. Of specific interest in relation to the following empirical analysis is the situation where the traditional policy alternative is expected to result in non-revenue raising regulation. Then the VA-set becomes a vertical line (figure 1b), i.e. becomes independent of the λ since $T^* = 0$. When VAs reduce efficiency (the situation illustrated in figure 1b) Governments that are more concerned with the environment will no longer be able to induce firms to attain higher environmental goals than under the traditional process because there is no credible threat of tax payment. Thus if the policy disagreement explanation applies (and many VAs are inefficient) only Governments that are less concerned with the environment than Congress will find VAs advantageous since VAs will achieve lower environmental goals than the traditional alternative. If on the other hand VAs generally improve efficiency (corresponding to the VA-set border moving to the right in figure 1b) it becomes possible for VAs to increase environmental performance relative to traditional regulation (because of the efficiency gain) and we would expect Governments that are less as well as more concerned with the environment than Congress to find VAs advantageous.

Irrespective of the VA's relative efficiency, the model implies that environmental performance increases relative to traditional regulation as Government's concern with the environment increases relative to Congress (corresponding to a move to a point in the figure with a higher δ-value and so to a higher $\tilde{R}$ -isoquant).

4. EXTENDING THE MODEL WITH FIRM IO SIGNALLING POWER

This augmentation of the model attempts to capture the *responsibility shifting explanation* suggested in the introduction. Presumably interest groups have the ability to do quiet lobbying under the traditional process as well as under the VA process. However, interest groups often participate in the public debate as well in order to sway public opinion in their favour thus putting indirect pressure on policy makers. The shift in responsibility that VAs cause may be important with respect to this part of the interest group's activity. It will probably reduce the effectiveness of an interest group's public criticism of Government, if Government is not perceived as responsible for the criticized act or if the interest group is perceived as sharing responsibility for the criticized act.

We will present the augmented model formally, but only give an intuitive explanation of the model results (the reader interested in the formal model derivation is referred to Hansen 1999).

In the augmented model it is assumed that interest groups may affect public opinion by signalling the utility affect that policy has on their constituents. If the signal is credible the public may punish/reward those perceived as responsible (e.g. through the ballot box or through consumer demand decisions). If the interest group is powerful (i.e. the public reacts to the signal that is sent by the interest group) this will affect the result of the traditional policy process by making it less attractive for Congress and Government to reduce the powerful interest group's utility.

Now consider the utility functions when the firm IO has signalling power under the traditional policy process. The IO is able to signal the utility effect of policy on firms to the public who in turn may punish Government and Congress that are responsible for policy, e.g. at the next election. The risk of punishment is assumed to be proportional to the utility effect signalled by the IO. Augmenting the previous model in this way we have the following agent utilities under the traditional policy process:

$$U_c^* = U_f^* + \lambda_c T^* + \delta_c R^* + s^f U_f^* \tag{9}$$

$$U_g^* = U_f^* + \lambda_g T^* + \delta_g R^* + s^f U_f^* \tag{10}$$

$$U_f^* = -T^* - C(R^*, T^*) \tag{11}$$

where s^f is the marginal utility effects of firm interest group criticism.

When responsibility for setting the environmental goal is shared with the IO through a voluntary agreement, it is no longer possible for the IO to credibly criticize policy and so agent utilities under the voluntary agreement process are as in the previous subsection:

$$\tilde{U}_f = -C(c\tilde{R}, 0) \tag{12}$$

$$\tilde{U}_c = \tilde{U}_f + \delta_c \tilde{R} \tag{13}$$

$$\tilde{U}_g = \tilde{U}_f + \delta_g \tilde{R} \tag{14}$$

In the model the effect of interest group signalling on Government and Congress utility is assumed to be the same so that only the policy priority weights can differ between the two utility functions. This means that the effect of signalling on the VA-set can be illustrated in the same type of diagram as in figure 1. The result of firm signalling power is that the VA-set moves up in the graph as shown in figure 2.

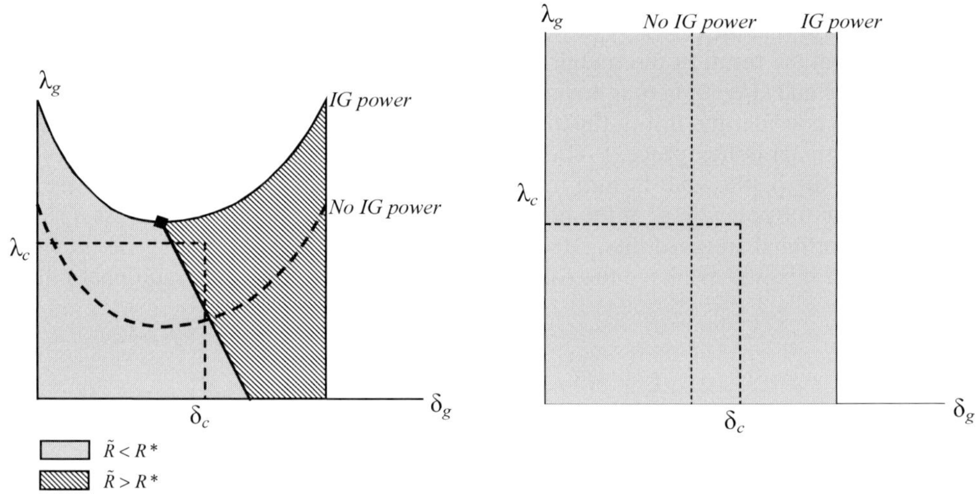

a. Revenue raising alternative b. Non-revenue raising alternative

Figure 2. *Illustration of the VA-set for a model with interest group signalling power*

The intuition is that when the firm IO has signalling power, it will criticize policy under the traditional process thus reducing Government utility. When responsibility for setting the environmental goal is shared with the IO through a voluntary agreement, it is no longer possible for the IO to credibly criticize policy. This gives an extra benefit to Government of entering into a VA vis-à-vis the situation without firm IO signalling power so that for all parameter sets the VA becomes more attractive (the outward shift of the border in figure 2a). At the same time firm bargaining power increases so that the resulting VA will have lower environmental performance than if the firm did not have signalling power (we see that isoquants and thus the area of reduced environmental performance also shift out in figure 2a). If the benefit of eliminating firm IO criticism is large enough it will make VAs that increase implementation costs attractive to Government even though there is no disagreement with Congress policy priorities. The utility benefit of eliminating firm IO criticism may be greater than the utility loss from increased implementation costs.

Model Predictions. When the traditional policy alternative is expected to result in non-revenue raising regulation (figure 1b) firm signalling power will also result in an outward shift of the vertical VA-set border and the underlying $\tilde{R}$-isoquants. Thus as in the previous section the model implies that efficiency reducing VAs will lower environmental performance relative to traditional non-revenue alternatives (because Government does not have a credible threat of tax payment as an inducement). However, even though environmental performance af the VA falls further as IO signalling power increases (because of increasing IO bargaining power) it may become advantageous even for Governments that are more concerned with the environment than Congress to choose the VA process. Governments that are more concerned with the environment may be willing to accept a lower environmental target than under traditional regulation in order to avoid harmful IO criticism.

Irrespective of the VA's relative efficiency and IO signalling power, the model still implies that environmental performance increases relative to traditional regulation as Government's concern with the environment increases relative to Congress (corresponding to a move to a point in the figure with a higher δ-value and so to a higher $\tilde{R}$-isoquant).

5. EXTENDING THE SIMPLE MODEL WITH ENVIRONMENTAL INTEREST GROUP SIGNALLING POWER

Now consider environmental interest group signalling power, but in a model allowing responsibility for goal setting to be decoupled from responsibility for policy implementation. Let R_g denote the goal set at the time of policy implementation and R the damage reduction actually attained. The environmental interest group can criticize goal setting as well as goal attainment and we assume that non-attainment of a set goal in itself increases the risk of punishment. The following simple specification catches this. At the time of goal setting the utility effect of criticism is:

$$s(R_g - \overline{R})$$

where $\overline{R}$ is the level of emission perceived by the public as acceptable. At the time of goal attainment the utility effect of criticism is:

$$2s(R - R_g)$$

so that non-attainment of a set goal is costly, while the net effect of criticism of goal setting and goal attainment when goals are reached is that the actually attained emission reduction is criticized (the net effect is $s(R - \overline{R})$).

Agent utilities under the traditional policy process become:

$$U_f^* = -T^* - C(R^*, T^*) \tag{15}$$

$$U_c^* = U_f^* + \lambda_c T^* + \delta_c R^* + s^e(R^* - \overline{R}) \tag{16}$$

$$U_g^* = U_f^* + \lambda_g T^* + \delta_g R^* + s^e(R^* - \overline{R}) \tag{17}$$

where s^e is the marginal utility effects of environmental interest group criticism on Congress and Government. Under the traditional process set goals are attained in order to avoid the extra criticism of goal non-attainment.

Under the voluntary agreement process the Congress has no responsibility. Government shares responsibility for goal formulation with the firm, while implementation is the sole responsibility of the firm. We then have

$$\tilde{U}_f = -C(c\overline{R}, 0) + s_f^e(\tilde{R} - \overline{R}) + 2\, s_f^e(\tilde{R} - \tilde{R}_g) \tag{18}$$

$$\tilde{U}_c = \tilde{U}_f + \delta_c \tilde{R} \tag{19}$$

$$\tilde{U}_g = \tilde{U}_f + \delta_g \tilde{R} + s^e(\tilde{R}_g - \overline{R}) \tag{20}$$

where s_f^e is the marginal utility effects of environmental interest group criticism on the firm.

Although the intuition is somewhat different, the result of environmental interest group signalling power is that if $s_f^e < s^e$ the VA-set moves up in the graph just as shown in figure 2a.

The intuition is that when the environmental interest group has signalling power, it will criticize policy (goal setting and goal attainment) under the traditional process thus reducing Government utility in proportion to the actual emission reduction achieved. When responsibility for attaining the goal is shifted to the firm IO through a voluntary agreement, Government is no longer susceptible to criticism of goal attainment. If Government can persuade the firm IO to set high goals then the VA entails an extra utility benefit for Government even though the higher goals are not met since Government is not responsible for goal attainment. However, the firm IO must be willing to accept responsibility and the utility loss associated with environmental interest group criticism of non-attainment of the set goal. Thus only if firms are less sensitive to environmental interest group criticism than Government ($s_f^e < s^e$) will there be potential gains from trading responsibility for reduced tax payment and reduced realised emission reductions. If $s_f^e < s^e$ this gives an extra benefit of entering into a VA which, like in the case of firm IO signalling, makes VAs more attractive to Government. If the benefits of trading responsibility are large enough it will make VAs that increase implementation costs attractive to Government even though there is no disagreement with Congress policy priorities. Actual emission reductions may be lower or higher than under traditional regulation, but in all cases goals are set higher than attained emission reductions (i.e. we will always have $\tilde{R}_g > \tilde{R}$). It is important to stress that this non-attainment of the set goal is an implicit part of the agreement with Government.

Model Predictions. When non-revenue raising traditional policy alternatives expected the situation can parallel to above be illustrated by figure 2b. EG-signalling power will just as IO-signalling power cause an outward shift of the vertical VA-set border and the underlying $\tilde{R}$-isoquants. Efficiency reducing VAs will also in this case lower environmental performance relative to traditional non-revenue alternatives (because of Government's lack of a credible tax alternative). In this case, however, environmental performance af the VA falls as EG-signalling power increases because this increases the IO's bargaining power (i.e. it increases the value of the asset supplied by the IO: taking responsibility). Despite this it may become advantageous for environment biased Governments to choose the VAs and accept lower environmental performance in order to reduce the effect of harmful EG-criticism by shifting responsibility to the IO.

Essentially then the model predicts that EG-signalling power has the same effect on the propensity to enter VAs and on the resulting environmental performance as IO-signalling power and that this effect is stronger the less susceptible firms are to public criticism from the EG. The key difference is the mechanism of responsibility transfer which in this case is through (planned) non-attainment of the environmental goals set in the agreement. Thus the model would predict increasing non-attainment of goals as EG-signalling strength increases (while this would not be expected if the agreement were driven by IO signalling strength).

Finally, we again note that irrespective of the VA's relative efficiency and EG-signalling power, the model implies that environmental performance increases relative to traditional regulation as Government's concern with the environment increases relative to Congress (corresponding to a move to a point in the figure with a higher δ-value and so to a higher $\tilde{R}$-isoquant).

6. THE META STUDY

The idea behind conducting a meta study was to utilize the knowledge gathered by researchers through case studies on about specific voluntary agreements. Our hope was to generate a sufficient number of observations of combinations of key model variables from studied VAs to allow testing of model predictions.

The meta study (see Chidiak et al. (1999) for details) was conducted by telephone interviewing researchers who have published case studies of voluntary agreements with industrial organizations. A questionnaire with multiple choice questions was used as an interview guide. Choices among the listed answers were elicited and explanatory comments noted by the interviewers. Researchers and case studies were located through the so-called snowball method (starting with case studies we were aware of, each researcher was asked if he/she had knowledge of other VA case studies).

The questionnaires covered most key variables of the model presented in the previous section (in addition to questions on information and cost structure and on implementation of the VA). In order to avoid arbitrary reference standards where possible, questions elicited rankings rather than ordinal evaluations. The questionnaire is reproduced in Chidiak et al. (1999), while the analysed VAs along with a list of interviewed researchers and referenced case studies are listed in appendix 1.

All in all 20 interviews were completed of which 19 turned out to be on VAs with industrial organizations. The number of answers to specific questions was in many cases reduced further since researchers were often unable to give qualified answers to all questions.

In the next subsection we present descriptive statistics for variables generated from answers to those questions that are relevant for testing model predictions (tables indicate the source question from which the variable is derived and where this is not apparent how the variable is derived). In the following subsection we formulate and attempt to test model predictions in terms of correlation patterns between these variables.

6.1. Descriptive Statistics

Initially we elicited the researcher's evaluation of the type of traditional regulation that would have been implemented without the VA and the credibility of this threat. In table 1 the background threat for the VA is summarised for the 19 relevant cases.

Table 1. *Strength and type of alternative policy if VA had failed*

Regulatory threat	*Number of cases*
Tax:	1
Direct regulation:	9
Clear that traditional regulation would have been implemented, but unclear which type:	3
Unclear whether traditional regulation would have been implemented :	3
No response:	3

It is notable that in most cases there was a clear threat of traditional regulation if the VA failed which suggests that the VAs probably have affected firm behaviour. It is also notable that the regulatory threat in most cases is direct regulation, i.e. non-revenue raising regulation. Thus our data should be interpreted using the model with a non-revenue raising traditional policy threat.

Then we asked the researcher for his opinion of Government's environmental and firm bias relative to opposition parties. In table 2 Government bias variables are summarised.

Table 2. *Government bias relative to opposition parties*

	Biased against	*No bias*	*Bissed in favour of*	*No response*
Gov. bias vis-à-vis firms (MGFB):	5	6	5	3
Gov. bias vis-à-vis environment (MGEB):	5	11	2	1

These questions try to get at the model bias parameters (δ_g, λ_g) and (δ_c , λ_c). There seems not to be any dominating bias pattern for the sampled VAs. However, it is notable that the variables are highly (negatively) correlated, i.e. governments that are biased in favour of firms also tend to be biased against the environment.

We also elicited the researcher's opinion of interest group signalling strength and firms' concern with their environmental image. In table 3 these variables are summarised.

Table 3. *Interest group signalling power*

	Below average	*Average*	*Above average*	*No response*
IO signalling power (MIOS):	2	5	12	0
EG signalling power (MEGS):	5	3	11	0
Firms' concern with environmental image (FIMI):	3	4	9	3

There seems to be a tendency for EGs and IOs in the sample to have above average signalling strength and for the sampled VA firms to be more concerned with their environmental image than other firms.

In table 4 the researchers' opinion of whether EG and opposition parties approve of the use of a VA is summarised.

Table 4. *Approval of the VA instrument*

	Disapprove	*Non-committed*	*Approve*	*No response*
EG approval of the use of the VA instrument (EGVA):	9	8		2
Opposition approval of the use of the VA instrument (OPVANY):	3	7	6	3

We note that while EGs are critical of the use of VAs in a majority of cases opposition parties are only critical in one out of five cases.

Finally, the interviewed researchers found that EGs generally are more critical of the use of VAs as the way to reach environmental goals than they were of the actual goals set in the studied VAs. We interpret this difference as a lack of confidence in the VA's ability to actually reach the goals set in the agreement. In table 5 we tabulate the variable 'EG goal confidence' constructed in the following way: If the EG has a better valuation of the VA goal than of the VA instrument the EG is placed in the 'Not Confident' category. If the EG has a lower evaluation of the goal than of the VA instrument than the EG is placed in the 'Confident' category. If the EG has the same evaluation of the goal as of the VA instrument then the EG is placed in the 'Non-committed' category.

Table 5. *EG goal confidence*

	Not Confident	*Non-committed*	*Confident*	*No response*
EG goal confidence (DVAGO):	4	12	1	2

Table 5 indicates a tendency for EGs not to be confident in the VA's ability to reach the goals set in the agreement.

6.2. Test of Model Predictions

We see from table 1 that the traditional policy threat was non-revenue raising regulation in almost all cases covered by the meta-study. In table 6 we summarize the empirical findings that the theoretical model predicts for the situation with non-revenue raising regulatory threats. In the first four rows the predicted findings (derived in sections 3, 4 and 5 in terms of questionnaire variable correlation patterns etc.) are summarised for each of the different VA-explanations allowed for by the model, i.e.:

- the efficiency explanation
- the disagreement explanation
- the responsibility shifting explanation driven by IO signalling power and
- the responsibility shifting explanation driven by EG signalling power

In the last row we present the actual correlation patterns etc. found in the data.

Columns 1 and 2: The key prediction for the disagreement explanation (i.e. that VAs are inefficient and driven by disagreement over policy priorities between Government and Congress) presented in section 3 was that we would only expect firm biased governments to enter VAs and that we would expect Congress to be opposed to the use of VAs. This pattern is not expected if any of the other three explanations apply. We do not find this pattern in our data.

Column 3: If VAs are explained by IO signalling power (i.e. that VAs are inefficient and driven by Government's desire to avoid being publicly criticized by powerful IOs) we would expect EG approval of the use of VAs (variable EGVA) to be negatively correlated with firm IO signalling power (variable MIOS). The reason being (as shown in section 4) that environmental performance falls as IO signalling strength increases. If VAs are not explained by firm signalling we would on the other hand not expect EG approval to be correlated with IO signalling power. There is no such correlation pattern in our data, rather there is a small highly insignificant positive correlation coefficient.

Column 4: If on the other hand VAs are explained by EG signalling power (i.e. that VAs are inefficient and driven by the Government's desire to avoid being publicly criticized by powerful EGs) we would expect EG approval of VAs (variable EGVA) to be negatively correlated with EG signalling power (variable MEGS) instead. In this case (as shown in section 5) environmental performance falls as EG signalling power (and thereby IO bargaining strength) increases. We do in fact find a highly significant negative correlation coefficient in our data.

Column 5 and 6: Further, if VAs are explained by EG signalling power 'planned' non-attainment of environmental goals increases with EG signalling power but falls with the firms' concern with their environmental image. With this explanation we would therefore also expect EG goal confidence (variable DVAGO) to be negatively correlated with EG signalling power (variable MEGS) and positively correlated with the firms' concern with their environmental image (variable FIMI). Though the coefficients are not significant at the 5% level we also find this pattern in our data.

Column 5 and 6: Finally, for all four explanations generated by the model (including the efficiency explanation) we expect EG approval of VAs (variable EGVA) to be positively correlated with government environmental bias (variable MGEB) as well as with the firms' concern with their environmental image (variable FIMI) since this in all cases increases environmental performance. We find a significant positive correlation coefficient for EGVA*MGEB in our data while the corresponding coefficient for EGVA*FIMI is highly insignificant.

Table 6. *Model predictions and meta study results*

	Dominating Government bias pattern	*Dominating Government-Congress agreement pattern on use of Vas*	*EGVA-MIOS correlation*	*EGVA-MEGS correlation*	*DVAGO-MEGS correlation*	*DVAGO-FIMI correlation*	*EGVA-MGEB correlation*	*EGVA-FIMI correlation*
Model predictions with:								
Efficiency explanation:	none	None	none	none	none	none	positive	positive
Disagreement explanation:	biased in favour of firms	Government-Congress disagreement	none	none	none	none	positive	positive
Responsibility shifting explanation (driven by IO signalling power):	none	none	negative	none	none	none	positive	positive
Responsibility shifting explanation (driven by EG signalling power):	none	none	none	negative	negative	positive	positive	positive
Meta study:								
Findings:	none	none	none	negative**	(negative)	(positive)	positive*	none
Spearman's rho:			0.170	-0.862**	-0.368	+0.309	+0.603*	+0.097
(significance level)			(0.513)	(0.000)	(0.146)	(0.283)	(0.013)	(0.741)

* significant at 5% level.
** significant at 1% level.

In conclusion no correlations of unexpected signs were found and of the two correlations that were predicted by the model regardless of VA explanation one was found. Given the small number of cases not finding all the predicted correlations is not surprising. On the other hand, it is also clear that applicability of the model has not been tested vigorously.

Taking the model for granted we do not find the variable patterns that would be generated if *the policy disagreement explanation* or *the IO signalling explanation* dominated in the studied cases. Given the small number of cases studied this should not in itself be be over interpreted. However the pattern expected if *the EG signalling explanation* applied is seen in the meta study data though it is not significant for all variables. Finding such a pattern even though the number of cases is small is an indication that *the EG signalling explanation* may apply in many of the cases studied.

Thus, if we take the model for granted our data suggests that EG signalling may explain the use of VAs in many sample cases. The model predicts that VAs in these cases are less efficient than the traditional alternative regulation that would have been adopted without the VA. Further, the model predicts that VAs in these cases achieve a lower environmental performance than the traditional alternative would have achieved which may reduce social welfare further.

When interpreting this study it should, however, be stressed that we can not expect the studied sample of VAs to be representative af all VAs (in fact the way researchers select VAs to be studied is probably far from generating a random selection as is the way we have located the subset of case studies included in the meta study).

7. CONCLUSIONS

The study reveals that in most of the analysed cases VAs were backed by a clear threat of regulatory action if the VA failed. Thus we expect that most of the VAs covered in the study have been able to induce changes in firm behaviour. The focus of the study has been whether the VA is more or less cost effective than the traditional regulatory alternative.

We have presented a theoretical model offering several explanations of VAs and confronted it with empirical evidence from case studies of voluntary agreements.

The empirical evidence does not falsify the model, however, the study only covers a small number of cases and testing is only attempted in a few dimensions. Thus even though the results are mildly encouraging with regard to the model's explanatory power, the theoretical model is by no means tested vigorously. Further generalisation beyond the specific sample of VA's studied is not possible because of representativity problems.

If the theoretical model is taken for granted, the data tends to support the hypotheses that Governments in many of the sample cases choose VAs in order to shift the responsibility for implementation to industrial organizations that are less sensitive to criticism from powerful environmental interest groups. When this explanation of a VA applies, the model predicts that the VA will be less cost effective and achieve lower environmental performance than the traditional regulatory alternative which would otherwise have been adopted.

Although the presented findings are not conclusive, nor necessarily representative, they do suggest the worrying possibility that some - possibly many - of the VAs being established today achieve lower environmental performance less cost effectively than the most likely traditional regulatory alternative.

8. ACKNOWLEDGEMENTS

The research leading to this chapter is a product of the VAIE-project and has been funded by The European Commission, DG XII for science, research and development through the JOULE programme.

9. REFERENCES

Chidiak, M., M. Glachant and L.G.Hansen, (1999), *VAIE project task A final report - Theoretical Perspectives on the Efficiency of Voluntary Approaches to Promote Energy Efficiency*, CERNA, Paris.

Glachant, M., (1994), The setting of Voluntary Agreements Between Industry and Government: Bargaining and Efficiency, *Business Strategy and The Environment* 3 (2), 43-49.

Hansen, L.G. (1999) Environmental Regulation through Voluntary Agreements, in *Voluntary Approaches in Environmental Policy,* eds: C. Carraro and F. Leveque, Kluwer academic publishers, Dordrecht.

IEA, (1995), *Survey on Voluntary Agreements: Status*, Interim Report and Next Steps, IEA note.

10. APPENDIX

10.1. VAs Covered in the Meta Study Sample

The meta study was conducted by telephone interviewing researchers who have published case studies of negotiated voluntary agreements. A questionnaire with multiple choice questions was used as an interview guide. Choices among the listed answers were elicited and explanatory comments noted by the interviewers. Researchers and case studies were located through the so-called snowball method (each researcher was asked if he/she had knowledge of other VA case studies). 20 interviews were completed of which 19 were negotiated VAs. Coverage of the studied cases is illustrated in Table 1, while the VAs in the sample are listed in Table 2. Finally, the full list of researchers and the case study references are listed at the end of the appendix.

Table 1. *Countries and environmental areas covered by the VAs of the study*

	Air	*Water*	*Waste*	*All*
Finland	1			
Norway	1			
Denmark	2			
Holland	2		2	2
Belgium	1		1	
Germany		1	2	
France		1	2	
Italy	1			

Table 2. *VAs in the meta study sample*

Belgian VA for SO_2 and NO_x emissions from Electricity Suppliers (1991)
Belgian Convention with the Cement Industry for waste treatment (1995)
Finnish Agreement on energy conservation with the paper sector
Danish VA with the plastics industry over PVC use
Danish VA with the electric utilities over SO_2/NO_x (1996)
Dutch Packaging Covenant (1997 version)
Dutch Covenant with the Chemical Industry (all environmental concerns)
Dutch VA with spray can producers (on CFC use in spray cans)
Dutch packaging covenant (1991 version)
Dutch VA with the Basic Metal Industry (on all environmental issues)
Dutch LTA (on energy efficiency) with the Chemical Industry
French VA on Packaging waste recycling: ECO-EMBALLAGES (1992)
French VA on soaps and detergents (over phosphates) (1986-89-90)
French VA on end-of-life vehicle recycling/reuse (1993)
German VA on Packaging waste recycling: DSD (1991)
German VA on end-of-life vehicles (recycling) (1996)
German VA on EDTA (Chemical gelatine agent) (1991)
Italian VAs on the quality of fuels (1989-92)
Norwegian VA on GHG emissions with the aluminium industry

(The interviewed researchers were: 1) Franck Aggeri, CGS, Ecole des Mines de Paris, Paris, France, 2) Steven Baeke, University of Ghent, Belgium,3) Peter Börkey, CERNA, Ecole des Mines de Paris, Paris, France, 4) Karl Brockmann, Centre for European Economic Research, Mannheim, Germany, 5) Jacco Farla, University of Utrecht, The Netherlands, 6) Matthieu Glachant, CERNA, Ecole des Mines de Paris, Paris, France, 7) Verina Ingram, University of Wageningen and IWACO, The Netherlands, 8) Katja Johannsen, AKF, Denmark, 9) Ralph Jülich, Öko Institute, Darmstadt, Germany, 10) Signe Krarup, AKF, Denmark, 11) Delphine Misonne, CEDRE, Bruxelles – Belgium, 12) Giulia Pesaro, IEFE, U. Bocconi, Milan, Italy, 13) Philippe Quirion, CERNA, Ecole des Mines de Paris, Paris, France, 14) Mikael Togeby, AKF, Denmark, 15) Asbjørn Torvanger, Center for International Climate and Environmental Research, Oslo, Norway)

11. CASE STUDY REFERENCES

Aggeri, F. & A. Hatchuel (1995), *A Dynamic Model of Environmental Policies. The case of innovation oriented voluntary agreements*, Nota di Lavoro 24.97, FEEM, February 1997.

Börkey, P. & M. Glachant (1997)' *Les engagements volontaires de l'industrie dans le domaine de l'environnement: nature et diversité*, final report to the Ministry for the Environment Ademe (French Environment Agency), CERNA.

Croci E. & G. Pesaro (1997)' Voluntary agreements in Italy: a new approach in environmental policy. in: *ENER Bullettin*, n. 20.97.

Croci E. & G. Pesaro (1997a), Gli accordi volontari in campo ambientale: evoluzione del quadro di riferimento' in: *Economia delle Fonti di Energia e dell, Ambiente*, n.3/97 [Environmental voluntary agreements: evolution of the reference framework].

Environmental Sociology and Social Methodology Group, Wageningen University (1998) *Joint Environmental policy-making*, Final Report, Wageningen, 1998.

Gebers, G. Biekart J-W.; Bizer, K.; Ingram, V. Jülich R., (1998), *New Instruments for Sustainability – The Role of Environmental Agreements*, Final Report and Annex Case Studies, Darmstadt.

Glachant, M. & T. Whiston (1997): Voluntary Agreements between Industry and Government – the case of Recycling Regulations' in: *Environmental Policy in Europe: Industry, Competition and the Policy Process*, edited by F. Lévêque, Edward Elgar, Cheltenham, UK.

Glachant, M. (1995), *DSD contre Eco-Emballages: Analyse comparée de l'efficacité des consortia de recyclage allemands, français et italiens*, study for ADEME, CERNA.

Ingram, V. (1998), *The Long Term Energy Efficiency Agreement with the Fine Ceramic Industry, (The Netherlands)*, in Öko Institut (1998) – Annex Case Studies, Darmstadt.

Ingram, V. (1998), *The SUBAT Agreement (The Netherlands)*, in Öko Institut (1998) – Annex Case Studies, Darmstadt.

Jülich, R. (1998), *The EDTA Agreement (Germany),* in Öko-Institut (1998) – Annex Case Studies, Darmstadt.

Jülich, R. (1998), *The Battery Agreement (Germany),* in Öko-Institut (1998) – Annex Case Studies, Darmstadt.

Krarup, S., M. Togeby and *K. Johannsen (1997), De* første aftaler om energieffektivisering – Erfaringer fra 30 aftaler indgået i 1996, AKF. Working paper.

Misonne, D. (1998): The Agreement between the Walloon Region and the Cement Industry (Belgium), in Öko-Institut (1998) – Annex Case Studies, Darmstadt.

Misonne, D. (1998), *The River Contract of the Upper-Meuse (Belgium*), in Öko-Institut (1998) – Annex Case Studies, Darmstadt.

Öko Institut (1998), *New Instruments for Sustainability*, edited by B. Gebers, Final Report, Darmstadt 1998.

Quirion, P. (1994),*La gestion des déchets d'emballages ménagers en France et en Allemagne: élements d'evaluation économique*, DEA thesis, Université de Paris 1 Pathéon-Sorbonne.

Rennings, K., K.L. Brockmann & H. Bergmann (1997),*Voluntary Agreements in Environmental Protection – Experiences in Germany and Future Perspectives*, published as ZEW Discussion paper.

Rietbergen M., Farla J., Blok K. (1998), *Quantitative Evaluation of Voluntary Agreements on Energy Efficiency*, paper for the CAVA workshop in Ghent on November 26-27, 1998, 15 p.

Seyad, A., S. Baeke & M. De Clercq (1998), Success determining factors for negotiated agreements: A Comparative Case Study of the Belgian Electricity Supply Industry and the Packaging Sector, to be published in "*Co-operative Environmental Governance – Public-Private Agreements as a Policy Strategy*", edited by P. Glasbergen, Utrecht University.

CHAPTER 3

THE EUROPEAN AND THE AMERICAN APPROACH TO ENVIRONMENTAL VOLUNTARY AGREEMENTS

THE EVOLUTION OF ENVIRONMENTAL AGREEMENTS AT THE LEVEL OF THE EUROPEAN UNION

G. SCHNABL*

European Commission, Directorate-General Environment, Brussels, Belgium

Abstract. The European Community starts using the instrument of voluntary environmental agreements in the late 1980s. In 1996, a Commission Communication takes stock of existing experiences and outlines ideas for a more consistent Community approach. In 2002, the Commission issues a second Communication, this time dedicated exclusively to environmental agreements at Community level. It builds on comments and requests put forward by the Council and the European Parliament. The Communication takes account of the specific legal and institutional constraints at EU level and proposes a set of criteria as well as procedural steps. It also details the concepts of self-regulation and co-regulation. Together with an interinstitutional agreement, it paves the way for the further use of the instrument at Community level. As of today experiences with environmental agreements at Community level are still limited, but encouraging.

1. INTRODUCTION

Voluntary agreements made their way as an instrument of environment policy in the course of the 1980s. Even if the first examples could be found in earlier years, as far back as the 1960s and 1970s, 1980-1990 was the key decade for their development on a larger scale (OECD,1999, p.45). It was also the decade of intensive debates on the benefits and shortcomings of voluntary agreements - debates that frequently set passionate "believers" against equally convinced "non-believers". This is not the place to re-open a discussion that will most probably go on for some more years. It is worth noting, however, that despite various assessment attempts and numerous analytical publications on voluntary approaches over the last fifteen years, the question of *environmental effectiveness and economic efficiency* of voluntary approaches has yet to be answered in a definitive way (OECD,2003, p.13).

The European Commission recognised the first environmental agreement at Community level in 1989. It concerned an engagement of the aerosol industry to limit the use of chlorofluorocarbons to applications for which these substances were essential. Three more agreements followed in 1989 and 1990. The complete list is given in the annex.

The modality chosen for the "recognition" of these agreements by the Commission – the adoption of a Commission Recommendation confirming the

E. Croci (ed.), The Handbook Of Environmental Voluntary Agreements, 93 - 106

content of the industry's engagement – proved to be a sound approach. It was taken up again in later years. Chapters three and four will deal in more detail with the procedural aspects.

Only the term "environmental agreement" is used throughout this text, in line with the wording of the official Commission Communications. This does not exclude that the agreement in question also responds to a challenge from other policy fields, for example energy policy.

These agreements are fundamentally different from voluntary *programmes*, developed by public authorities to which individual firms are invited to participate. If participation as such is voluntary, the decision to adhere to the programme implies strict respect of the set rules. At Community level, the eco-management and audit scheme EMAS[1] and the eco-label award scheme[2] are well-known examples of such voluntary programmes.

2. THE 1996 COMMUNICATION AND ITS FOLLOW-UP

In 1996, the Commission adopted its first "Communication on Environmental Agreements"[3]. This document took stock of experiences with agreements as environmental policy instruments in Member States, developed guidelines for their effective use and set out conditions under which agreements could be used for the purpose of implementing certain provisions of Community Directives. It proposed a "checklist" for their assessment in terms of reasons for choice, content and compliance with the EC Treaty.

It is evident that agreements at Member States level have to comply with the EC Treaty, in particular the provisions relating to the internal market and competition, and its derived legislation. Their potential role in implementing certain provisions of Community Directives is also limited: as a general rule, the binding nature of Directives requires equally binding transposition measures. The Communication stated that

> "where Directives intend to create rights and obligations for individuals, it is generally not possible to implement the relevant provisions through agreements … on the other hand, where a provision of a Directive provides for the setting up of general programmes or for the achievement of general targets, the full achievement of the set objectives or targets does not necessarily require regulatory action".

A Directive may of course explicitly provide for voluntary implementation measures. Examples in environmental Community legislation are so far Directive 2000/53/EC on end-of-life vehicles (Article 10(3))[4], Directive 2002/96/EC on waste electrical and electronic equipment (Article 17(3))[5], Directive 2004/12/EC amending Directive 94/62/EC on packaging and packaging waste (Article 22(3)a)[6] and the recent Proposal for a Directive on batteries and accumulators and spent batteries and accumulators (Article 33)[7]..

The Communication did not establish any reporting or notification system for environmental agreements at Member States level. The Commission never established a comprehensive register of such agreements, nor tried to monitor them in a systematic way. Only draft agreements containing technical specifications for products were subject to screening, in line with the general procedures for the provision of information in the field of technical standards and regulations[8].

The Communication sent forward two unambiguous messages:

- Yes, environmental agreements can play a beneficial role and they offer certain advantages: they can promote a pro-active attitude on the part of industry, they can provide cost-effective, tailor-made solutions and allow for a quicker and smoother achievement of objectives.
- No, voluntary approaches are not meant to replace legislation. Legislation will remain the necessary backbone of Community environmental policy, but it needs to be supplemented by market-based elements.

The document focussed on agreements concluded at the level of Member States. A short chapter was dedicated, however, to environmental agreements at Community level.

It stated *inter alia* that:

> "For the time being, the Commission has … to resort to non-binding agreements as the available instrument to encourage a pro-active approach from industry and as an incentive for effective environmental action … the Commission will consider on a case by case basis whether commitments of industry can be used as an effective environmental measure … as regards the institutional framework, it is necessary to base non-binding agreements on objectives already endorsed by the Community institutions … keeping the institutional balance is not only a guarantee for equitable results, it also provides industry with an element of stability".

The Communication hereby touched on two major issues linked to the use of environmental agreements at Community level:

- their legal character on the one hand, and
- the role of the Community legislator – European Parliament and Council – on the other.

In reply to the Communication, the Council adopted on 7 October 1997 a short Resolution[9], stating its belief that

> "environmental agreements can play an important role within the mix of instruments by encouraging a pro-active approach from industry, particularly in sectors with representative organisations or with a limited number of companies"

and inviting the Commission to do further work in this field.

The European Parliament's Resolution[10] of 17 July 1997 was more exhaustive. Quoting some of the text's key elements, it affirmed

> "that binding legislation will continue to form the backbone of Community environmental policy, which may be supplemented by voluntary agreements".

It further noted that

> "when EAs are used, the same level of legal protection … as is offered when legislation is implemented must be guaranteed"

and expressed its concerns that

> "the increasing shift of environmental protection tasks from parliamentary legislation, backed by democratic legitimacy, to self-regulation by industry, is leading to reduced legal certainty."

The European Parliament was to return to the issue. In its Resolution on the Commission Green Paper on environmental issues of PVC of 3 April 2001[11], it called on the Commission

> "to present as soon as possible a proposal for framework legislation on environmental agreements, which lays down the relevant criteria with regard to conditions, monitoring arrangements and penalties".

This added to the key issues already identified by the Commission the questions of
- representativeness of an agreement and of
- the inherent conflict between the voluntary nature of the instrument and the wish to make it as "binding" and legally controllable as possible.

Between 1998 and 2000, the Commission recognised eight more environmental agreements. In parallel, it took part in the continuing international debate on the potential benefits and shortcomings of the instrument. The work of OECD, resulting in two reports published in 1999 and in 2003 (see footnotes 1 and 2), and the CAVA project ("Concerted Action on Voluntary Approaches") finalised in 2001 are outstanding in this respect.

In response to the growing concerns of the European Parliament, the Commission committed itself not to acknowledge any additional agreements before having presented a comprehensive framework for the further use of the instrument. In line with this commitment, the Commission came forward with the "Communication on Environmental Agreements at Community Level within the Framework of the Action Plan on the Simplification and Improvement of the Regulatory Environment" on 17 July 2002[12].

This Communication deals *exclusively* with the ways and means of using environmental agreements as a policy tool at the Community level. Contrary to the 1996 Communication, it does not contain guidance or recommendations for the application of the instrument in Member States. Environmental agreements used, at national level, for the implementation of provisions laid down in European Directives are not covered by this paper neither.

3. LEGAL CONSTRAINTS FOR ENVIRONMENTAL AGREEMENTS AT COMMUNITY LEVEL

Before going into the details of the 2002 Communication, it is necessary to clarify some basic legal considerations that apply to environmental agreements at Community level. Given the basically positive evaluation of the instrument, the limited number of agreements that have actually been acknowledged might surprise, in particular when compared with the much more widespread use of agreements in some Member States. Legal constraints may be partly responsible for this. Whereas national authorities normally have a wider range of possibilities at their disposal – for example, civil as well as administrative law might be used, under control of courts or arbitration bodies - the Treaty establishing the European Community (TEC) is much less permissive (for an overview on the diversity of voluntary approaches, see OECD, 1999, p.45 ff).

Article 249 TEC lists the instruments to be used by the institutions "to carry out their task and in accordance with the provisions of this Treaty": regulations, directives, decisions, recommendations and opinions. This list is exhaustive. It is not possible to create additional policy instruments at the level of secondary legislation, thereby bypassing the Treaty.

Furthermore, the wording of Article 249 confirms the link between instruments and *tasks*, in other words the policy provisions of the Treaty. In the case of environmental policy, Article 174 TEC defines the scope of the Community policy and its objectives. The conclusion, in the view of the Commission's legal experts, is that the instrument must name and contain the environmental objective. It is not possible to adopt a "void box", an instrument only setting modalities and no objectives, the latter being brought in at a later stage and by a third party.

In the light of the above, the Commission was unable to propose "framework legislation" for the use of environmental agreements, as requested by the European Parliament.

The legal options for environmental agreements remain fairly restricted. In the absence of any tailor-made TEC provision, "environmental agreements" are unilateral commitments submitted by enterprises and/or their associations, recognised by the Commission by means of a Recommendation or another formalised act. It is worth noting that such a special provision exists in the field of social policy, Article 139 TEC.

The Recommendation prevails as the most commonly used approach. Only occasionally, has an exchange of letters been used (see the last two agreements listed in the annex). It is worth noting, however, that for some agreements the "recognition" took the form of a notice pursuant to Article 19(3) of Council Regulation N° 17[13], the first Regulation implementing (former) Articles 85 and 86 of the Treaty (now Articles 81 and 82 TEC). As the Commission, when overhauling its system of competition control, has recently abandoned the systematic *ex ante* evaluation of agreements, this model is no longer applicable in the future.

These legal considerations determined the choice of approaches laid down in the 2002 Communication on environmental agreements at Community level.

4. THE 2002 COMMUNICATION

4.1 The wider context

The Communication is explicitly linked to the Action Plan "Simplifying and improving the regulatory environment"[14], itself an offshoot of the wider "European Governance" issue ("European Governance – A White Paper", COM(2001)428 of 25.7.2001, and the Communication from the Commission "European Governance: Better Lawmaking", COM(2002)275 of 5.6.2002). These initiatives are not limited to environmental policy, but apply to all areas covered by Community competence.

In the chapter of the Action Plan entitled "Making more appropriate use of legislative instruments", the Commission stresses

> "that appropriate use can be made of alternatives to legislation without undermining the provisions of the Treaty or prerogatives of the legislator. There are several tools which ... can be used to achieve the objectives of the Treaty while simplifying lawmaking activities and legislation itself".

The text continues by identifying such tools: co-regulation, self-regulation, voluntary sectoral agreements, open co-ordination method, financial interventions and information campaigns.

Quoting again from the Action Plan,

> "voluntary agreements constitute one form of self-regulation. Voluntary agreements can also be concluded on the basis of a legislative act, i.e. in a more binding and formal manner in the context of co-regulation, thereby enabling parties concerned to implement a specific piece of legislation".

Co-regulation and self-regulation are again highlighted in a specific chapter in the Communication on environmental agreements. The basic difference between the two concepts lies in the fact that self-regulation, which covers a large number of more or less informal practices, does not involve any legislative act, whereas co-regulation does.

Whereas the notion of self-regulation is rather easy to understand, co-regulation has triggered many questions and even some perplexity. It is an attempt to combine the flexibility of a purely voluntary instrument with the added value of a regulatory approach in terms of control and enforcement. The Action Plan enumerates a series of criteria for the use of co-regulation: obligatory advance information of the legislator, compliance with the general interest, limitation of the legislation to "essential aspects", transparency, and representativeness of the parties concerned. Of course,

> "in cases where using the co-regulation mechanism has not produced the expected results, the Commission reserves the right to make a traditional legislative proposal".

The Communication develops the concept in more detail. The "essential aspects" to be addressed in the legal act (typically a Directive) are defined: the objectives to be achieved (potentially also interim objectives), deadlines and implementation mechanisms, methods of monitoring the application of the legislation, review clauses and sanctions. Detailed provisions on how to reach the set objective in time are left to an environmental agreement proposed by the economic actors concerned.

Co-regulation has as yet not passed beyond the conceptual stage and there is no practical experience with the instrument. The appropriate design of the interface between the elements covered by the Directive and the aspects left to the agreement might well prove difficult, in particular when it comes to safeguard provisions in the case of the agreement's failure to deliver the expected results.

4.2 The main content of the Communication

The Communication covers more than just the concepts of self-regulation and co-regulation. It also addresses

- basic legal conditions for the use of environmental agreements,

- criteria for their assessment and
- procedural requirements.

Environmental agreements should be fully compatible with all provisions of the TEC. The internal market and competition rules are key issues in this respect. It is worth noting that, in the absence of a full *ex ante* assessment by the Commission (see chapter 3), the proponents of the agreement are responsible for its appropriate design. The recognition of the agreement by the Commission does not constitute a failsafe protection against competition complaints and subsequent legal procedures.

Proposed agreements must also be compatible with the international commitments of the European Community and multilateral trade rules. Finally, the UN-ECE Convention on access to information, public participation in decision-making and access to justice in environmental matters[15] grants an extended "right to know" to a very broadly defined "public". Environmental agreements are explicitly included in the definition of "environmental information" to be made available to the public (Article 2(3)b).

The *assessment criteria* are derived from the checklist of the 1996 Communication, but they focus on aspects of particular relevance for agreements at Community level. It goes without saying that the agreement should, first and foremost, aim at a high level of environmental protection. "Business as usual" commitments are thus unwelcome. The other criteria listed are:

- Cost-effectiveness, in terms of comparative administrative costs for the Community institutions: Whereas it can be expected that agreements prove to be cost-effective for their proponents when compared with "command and control" regulation, it is by no means certain that they will be less resource-consuming for the institutions. The Commission will look very closely at the workload resulting from an agreement, in particular when it comes to monitoring and reporting.
- Representativeness: the proponents of an agreement must be sufficiently representative, organised and responsible. This should ensure that the agreement is effective for the larger part, if not the totality, of the sector concerned at European level. If representativeness is missing, the desired environmental added value is likely to be absent, too.
- Setting of quantified and staged objectives: this is considered as a must, as unclear objectives and the absence of a binding timetable increase the risk that the instrument be disqualified as inefficient window-dressing.
- Involvement of civil society: beyond the fact that this is already a legal requirement under the Aarhus Convention, transparency and openness to the public are likely to enhance the credibility and acceptance of the instrument. The public should also have the possibility to give comments on proposed agreements.
- Monitoring and reporting: these are crucial issues for the success or failure of an environmental agreement. The lack of appropriate mechanisms makes it impossible to assess whether the agreement has really reached its objective or not. On the other hand, monitoring and reporting obligations, if designed in an ambitious way, might become so burdensome that they conflict with the desirable flexibility of the instrument as well as with cost-efficiency considerations. A balanced approach and a well thought out sharing of the burden between self-control and external verification will be needed, but might be difficult to obtain.

- Sustainability: environmental policy is part of the sustainability triangle, together with economic and social development. Agreements cannot neglect this overall setting.
- Incentive compatibility: the success of an agreement is likely to be hampered if other elements – market trends, legislation or other policy instruments – influence developments in an opposite way. The agreement should therefore be consistent with other applicable instruments.

The *procedural requirements* proposed by the Commission reflect the wish to grant transparency and to respect the institutional balance. The steps to be observed are slightly different for agreements under the self-regulation and under the co-regulation approach, but in both cases, the "right to know" of the Community legislators and of the public at large acts as the guiding principle. Council and the European Parliament will be systematically informed on the intention to make use of an agreement. The agreements, as well as the monitoring of results and reports will be made accessible to the public by means of information technology.

Nothing new is foreseen for the formal recognition of agreements. In the case of co-regulation, no specific act is required, whereas for self-regulation, the Communication endorses the two already well-known approaches: an exchange of letters or the adoption of a Commission Recommendation.

These procedural rules will be complemented by provisions of the Interinstitutional Agreement on better law-making[16]. They reinforce the obligation on the Commission to inform Council and Parliament well before recognising any agreement. In the case of co-regulation, the basic legislative act may also include a provision for a two-month "*period of grace*" (sic) during which the legislators can suggest amendments to a draft agreement, or object to its entry into force.

It is worth noting that the Commission does not expect all environmental agreements at Community level to comply with these assessment criteria and procedural requirements. The Communication explicitly encourages so-called "spontaneous agreements", initiated by stakeholders in areas where the Commission has neither proposed legislation so far, nor expressed an intention to do so. Such agreements must not conflict with TEC rules or other Community legislation. Apart from this requirement, they are not subject to any constraint from the side of the institutions. On the other hand, of course, there is no formal "recognition" of spontaneous agreements as policy instrument at the European level.

The Communication finally proposes some policy fields where environmental agreements are likely to be considered first: the planned PVC Strategy, integrated product policy, waste management and climate change.

All in all, the 2002 Communication builds on the ideas and concepts already laid down in 1996 and on the experience gained so far with environmental agreements. The Commission's attitude towards this instrument is characterised essentially by continuity.

4.3 Follow-up: reactions from the EESC and the EP

Whereas the Committee of the Regions and – surprisingly – also the Council refrained from issuing any formal position on the Communication, both the European

Economic and Social Council (EESC) and the European Parliament (EP) reacted formally.

The EESC adopted its Opinion on 18 September 2002[17]. The overall tone is positive towards the Commission's approach. The ten EESC proposals address four aspects: compatibility of agreements with competition law, procedural suggestions, the development of more detailed criteria and – last but not least – the risk that the participation of the European Parliament and the Council could

> "make the process extremely complex and costly ... this could have a very damaging impact on cost-efficiency for partners in the voluntary agreement".

The European Parliament's Resolution, adopted on 13 May 2003[18], is equally supportive in its overall message. Divergent views appear, however, on some points. The EP

> "deplores ... the form of a non-binding communication",

expresses a preference for co-regulation, considers that

> "the results of spontaneous agreements ... should be systematically closely monitored"

and sees a comparative environmental impact assessment as essential prerequisite for the use of an environmental agreement. It stressed again, as it did already in 1997,that

> "traditional legislative instruments must continue to be the normal means of achieving the environmental policy objectives laid down in the Treaties".

A certain conflict between the two positions is evident: whereas the EESC is concerned by a potential "overburdening" of the instrument by procedural requirements, the European co-legislator is keen to keep a closer eye on all kind of environmental agreements, including those not recognised by the Commission. Industry is likely to share the EESC's position.

Both the Opinion and the Resolution endorse the Commission's choice of candidate policy fields. The EP suggests adding the Action Plan approved by the World Summit on Sustainable Development to the list.

The Commission signalled that such an extension of the field of application could be envisaged. When it comes to procedural requirements and criteria, it seems however preferable not to make the system designed in the Communication and the Interinstitutional Agreement even stricter and more demanding. The balance between the contradictory requests appears to be right. Priority will be given to testing these arrangements on a case-by-case basis before concluding on any need for conceptual or procedural changes.

5. FOLLOW-UP OF ENVIRONMENTAL AGREEMENTS

What happens after the formal recognition of an agreement? As outlined above in chapter 4.2, monitoring and reporting arrangements are crucial for the evaluation of the agreement's performance and the achievement of the objective. The credibility of the voluntary approach as such is at stake in this phase.

The modalities are normally tailor-made to the individual agreement, depending on its objective, scope, participants and timeframe. Some of the already existing environmental agreements indicate possible solutions. The follow-up modalities under the three agreements of the European, Japanese and Korean carmakers on CO2 emission reductions from new passenger cars may serve as first example.

The Recommendations recognising these agreements (see annex) have already established the principle of joint monitoring. In addition, Council and EP adopted a Decision establishing a scheme to monitor the average specific emissions of CO2 from new passenger cars[19]. As the implementation of the Decision took some time, the carmakers provided monitoring data in the beginning and for the base years (1995 to 2001). Post 2002, data provided by the Member States have been used. However, the car industry continues to collect its own data for comparison purposes. This ensures the availability of a two-track data stream, both from industry and from Member States.

Commission and industry representatives meet regularly – at least twice a year - at expert level (steering group) in order to prepare joint monitoring reports. In addition, they meet at least once per year on Director-General level (supervisory group) in order to adopt these joint reports and to discuss progress, the underlying assumptions as well as other relevant factors that might influence the commitment. Member States also meet with the Commission and in an informal expert group established under the above mentioned monitoring Decision.

The results of the whole monitoring process are published annually as formal Communications to the Council and the European Parliament. These Communications bring the three environmental agreements into the overall context of the Community Strategy to reduce CO2 emissions from cars, a strategy also covering the fuel-economy labelling of cars and the promotion of fuel efficiency by fiscal measures.

Four annual reports have been issued so far[20]. They are freely available on the Commission's website to anybody who is interested[21]. It is evident that this mechanism, put in place well before the adoption of the 2002 Communication on environmental agreements at Community level, already ensures a considerable degree of transparency.

Another example is the implementation of the 1998 agreement concerning good environmental practice for household laundry detergents. The proponent of the agreement, AISE (*Association internationale de la savonnerie, de la détergence et des produits d'entretien*), committed itself to a set of reduction targets in terms of energy consumption per wash cycle, weight of consumed detergents, weight of packaging and poorly biodegradable organic ingredients.

Under this agreement, an independent private organisation monitored progress towards the targets every two years. These reports, complemented by additional information given by AISE, formed the basis for a consultation process organised by the Commission with Member States, consumer organisations and environmental NGOs. The overall assessment was published in an official Commission document reporting on the results reached for the period 1996-2000[22].

As in the case of the CO2 reduction agreements, the monitoring system builds on a "two-track approach" (reporting by an independent organisation and by the Commission) and ensures stakeholder participation.

6. RECENT DEVELOPMENTS AND PERSPECTIVES

Communication on environmental agreements in July 2002, the Commission has not recognised any new agreement. This is simply due to the fact that no new draft agreement has been proposed, neither in the priority fields identified in the Communication nor in other areas. It is difficult to say why the economic actors are apparently reluctant to commit themselves to environmental objectives, despite their widespread interest in "good governance" and corporate social responsibility. It may be that another "new" instrument, namely emissions trading, is absorbing all attention for the time being.

Outside the field of environmental policy, a commitment by the European automobile industry to increase the protection of pedestrians and other road users from injury as a result of collision with a motor vehicle had an interesting fate. Debate on this commitment started at the same time that the 2002 Communication on environmental agreements was under development. The Commission presented its assessment of the terms of the commitment to the Council and the European Parliament by means of a formal Communication[23]. In the light of the legislators' reaction, the Commission finally decided to write the essential parts of the initial commitment into a proposal for a Directive[24].

Issues like health and safety of human beings are apparently considered as too crucial to be left to voluntary approaches. The European Parliament had left no room for doubt on this, notwithstanding its diplomatic language, when stating in its Resolution of 13 June 2002[25] that "the simple conclusion of a negotiated commitment does not seem to be a convincing means which will necessarily contribute to reaching the proposed goal" and asking the Commission to come forward with a framework Directive.

The use of *environmental* agreements will certainly be subject to the same restriction. On the procedural side, it is worth noting that the formalised information of the Community legislators already anticipated the approach laid down in the 2002 Communication.

Environmental agreements proposed to, but not recognised by the Commission – normally on grounds of insufficient objectives - do not simply go to the dustbin. They tend to be implemented by their proponents like spontaneous agreements, without any formal involvement of the Community institutions. This is notably the case for the "European Declaration on Paper Recovery", a commitment launched in November 2000 by three major European players in the field of paper and board manufacturing and recycling, and of "Vinyl 2010", the voluntary commitment of the PVC industry formalised in October 2001. The latter was initially proposed as a reaction to the Commission's intention to propose legislation in this field. It failed, however, to convince in terms of ambition. As the adoption of a Community wide PVC strategy is still pending, Vinyl 2010 now continues its career as a stand-alone instrument.

It is difficult to say, for the time being, at what point in time or for which economic sector the next environmental agreement will be proposed. It is most likely that the instrument will find its place in broader policy packages like the "thematic strategies" developed by the Commission under the Sixth Environmental Action Programme. Embedded in such strategies, environmental agreements will play a role as complement to "traditional" legislation and possibly also to economic instruments. A radical switch away from legislative instruments, towards a systematic and widespread use of agreements in all environmental policy fields, is unlikely to happen.

7. ANNEX: ENVIRONMENTAL AGREEMENTS ACKNOWLEDGED AT COMMUNITY LEVEL (IN CHRONOLOGICAL ORDER)

►Agreement on the reduction of chlorofluorocarbons by the aerosol industry (Commission Recommendation 89/349/EEC of 13 April 1989, OJ L 144, 27.5.1989, p.56)

►Agreement on the labelling of detergents and cleaning products (Commission Recommendation 89/542/EEC of 13 September 1989, OJ L 291, 10.10.1989, p.55; repealed by Regulation (EC) n° 648/2004 of 31 March 2004 on detergents with effect from 8 October 2005, OJ L 104, 8.4.2004, p.1)

►Agreement on the reduction of chlorofluorocarbons used by the Community's foam plastics industry (Commission Recommendation 90/437/EEC of 27 June 1990, OJ L 227, 21.8.1990, p.26)

►Agreement on the reduction of chlorofluorocarbons used by the Community's refrigeration industry (Commission Recommendation 90/438/EEC of 27 June 1990, OJ L 227, 21.8.1990, p.30)

►Agreement to reduce energy consumption by televisions and video recorders in standby mode (notice in OJ C 12, 16.1.1998, p.2)

►Agreement concerning good environmental practice for household laundry detergents (Commission Recommendation 98/480/EC of 22 July 1998, OJ L 215, 1.8.1998, p.73)

►Agreement on the cessation of production and imports of several models of washing machines with low energy efficiency (notice in OJ C 382, 9.12.1998, p.6)

►ACEA (European Automobile Manufacturers Association) agreement on CO2 emissions reductions from new passenger cars (Commission Recommendation 1999/125/EC of 5 February 1999, OJ L 40, 13.2.1999, p.49)

►KAMA (Korea Automobile Manufacturers Association) agreement on CO2 emissions reductions from new passenger cars (Commission Recommendation 2000/303/EC, OJ L 100, 20.4.2000, p.55)

►JAMA (Japan Automobile Manufacturers Association) agreement on CO2 emissions reductions from new passenger cars (Commission Recommendation 2000/304/EC, OJ L 100, 20.4.2000, p.57)

►Agreement CEFIC (European Chemical Industry Council) concerning the implementation of certain provisions of the Rotterdam Convention (Convention on the Prior Informed Consent Procedure for certain hazardous chemicals and pesticides in international trade) before its entry into force (exchange of letters CEFIC 27.11.2000/ Commissioners Liikanen and Wallström 22.12.2000; not published in the OJ)

►Agreement FECC (European Chemical Distributors' Association) concerning the implementation of certain provisions of the Rotterdam Convention (Convention on the Prior Informed Consent Procedure for certain hazardous chemicals and pesticides in international trade) before its entry into force (exchange of letters FECC 27.11.2000/ Commissioners Liikanen and Wallström 22.12.2000; not published in the OJ)

8. NOTES

* This text reflects the personal opinion of the author. It is not an official position paper of the European Commission.

[1] Regulation (EC) N° 761/2001 of the European Parliament and of the Council of 19 March 2001 allowing voluntary participation by organisations in a Community eco-management and audit scheme (EMAS), OJ L 114, 24.4.2001, p.1 (replaces Council Regulation N° 1836/93, OJ L 168, 10.7.1993, p.1)

[2] Regulation (EC) N° 1980/2000 of the European Parliament and of the Council of 17 July 2000 on a revised Community eco-label award scheme, OJ L 237, 21.9.2000, p.1 (replaces Council Regulation N° 880/92, OJ L 99, 11.4.1992, p.1)

[3] COM(96)561 final of 27.11.1996

[4] OJ L 269, 21.10.2000, p.34

[5] OJ L 37, 13.2.2003, p.24

[6] OJ L 47, 18.2.2004, p.26

[7] COM 2003(723) final of 21.11.2003

[8] Council Directive 83/189/EEC, OJ L 109, 19.4.1983, as amended by Council Directive 88/182/EEC, OJ L 81, 26.3.1988, and by European Parliament and Council Directive 94/10, OJ L 100, 19.4.1994, p.30

[9] OJ C 321, 22.10.1997, p.6

[10] OJ C 286, 22.9.1997, p.254

[11] OJ C 21E, 24.1.2002, p.112

[12] COM(2002)412 final

[13] OJ P 13, 21.2.1962, p.204

[14] COM(2002)278 final of 5.6.2002

[15] "Aarhus Convention", signed on behalf of the European Community on 25 June 1998. Formal approval still pending (Proposal for a Council Decision COM(2003)625 final of 24 October 2003).

[16] OJ C 321, 31.12.2003, p.1

[17] Doc. CES 1029/2002
[18] EP document P5_TA-PROV(20030513)0205. Not yet published in the OJ.
[19] Decision n° 1753/2000/EC, OJ L 202 of 10.8.2000, p.1
[20] COM(2000)615 final of 4.10.2000, COM(2001)643 final of 8.11.2001, COM(2002)693 final of 9.12.2002 and COM(2004)78 final of 11.2.2004
[21] Website address: http://europa.eu.int/comm/environment/co2/co2_monitoring
[22] COM(2002)287 final of 10 June 2002
[23] COM(2001)389 final of 11 July 2001 "Pedestrian protection: Commitment by the European automobile industry"
[24] COM(2003)67 final of 19 February 2003, proposal for a Directive of the European Parliament and the Council relating to the protection of pedestrians and other vulnerable road users in the event of a collision with a motor vehicle and amending Directive 70/156/EEC. Directive adopted on 17 November 2003 (2003/102/EC, OJ L 321, 6.12.2003, p.15)
[25] EP DOCUMENT P5_TA-PROV(20020613)0323. NOT YET PUBLISHED IN THE OJ.

9. REFERENCES

OECD, (1999), *Voluntary Approaches for Environmental Policy – an assessment*. OECD Publications, ISBN 92-64-17131-2

OECD, (2003), *Voluntary Approaches for Environmental Policy – effectiveness, efficiency and usage in policy mixes*. OECD Publications, ISBN 92-64-10177-2

THE USE OF VOLUNTARY APPROACHES FOR ENVIRONMENTAL POLICYMAKING IN THE U.S.

K. BROUHLE[#], C. GRIFFITHS[##], A. WOLVERTON[##]

[#]*University of Alberta, Canada*

[##]*U.S. Environmental Protection Agency, Washington, D.C., USA*

Abstract. The use of voluntary approaches to achieve environmental improvements has grown dramatically in the United States (U.S.) since they were first introduced thirteen years ago. As of 2004, there are over 50 voluntary programs in the U.S. at the federal level alone. These programs take a variety of forms, from large, cross-industry efforts to reduce global climate impacts to smaller, "boutique" efforts aimed at specific industrial sectors. Other voluntary approaches used in the U.S. include negotiated agreements, industry-initiated unilateral commitments, and state and regional voluntary initiatives, but these tend to be used less regularly.

Despite the diversity of voluntary approaches in the U.S., they often pursue common, and sometimes overlapping environmental objectives and use similar methodologies to achieve such goals. While most voluntary initiatives in the U.S. state an explicit environmental goal, they may also have less direct policy objectives such as enhancing innovation or increasing awareness of environmental issues.

Many argue in favour of the increased use of voluntary approaches in environmental policymaking on the basis of environmental effectiveness, economic efficiency, reductions in government administrative, monitoring and enforcement costs, increases in environmental awareness, and encouragement of innovation. Few programs have been evaluated properly on the basis of these objectives, however. The empirical literature sheds little light on the value of voluntary approaches in achieving goals set by U.S. environmental policy. The difficulty in evaluating voluntary approaches lies in sorting through the myriad of programs, identifying a discernible environmental goal, gathering adequate data for analysis, and measuring achievement of the environmental goal relative to a reasonable baseline scenario.

1. INTRODUCTION

The number of voluntary approaches used to address environmental issues in the U.S. has grown dramatically since the introduction of the Environmental Protection Agency's 33/50 program in 1991. In 1999, the OECD identified 42 voluntary initiatives in the U.S. with an estimated 13,000 participants. The U.S. EPA, the primary environmental regulatory agency in the U.S., administered 33 of these initiatives (OECD, 1999). Currently, the U.S. EPA's Partners for the Environment website lists 40 voluntary initiatives (U.S. EPA, 2004a). At the federal level alone, we identify over 50 voluntary initiatives in the U.S.

The majority of U.S. voluntary efforts are what the OECD (2003) describes as public voluntary programs. That is, the federal, state, or regional regulatory authority designs the programs, and individual firms are invited to participate. The 33/50, Energy Star™, and WasteWise programs are all examples of public voluntary

E. Croci (ed.), The Handbook of Environmental Voluntary Agreements, 107 - 134

programs in the U.S. This type of approach is in contrast to negotiated agreements between government and industry that are more commonly found in Europe, but are relatively rare in the U.S. (Croci, 2003). A third type of voluntary effort sometimes found in the U.S. is a unilateral commitment, also know as an industry-led initiative. These tend to be proposals generated by industry or trade associations with little or no involvement from government regulators. Examples of this type of voluntary initiative include the chemical industry's Responsible Care® program, and an industry-led carbon emissions trading system called the Chicago Climate Exchange.[1]

Voluntary initiatives in the U.S. are typically designed to enhance the efficacy and scope of existing regulations. Benefits of participation in these programs generally include technical assistance, information subsidies, and public recognition, but firms usually do not garner regulatory exemption. While participating firms generally sign an agreement to adhere to a specified environmental goal and sometimes provide self-reported information on pollution levels or achievement of goals, voluntary initiatives are typically non-binding and do not penalize firms for non-attainment beyond possible revocation of membership (Darnall and Carmin, 2004). This is largely due to the institutional framework in which environmental policymaking occurs. In the U.S., regulatory agencies are held at arms length from the policy process by Congress and therefore lack sufficient credibility to issue threats of new regulation (Delmas and Terlaak, 2002a). In some cases, voluntary approaches are used to reduce emissions in sectors or for pollutants for which political will to pass formal regulation is lacking (e.g., in the area of climate change). In general, voluntary programs are not a central component of the U.S. EPA's regulatory activities (Mazurek, 1999). In contrast to the experience in some European countries, U.S. agencies continue to rely on more traditional forms of environmental regulation and in particular on standard-based regulation.

That said, voluntary approaches are not necessarily unimportant or ineffective. The U.S. EPA's 33/50, Green Lights, and Energy Star™ programs have achieved moderate levels of success in furthering environmental improvements (Decanio, 1994; Khanna and Damon, 1999; Horowitz, 2001; Nadeau, Cantin, and Wells, 2003). Some researchers, however, question the ability of voluntary approaches to produce environmental benefits beyond what would occur if the initiatives were not in existence, particularly given the potential for firms with a predisposition towards environmentally-responsible behaviour to join these programs (Hartman, 1988; Ozog and Waldman, 1994), the temptation for some firms to free ride (King and Lennox, 2000; Welch, Mazur, and Bretschneider, 2000), and other constraints (Maitland, 1985). Given the growing number of voluntary approaches being implemented in the U.S., measuring the success of voluntary programs has become increasingly important. The largest difficulty in conducting an evaluation is gathering adequate data for the analysis. Available research on voluntary programs generally evaluates older programs that have been in existence long enough to produce potentially measurable results. Current voluntary programs may have learned from past mistakes but are still too new to be evaluated. Evaluation is further complicated by the task of sorting through the myriad of voluntary programs, identifying a discernible environmental goal, and measuring achievement of the environmental goal relative to a reasonable baseline scenario.

This paper examines U.S. voluntary initiatives, their key similarities and differences, evidence on their effectiveness, and some of the difficulties in evaluating them. While pollution prevention and other environmental activities are worthwhile for firms to undertake--firms clearly recognize that there are benefits from engaging in such activities--our key interest is in determining whether third-party or government voluntary initiatives induce firms to reduce emissions or incorporate pollution prevention activities into production beyond what they would have undertaken without such a program in place. Section II describes some of the common methods used in the design of U.S. public voluntary programs at the federal level; Section III discusses the types of voluntary approaches typically found in the U.S.; and Section IV discusses the arguments and evidence for how well U.S. voluntary initiatives meet five possible policy objectives. Section V describes the main difficulties faced by researchers and policymakers alike when attempting to evaluate U.S. voluntary programs. Section VI concludes.

2. THE METHODOLOGY OF DESIGNING PUBLIC VOLUNTARY PROGRAMS

Broadly speaking, there are three basic methods commonly used by U.S. public voluntary programs at the federal level to achieve environmental improvements: (1) require firms to set specific environmental goals; (2) promote firm environmental awareness and encourage process changes; and (3) supply the public with information on firm participation or environmentally-responsible products. No one approach is used to the exclusion of others; most U.S. voluntary programs use a combination of several methods.

By far, the most common method used in the design of U.S. voluntary environmental programs is goal setting. These may be implementation-based goals, where participants meet specific targets set by the U.S. EPA, or they may be target-based goals, where the U.S. EPA specifies a qualitative or process-oriented goal and firms individually set and then meet a specific target (OECD, 1999). Implementation-based goals provide consistency in objectives across firms, while target-based goals allow firms increased flexibility.

Since U.S. programs tend to augment existing regulation, implementation-based goals make it easier to design approaches that go beyond standards specified in formal regulation. They also make it simpler to monitor and measure whether participants in the program are meeting the goal. The 33/50 program is an example of a voluntary program that sets an implementation-based goal. Firms were challenged to reduce their emissions of 17 priority toxic chemicals by 33% by 1992 and by 50% by 1995. While firms were allowed to decide how to accomplish these goals, the U.S. EPA dictated the percentage reductions.

With target-based goals, participants choose (and sometimes publicly announce) their own targets. This allows firms the flexibility to set a goal that meets the qualitative objective specified by the voluntary approach and allows for heterogeneity in costs across firms. The U.S. EPA's WasteWise program is an example of this type of program. Partners are required to implement three waste prevention activities, improve the collection of recyclable material, and increase the purchase of recycled material. However, individual firms are allowed to determine the degree to which these broad goals are achieved and how they should be implemented. Another example is the Climate Challenge program where firms set their own goals on emissions reductions.

Voluntary initiatives with target-based goals require the regulator to know much less about the ability of individual firms to respond to a particular goal, since the firm selects a goal it feels is appropriate given its particular cost structure. The problem with this approach is that it is harder to measure achievement of the goal because of the host of possible goals a firm could set and the need for firm-specific information to evaluate success or failure. This goal-setting approach has also been criticized because firms are unlikely to set goals that are a challenge for them to meet. This is one of the criticisms of the Responsible Care® program (Howard, Nash, and Ehrenfeld, 2000; King and Lennox, 2000). Firms may simply set goals that require actions they would have taken anyway. In other words, firms pre-disposed to undertake environmentally beneficial actions self-select into the program to earn an added benefit such as public recognition.

A second method used in the design of U.S. public voluntary programs is to promote environmental awareness of participating firms and/or to encourage environmentally beneficial process changes. The Green Lights program encouraged firms to adopt energy efficient changes that translate into savings in firm electricity costs. The SF6 Emissions Reduction Partnership program attempts to raise environmental awareness by providing information on the contribution of sulfur hexafluoride to greenhouse gases and to encourage the industry to adopt new production processes that significantly reduced the use of SF6. The Design for the Environment and Green Chemistry programs work with industry to research environmentally friendly processes and technologies and then encourage firms to adopt them.

The difficulty with this design method is that it requires detailed knowledge of a firm's ongoing operations for evaluation. Also, firms may not be comfortable with the government in the role of directing technology investment decisions toward particular technologies and away from others based on the information they provide. Critics argue that the market is better at choosing environmentally friendly process changes than is government. Advocates of this approach counter that the market may not be better at guiding technology decisions if there is a lack of information about the technologies available (Office of Technology Assessment, 1991), especially if the technologies are in an area unrelated to the firm's primary revenue activity. Even profitable process changes may not be known if the firm is not able to justify the cost of researching it. A government agency, on the other hand, may have economies of scale advantages in researching and disseminating information.

A third method used in the design of U.S. voluntary programs is to publicize the environmental responsibility of program participants or products. The idea is to supply this information so that green consumers and investors can alter consumption and investment patterns in favour of cleaner firms or products (Arora and Gangopadhyay, 1995; Cohen and Konar, 1997, 2001). This is the intent behind programs such as Energy Star™, Green Power Partnership, and It All Adds Up to Cleaner Air. Two common methods of conveying information on participants or products are (1) to simply publicize the membership in the program, and (2) to establish a set of criteria that when met earn the firm or product a label. The easiest type of information to publicize is general in nature. Energy Star™ publicizes participant names, highlights companies that excel at energy efficiency and provides results for particular energy-efficient buildings or products that have earned the label. The National Environmental Performance Track and Climate Leaders programs also highlight outstanding performers.

The effectiveness of this information remains an open question (Bui, 2003). If the provision of these data truly provides a business advantage, then one wonders why firms do not self-report this information to the public directly. One answer may have to do with credibility. The U.S. EPA serves as a third party, offering neutrality and potentially some control as verifier and presenter of the information. Firms may use their goal achievements as a way to potentially differentiate their product from their competitors' products (Arora and Cason, 1995; Videras and Alberini, 2000). Even if the publicity does not attract environmentally-minded consumers, it may still prove useful to the firm to be recognized by the U.S. EPA because it adds to a company or product's overall name recognition in the marketplace.

3. DESCRIPTION OF U.S. VOLUNTARY APPROACHES

Since the start of the 33/50 and Green Lights programs in 1991, over 50 other voluntary approaches have been initiated. Maxwell and Lyon (1999) argue from a political economy perspective that four forces in the late 1980s set the stage for the growth of voluntary initiatives in the U.S.: mounting and increasingly complex legislation, technological innovation and scientific discoveries, regulatory budget cuts, and the increasing use and effectiveness of "citizen lawsuits." Another reason voluntary initiatives have become increasingly popular methods for environmental regulation is that they allow policy-makers to address environmental issues that may lie outside the existing regulatory framework or issues for which regulation may be difficult to pass due to political considerations. For instance, due to limited authority under the Clean Air Act to control greenhouse gas emissions and lack of political will to impose strict standards, the U.S. government has increasingly turned to voluntary initiatives as a mechanism to address climate change concerns: Over one third of U.S. voluntary initiatives are designed to address global warming issues (see Table 1).

Most voluntary environmental initiatives in the U.S. are conceived and designed by the U.S. EPA. Other U.S. federal agencies that have designed voluntary environmental initiatives, either on their own or in conjunction with the EPA, include the Department of Energy, Department of the Interior, Department of Agriculture, Department of Transportation, and Occupational Safety and Health Administration. As noted in section 1, most voluntary initiatives in the U.S. are public voluntary programs. We separate this category of voluntary initiatives into multi-sector federal voluntary programs, single-sector federal voluntary programs, and state and local voluntary programs. Negotiated agreements and unilateral commitments also are discussed as separate categories in this section. Table 1 lists the U.S. EPA's main environmental voluntary initiatives at the federal level and a select group of voluntary initiatives at the state and regional level in the U.S., organized by major category, and also describes the targeted industries or sectors and the main environmental goal of the program.[2]

3.1. Multi-Sector Federal Voluntary Programs

Multi-sector federal voluntary programs include many of the most well known voluntary initiatives in the U.S., such as Green Lights, Energy Star™, National Environmental Performance Track, Sector Strategies, and WasteWise. Public authorities typically design these programs with possible consultation but no direct negotiation with affected industries and then invite firms to participate in the voluntary program.[3] Most of these public voluntary programs are designed to meet the goals of President Clinton's 1993 Climate Change Action Plan, which seeks to reduce greenhouse gas emissions, or to meet voluntary goals set out in the Pollution Prevention Act of 1990 (Mazurek, 1998). Nearly one third of multi-sector federal voluntary programs focus on energy efficiency and climate change issues. Another one third of these programs have pollution prevention as an environmental target (see Table 1).

A key characteristic of multi-sector federal voluntary programs is that they tend to target a wide variety of firms from different industries. Energy Star™ has more than 8,000 firms listed as partners and has awarded labels to items in over 40 different product categories (U.S. EPA, 2004b), while the WasteWise program has over 1,300 participating firms from 54 industry sectors (U.S. EPA, 2004c). To attract participants from different industrial sectors, multi-sector federal voluntary initiatives also tend to have general environmental objectives. Both Green Lights and Energy Star™, for example, focus on reducing energy consumption and improving energy efficiency. The WasteWise program encourages firms to reduce the amount of waste generated and increase recycling.

Increasingly, multi-sector federal voluntary programs promote the adoption of environmental management systems (EMS) as a way to manage and monitor participants' environmental responsibilities and achievements. The National Environmental Performance Track, Sector Strategies, and Design for Environment programs all encourage EMS implementation as a way to increase the likelihood that firms will meet environmental goals. Since firms in any industry can adopt an EMS, their use also encourages participation in voluntary initiatives across a variety of industries (Darnall and Carmin, 2004).

3.2. Single-Sector Federal Voluntary Programs

Single-sector federal voluntary programs in the U.S. differ from multi-sector federal voluntary programs in terms of both the scope of the environmental objectives and the number and type of participating firms that the programs target. While multi-sector federal voluntary programs have general environmental objectives that target firms from different industries, single-sector federal voluntary programs typically focus on more specific environmental problems that are often relevant to a particular industrial sector. For example, the 33/50 program targeted toxic emissions from the chemical industry; the Voluntary Aluminum Industry Partnership targets perfluorocarbon emissions from aluminium producers; and the AgSTAR program targets methane emissions from concentrated animal feeding operations (CAFOs). Many single-sector federal voluntary programs target global climate change problems associated with transportation-related issues and energy-producing sectors like coal mining and power generation (see Table 1). The focused aim of these programs attempts to provide targeted and effective technological expertise and assistance to participating firms.

3.3. State and Regional Voluntary Programs

A third type of public voluntary program is designed by state, regional, or local authorities, often to implement federal directives. Several regional offices of the U.S. EPA have established pollution prevention programs to help implement the federal Pollution Prevention Act of 1990. Regional U.S. EPA offices also help administer federal voluntary programs like National Environmental Performance Track in a given locality. More often, though, state and regional voluntary programs are employed to confront local environmental problems or issues. For example, the Southern Appalachian Mountains Initiative addresses air quality issues in the Appalachian Mountain region, while Project Loko I'a encourages restoration of coastline and the redevelopment of traditional Hawaiian fish ponds. Finally, some state and regional voluntary programs can be viewed as a testing ground for a national program. The San Francisco Bay Area Green Business Program, for example, is used to gauge the demand for a green business program that could be expanded in the future (U.S. EPA, 2004d). Table 1 lists a few of these state and regional programs.

3.4. Federal Negotiated Agreements

A fourth type of voluntary initiative in the U.S. is negotiated agreements. Unlike public voluntary programs, negotiated agreements are jointly designed by industry and federal regulators. The only two examples of negotiated agreements in the U.S. at the federal level are the Common Sense Initiative and Project XL (see Table 1). These agreements offer firms relief from existing regulations in return for demonstrating environmental performance above and beyond the existing status quo (Mazurek 1998). By involving firms in the regulatory process and by offering regulatory flexibility, negotiated agreements aim to reduce the adversarial relationship between firms and regulators. In addition, it is hoped that regulatory flexibility encourages firms to reduce pollution in the most cost-efficient manner. There has been concern, however, that firms may use negotiated agreements to attempt to influence regulatory authorities. In particular, firms may use negotiated agreements as a smokescreen for real environmental action, and hence delay or even pre-empt future regulations (Lutz, Lyon, and Maxwell, 2000; Maxwell, Lyon, and Hackett, 2000).

Negotiated agreements are not used frequently in the U.S. for several reasons. First, government-business relationships in the U.S. tend to be more adversarial, especially compared to those in Europe, and therefore do not create an environment conducive to negotiated agreements (Lyon and Maxwell, 2001). Second, questionable legal authority of regulatory agencies to exempt firms from existing regulations has prevented the adoption of negotiated agreements (GAO, 1997a; OECD, 1999). A third obstacle to more wide scale adoption of negotiated agreements is the potentially large cost of negotiations. For example, Intel's Project XL agreement was both time-consuming and expensive to negotiate. It took over 100 official meetings and involved 23 official representatives from different government agencies and the local community (Mazurek, 1998). It took 17 months of negotiation to put the agreement in place, and cost both firm and government a total of $588,000 (Blackman and Mazurek, 2001). Delmas and Mazurek (2001) find that by 2001 the average cost of negotiating an agreement for Project XL had fallen to $108,000. That said, the uncertain legal environment and substantial time and money involved in a negotiated agreement indicate that negotiated agreements are unlikely to be an important form of voluntary initiative in the U.S. in the future. In fact, the Common Sense Initiative is no longer active and Project XL is currently being phased out of operation (U.S. EPA, 2004e).[4]

Table 1. *U.S. Voluntary Initiatives*

Name of Voluntary Initiative (year established)	Industry / Sector	Environmental Target
Multi- Sector Federal Public Voluntary Programs		
Climate Leaders (2002)	multi-sector	global climate change
Climate Wise* (1993)	multi-sector	energy efficiency and global climate change
Consumer Labeling Initiative (1996)	multi-sector	labeling
Design for the Environment (1992)	multi-sector	green manufacturing; pollution prevention
Energy Star™ (1992)	multi-sector	energy efficiency and global climate change
Environmental Technology Verification (1995)	multi-sector	green manufacturing, labeling
Green Chemistry (1992)	multi-sector	pollution prevention
Green Engineering (1998)	multi-sector	pollution prevention
Green Lights* (1991)	multi-sector	energy efficiency and global climate change
Green Power Partnership (2001)	multi-sector	energy efficiency and global climate change
Green Suppliers Network (2001)	multi-sector	green manufacturing; pollution prevention
Improving Air Quality through Land Use (2000)	multi-sector	air quality
It All Adds Up to Cleaner Air (1999)	multi-sector	air quality
National Environmental Performance Track (2000)	multi-sector	regulatory innovation
National Waste Minimization Partnership (1994)	multi-sector	pollution prevention
Sector Strategies Program† (2003)	multi-sector	regulatory innovation
WasteWise (1992)	multi-sector	waste management
Water Efficient Product Enhancements Program† (1992)	multi-sector	water

** This program is no longer active † This program has undergone a name change at some time*

Table 1 (cont.)

Table 1. (cont.)

Name of Voluntary Initiative (year established)	Industry / Sector	Environmental Target
Single-Sector Federal Public Voluntary Program		
33/50 Program* (1991)	chemical	reduce toxic waste
Adopt Your Watershed (1994)	agriculture, public	water quality
AgSTAR (1993)	agriculture, CAFOs	global climate change
Best Workplaces for Commuters† (2000)	transportation	air quality and global climate change
Carpet America Recovery Effort (2001)	carpet	recycling
Clean Air Transportation Communities (2001)	transportation	air quality and global climate change
Clean School Bus USA (2003)	transportation	air quality and global climate change
Climate Challenge* (1994)	electric utilities	energy efficiency and global climate change
Coal Combustion Products Partnership (2003)	coal combustion	energy efficiency and global climate change
Coalbed Methane Outreach Program (1994)	coal mining	air quality and global climate change
Combined Heat and Power Partnership (2001)	power generation	energy efficiency and global climate change
Five-Star Restoration Program (1998)	wetlands	habitat restoration
Golf and the Environment (1995)	golf courses	land use
GreenScapes (2003)	landscaping	land use
High Production Volume Challenge Program (1998)	chemical	reduce toxic waste
HFC-23 Emission Reduction Program (1993)	chemical	global climate change
Hospitals for a Healthy Environment (1998)	hospitals	waste management

* *This program is no longer active* † *This program has undergone a name change at some time*

Table 1 (cont.)

Table 1 (cont.)

Name of Voluntary Initiative (year established)	Industry / Sector	Environmental Target
Single-Sector Federal Public Voluntary Program (continued)		
Indoor Air Quality Tools for Schools (1995)	schools	air quality
Landfill Methane Outreach Program (1994)	waste management	global climate change
Mobil Air Conditioning Climate Protection Partnership (1999)	automobile	energy efficiency and global climate change
Natural Gas STAR Program (1993)	oil and natural gas	global climate change
Pesticide Environmental Stewardship (1994)	agriculture; other users	reduce risk from use of pesticides
PFC Reduction/Climate Partnership (1996)	semiconductor	global climate change
Plug-In to e-Cycling (2003)	electronics	recycling
Reduced Risk for Conventional Pesticides (1993)	chemical	regulatory innovation
Ruminant Livestock Methane Efficiency* (1993)	livestock operations	global climate change
Seasonal Gas Use to Control Nitrous Oxide* (1993)	utilities; industrial facilities	energy efficiency and global climate change
SF6 Emission Reduction Partnership for Electric Power Systems (1999)	electric power	global climate change
SF6 Emission Reduction Partnership for Magnesium Industry (1999)	magnesium	global climate change
SmartWay Transport Partnership (2004)	freight	air quality and global climate change
SunWise School Program (2000)	schools	health improvement
Suppliers Partnership for the Environment (2001)	automotive	green manufacturing
Sustainable Future Initiative (2002)	chemical	pollution prevention

** This program is no longer active † This program has undergone a name change at some time*

Table 1 (cont.)

Table 1 (cont.)

Name of Voluntary Initiative (year established)	Industry / Sector	Environmental Target
Single-Sector Federal Public Voluntary Program (continued)		
Voluntary Aluminum Industrial Partnership (1995)	aluminum	global climate change
Voluntary Children's Chemical Evaluation Program (2000)	chemical	information provision
Voluntary Diesel Retrofit (2000)	transportation	air quality
State and Regional Public Voluntary Programs		
California Dairy Quality Assurance Program (1999)	dairy	water quality
Chemical Facility Security (2002)	chemical	pollution prevention
Hospital Compliance Environmental Initiative (2002)	hospitals	reduce toxic waste
Nevada Mining Partnership Program (2001)	mining	reduce toxic waste
Project Loko I'a (1999)	aquaculture	water
S.F. Bay Area Green Business Program (1996)	multi-sector	green goods; labeling
Southern Appalachian Mountains Initiative (1992)	multi-sector	air quality
Sustainable Agriculture Partnership (1993)	agriculture	agriculture
Negotiated Agreements		
Common Sense Initiative* (1994)	multi-sector	regulatory innovation
Project XL* (1995)	multi-sector	regulatory innovation

** This program is no longer active † This program has undergone a name change at some time*

Table 1 (cont.)

Table 1 (cont.)

Name of Voluntary Initiative (year established)	Industry / Sector	Environmental Target
Unilateral Commitments		
Chicago Climate Exchange (2003)	multi-sector	energy efficiency and global climate change
Coatings Care™ (1996)	paint	pollution prevention
Encouraging Environmental Excellence (1992)	textile	pollution prevention
Environmental Stewardship Program† (1990)	oil and natural gas	pollution prevention
Great Printer's Project (1992)	printing	pollution prevention
Green Seal (1992)	general	green goods; pollution prevention
International Standards Organization 14001 (1996)	multi-sector	pollution prevention
Responsible Care® (1988)	chemical	pollution prevention
Responsible Carrier (1994)	shipping	safety
Responsible Distribution Process℠ (1991)	chemical	pollution prevention
Sustainable Forestry Initiative® (1995)	forestry, paper	land use; labeling
Sustainable Slopes (2000)	ski resorts	land use

** This program is no longer active † This program has undergone a name change at some time*

3.5. Unilateral Commitments

A fifth type of voluntary initiative found in the U.S. is unilateral commitments. These programs typically do not involve federal regulators but are industry or trade association-led efforts to improve environmental performance (Table 1 lists a few examples). Motivations for participation in unilateral commitments vary. Some commitments may be a response to outside pressure from consumers and shareholders to improve the image of an industry. The Responsible Care® program, for example, is largely seen as a response by the chemical industry to several accidents and spills, including the Bhopal accident in 1984, which killed over 3,000 people. Other unilateral commitments may be undertaken in an attempt to differentiate members' products from others and to cater to green consumers (Arora and Gangopadhyay, 1995). Some unilateral commitments may attempt to pre-empt future regulation. Nash (2002) writes that the "chemical, petroleum, and forestry industries have used [voluntary initiatives] as defensive strategies to protect themselves from external interference in the form of public regulation." Finally, other programs may be viewed as a mechanism for firms to prepare themselves for future regulation (Lutz, Lyon, and Maxwell, 2000; Maxwell, Lyon, and Hackett, 2000). For example, with the possibility of global climate change regulation on the horizon, some firms are using the Chicago Climate Exchange to demonstrate that reductions in greenhouse gas emissions can occur in a cost-effective manner through the use of a cap-and-trade system. While emission targets are not demanding (members are to reduce emissions by 4% below baseline by 2006), participation in the Exchange since its inception in 2003 has grown from 13 to 51 members and includes a wide range of sectors such as chemical, pulp and paper, and electric power generation. Approximately 83,000 metric tons of CO2 were traded in January 2004, double the previous month's trading volume. In February 2004, over 400,000 metric tons were traded (Chicago Climate Exchange, 2004).

Unilateral commitments often are developed by industry trade associations or international organizations. The American Chemical Association, for example, started the Responsible Care® program to encourage safety in the handling of chemical products from inception through distribution and disposal. Other examples of industry-led unilateral agreements include Coatings Care™ by the National Paint and Coating Association and the Sustainable Forestry Initiative® by the American Forestry and Paper Association. Another source of unilateral commitments is international organizations. The International Standards Organization introduced ISO 14001 in 1996 to recognize and standardize the efforts of firms in incorporating environmental management systems.

Due to anti-trust concerns, industry-initiated voluntary approaches generally avoid prescribing specific actions, outcomes, or strategies (Kappas, 1997; Mazurek, 1998). Instead, these initiatives offer public recognition, information subsidies, and technical assistance in the pursuit of qualitative goals. They also rarely contain monitoring or sanction provisions, although some effort has been made by companies to appear credible through third-party verification. (OECD, 1999). Most unilateral agreements in the U.S. are best characterized as pollution prevention activities.

4. COMMON POLICY OBJECTIVES OF VOLUNTARY APPROACHES

The literature discusses a variety of reasons why voluntary approaches should be preferred to other more traditional forms of environmental regulations. However, there is a great disparity between the power of the arguments from a policy standpoint and the empirical evidence supporting such arguments. This section reviews five common policy objectives of voluntary approaches in the U.S. and elsewhere. These objectives include environmental effectiveness, economic efficiency, savings in administrative, monitoring, and enforcement costs, inducement of innovation, and increased environmental awareness. For each of these policy objectives, this section discusses the arguments for and against the achievement of such goals, and whether the available empirical evidence on U.S. voluntary approaches lends support to these objectives.

4.1. Environmental Effectiveness

The most commonly discussed objective of any environmental policy or regulation is how effective it is at improving environmental quality. Does the policy instrument accomplish a measurable environmental goal? It is argued that voluntary approaches are effective because they promote general environmental improvements or emission reductions beyond what is mandated by formal environmental regulation. It is also argued that voluntary approaches are instrumental in inducing firms to reduce emissions by greater amounts than they would have without the voluntary initiative in place. In other words, even the more environmentally responsible firms will be induced to take actions beyond what is defined as "business-as-usual" behaviour.

Critics of these arguments point to the fact that voluntary approaches often set easy-to-achieve targets that barely go beyond what is specified by existing regulation and that, compared to the alternative of establishing a more formal regulation mandating further reductions in emissions, the commitments made by firms to voluntary goals are less stringent and far less demanding (OECD, 1999). Also, once these targets have been set, most voluntary approaches in the U.S. are non-binding. This means that even if a firm agrees to a particular environmental target, it is not legally obligated to achieve it, and may merely use its membership in a voluntary initiative as a marketing tool. Finally, as touched on in section 3, a number of academics have pointed out that firms' incentives to over-comply with existing regulation through adherence to a voluntary goal may, in the long term, be designed to avoid future, more stringent regulations. These firms may, in fact, be co-opting the regulatory policy process by limiting regulators' options in the future. For instance, firms may invest in a particular technology to meet voluntary goals, knowing that it limits the practicality of mandating more stringent and therefore more costly emission reductions in the future (Lutz, Lyon, and Maxwell, 2000; Maxwell, Lyon, and Hackett, 2000).

In many cases, a lack of data does not allow for an evaluation of the effectiveness of a particular voluntary agreement in attaining its environmental goal (see Section 5). Thus, for these programs how environmental effective they are is still an open question. For those voluntary agreements that have been evaluated, the evidence of their environmental effectiveness is mixed. Horowitz (2004) finds that public voluntary programs promoting energy efficient products and services have been successful in achieving energy reductions. He estimates that the market effects due to the Green Lights program resulted in savings of over 5.6 million metric tons of carbon in the year 2000 (Horowitz, 2001, 2004). Other studies indicate less success. In many cases where noticeable emission reductions have occurred, factors other than the voluntary approach itself seem to have contributed to the reduction (GAO, 1997b; OECD, 2003). Investigators such as Morgenstern and Al-Jurf (1999) have found that voluntary agreements such as the U.S. EPA's Green Lights and 33/50 programs result in much more modest reductions than what is often claimed by regulators as attributable to the programs themselves. Hartman (1988) finds comparable results for voluntary energy conservation programs in the U.S. King and Lenox (2000) find that U.S. chemical companies that have signed on to the unilateral commitment Responsible Care® improve their environmental performance more slowly than non-participating firms, lending support to the argument that firms use the program as cover for less environmentally benign actions. Welch, Mazur, and Bretschneider (2000) find that participation of firms in the Department of Energy's Climate Challenge program has had no effect on CO2 reduction levels.

Is it possible to improve the environmental effectiveness of voluntary approaches? Alberini and Segerson (2002) demonstrate theoretically that the effectiveness of a voluntary approach increases when there is a credible regulatory threat and reliable monitoring of goal attainment. They also point out that the degree to which a voluntary approach results in real environmental improvements depends on the number of polluters that participate; the amount of abatement undertaken by each participant, and the effect that the voluntary approach has on the competitiveness of the market.

4.2. Economic Efficiency

A second objective by which environmental regulatory policies are evaluated is economic efficiency. In evaluating whether such an objective is reached, the relevant question is whether a particular approach reaches a given environmental goal at the lowest possible cost to firms and consumers. In particular, how close do voluntary approaches get to the most efficient outcome? Two slightly different arguments are made regarding the efficiency of voluntary approaches. The first argues that, to the extent that voluntary approaches help reduce uncertainties or supply information not provided in the marketplace, they bring to light new ways in which firms can reduce costs of production while also improving environmental performance. This is often referred to in the literature as a "win-win" argument (Lyon and Maxwell, 2001) and can be thought of as efficiency in absolute terms. The second argues for voluntary approaches from the perspective of relative efficiency. Voluntary approaches, while not as efficient as market-based instruments that attempt to internalise and price externalities appropriately in the market, are more efficient than command-and-control regulations because they allow firms greater flexibility to determine the best way to reduce emissions to meet the voluntary goal (OECD, 1999).

Many economists are sceptical of the "win-win" argument and often ask why a firm operating in a competitive market would leave such profit-making opportunities unexploited. Boyd (1998), through a set of case studies, finds that pollution prevention may not be undertaken by firms due to regulatory barriers; lack of consistent monitoring and enforcement by regulators; or regulatory, market or technical uncertainties and challenges that prevent the diffusion of new technologies. This suggests a potential role for government in offering firms greater regulatory flexibility, technical assistance, and increased pressure to comply with existing regulations, some of which could be provided through voluntary programs.

Regarding the second argument of relative efficiency, there is agreement that in some cases voluntary approaches may get closer to the efficient outcome than command-and-control regulations. However, they are unlikely to achieve full efficiency because, while they often allow firms the flexibility to choose the abatement technique for reaching the environmental goal, they do not establish incentives designed to minimize production costs. Equalization of marginal abatement costs across participating firms - a requisite for minimizing production costs - is not achieved since many voluntary agreements set a common goal for all participants or emission targets are set at the industry level, and these targets often are non-obligatory (Bizer, 1999; OECD, 2003). Finally, concerns have been raised over the effect of voluntary approaches on competitiveness. Participation in a voluntary agreement allows for the possibility of collusive behaviour through the creation of barriers to entry for non-participants and foreign firms, the phasing out of particular products, or through price setting (Brau and Carraro, 1999). This is particularly true in a negotiated agreement between government and industry, of which there are currently few examples in the U.S.

There is little empirical evidence on the efficiency or inefficiency of voluntary approaches. A survey of voluntary approaches reveals that, in many cases, abatement targets are not differentiated (i.e., each firm faces the same target); as such these agreements are not dissimilar to an emissions standard that does not specify the abatement technique. In other words, voluntary approaches are likely to suffer the same inefficiencies that result from mandating a uniform standard for all participating firms, amplified by the fact that not all firms participate. Evidence from the unilateral commitment Responsible Care® demonstrates that there is little likelihood that this program can ever achieve an efficient outcome because firms operate within a framework of existing regulations that mandate particular abatement methods. However, there may be some reduction in operating costs due to reduced insurance premiums and worker compensation costs made available to participants (Mazurek, 1998). A study by Nadeau, Cantin, and Wells (2003) implies some gain in economic efficiency for firms participating in Energy Star™. They find that energy efficiency measures undertaken as part of the program are responsible for a market return to member companies of approximately $16,000 per million dollars in asset value. They also find that the cost of not joining the program represents almost 10 percent of the asset value of the non-participants analysed.

Are there any factors that affect the efficiency of voluntary approaches? A number of publications have observed that voluntary approaches that allow for differentiated abatement targets across firms may increase the efficiency of a voluntary agreement. If that is not feasible, compensation to low-cost firms or the use of bubbles at the firm-level can partially offset this inefficiency (OECD, 1999). Glanchant (1996) demonstrates theoretically that voluntary approaches are efficient when there exists a high level of uncertainty surrounding the costs of abatement techniques for a concentrated industry with little heterogeneity in abatement activities and costs across firms. Segerson and Miceli (1998) also have observed that, at least theoretically, the efficiency of voluntary approaches depends on the allocation of bargaining power, the magnitude of the background regulatory threat, and the social cost of public funds.

4.3. Reductions in administrative and monitoring and enforcement costs

Proponents of voluntary approaches often claim that the government benefits from reductions in administrative costs as well as declines in monitoring and enforcement costs when compared with more traditional forms of environmental regulation. Government does face some costs when using a voluntary approach: the cost of preparing and sometimes negotiating an agreement, and the cost of implementing the agreement (OECD, 2003). However, government also may benefit from the shift of monitoring and enforcement costs to the private sector, if compliance is self-reported, or to a third-party, if audits are conducted. In the event that government requires submission of reports directly to the agency, it is still likely that monitoring costs are reduced because these requirements are often much less time-consuming than those required by traditional regulation (OECD, 1999).

How large are these cost savings compared to traditional forms of regulation? Alberini and Segerson (2002) point out that the argument of significant cost savings accruing to government depends on comparison to an inefficient and inflexible policy alternative. To the extent that the regulatory alternative under consideration is a more flexible command-and-control or market-based policy, such costs savings diminish or even disappear. Also, those voluntary approaches that have the lowest administrative costs are also the same approaches that risk being the least environmentally effective since the likelihood of free-riding increases when there is little oversight by a regulatory agency.

To the extent that firms are better informed about abatement activities than a government regulator, voluntary approaches are likely to result in some savings in both administrative and compliance costs. However, the empirical evidence is scarce. A number of studies have demonstrated that as a fraction of the total cost of federal environmental regulations in the U.S., voluntary approaches have contributed little in cost savings (NAPA, 1997, Mazurek, 1998). It has also been noted that administrative and implementation costs tend to be high when many parties are involved, the legal status of the agreement is ambiguous, and a detailed technical analysis of abatement options is needed prior to the implementation of the voluntary approach (OECD 2003). Project XL, a negotiated agreement in the U.S., had high transaction costs initially due to the negotiation of each agreement with an individual firm. For example, Intel's agreement cost the government $110,000 (Blackman and Mazurek, 2001). These costs have fallen substantially since the early agreements (Delmas and Mazurek, 2003). However, there is a bias built into the program of attracting larger firms to participate because they are more likely to be able to absorb the cost of joining (Blackman and Mazurek, 2001).

4.4. Environmental Awareness and Attitudinal Changes

It has also been argued that voluntary approaches are responsible for inducing long-term changes in the environmental awareness of industry, consumers, or both. In other words, through the act of going beyond existing regulations to meet particular environmental goals, firms are educating themselves on the nature of the environmental problem and ways in which it can be mitigated. To the extent that firms promote their membership in these voluntary initiatives to consumers, it may also affect environmental awareness or priorities of consumers and result in a demand for greater emissions reductions. Again, because voluntary approaches rarely set stringent environmental goals, they do not change the status quo of the industry, which allows for little change in the way an entire industry views particular environmental issues. For this reason, participation of firms in voluntary agreements often does little to convince consumers of the sincerity of a firm's environmental commitment and often is viewed largely as propaganda, falling into same category as advertising (Fierman, 1991).

The U.S. EPA (2002) claimed 11,300 participants in its programs as of 2000, up from 6,900 in 1996, an argument for the increased environmental awareness of industry. However, researchers have found that the Responsible Care® program appears to have done little to change consumers' view of the chemical industry. Why has it been so unsuccessful? Researchers point to the fact that the program narrowly promotes its membership to employees and people living near existing plants. The result is that consumer awareness of the program is low. However, individual companies have reported improvements in community relations and public perception; they have also reported an increased understanding of environmental issues by both industry and community (Mazurek, 1998).

4.5. Innovation and Dynamic Effects

A fifth objective of voluntary approaches is to induce innovation in abatement techniques that make the cost of compliance with environmental regulations decrease over time. It is argued that voluntary approaches induce such innovation because they signal to firms possible future regulatory requirements. In anticipation of meeting such requirements, firms look for ways to reduce the costs of compliance through new, better methods of emission reductions. The Chicago Climate Exchange promotes participation in its emissions trading program as a way to reduce regulatory risks and increase the potential reward for taking early action in the event of legislation (Chicago Climate Exchange, 2004). It is also possible that firms learn of better, more efficient ways of abatement through participation in voluntary programs, either through learning-by-doing that leads to technological improvements over time or information sharing among firms (OECD, 1999).

Critics have pointed out that voluntary approaches rarely set stringent, "technology-forcing" environmental targets and as such provide weak incentives for firms to innovate (OECD 1999). Instead of creating an incentive to innovate, voluntary approaches may be used to buy time and postpone regulation without any intention to seriously meet the voluntary obligations. Since firms are trying to avoid regulation, little innovation takes place (Bizer, 1999; Lyon and Maxwell, 2003). What little empirical evidence exists, suggests that voluntary approaches provide weak incentives for the development of new abatement technology (OECD, 2003).

While the ways in which voluntary approaches meet the five policy objectives discussed in this section have been explored in the literature on a theoretical basis, the validity of these arguments largely has not been tested empirically. To the extent that empirical research exists, it tends to focus on a narrow set of U.S. voluntary initiatives for which data are available and limits itself to examining the evidence for environmental effectiveness of the agreements. Evidence indicates that few voluntary approaches in the U.S. have resulted in anything greater than moderate reductions in emissions. One possible reason for the limited success of the programs examined is that the firms that tend to participate in the programs are already quite environmentally aware and as such would have reduced emissions beyond what is required by regulation even without the existence of the voluntary agreement. Also, many voluntary initiatives set easy-to-meet goals, which do not translate into large environmental improvements. Finally, programs that have been evaluated tend to be those that have been around the longest, those with which the government has "cut its teeth," and as such may not reflect government learning that has resulted in improvements in the way in which voluntary initiatives are structured. It remains largely unknown if any of the unstudied U.S. voluntary agreements attain the policy objectives highlighted here.

5. DIFFICULTIES IN EVALUATING VOLUNTARY INITIATIVES

To better understand why there has not been more empirical evaluation of voluntary agreements, we next turn to some of the challenges in evaluating the effectiveness of voluntary agreements in meeting these various policy objectives. One obstacle to providing a comprehensive evaluation of voluntary initiatives in the U.S. is that voluntary initiatives are a relatively new policy instrument. Voluntary initiatives started in the early 1990s and most began in the middle to late 1990s (see Table 1). Furthermore, voluntary initiatives often set target dates for environmental improvements several years into the future, which imply that it is too early to evaluate many voluntary initiatives. For example, the SmartWay Transport program aims to improve fuel efficiency standards and reduce emissions of carbon dioxide by

33 million metric tons and emissions of nitrogen oxide by up to 200,000 tons by the year 2012.

Another obstacle to the measurement of the effectiveness of voluntary initiatives is that many programs target general environmental objectives and therefore lack a measurable environmental output. While all programs focus on improvement to an existing environmental problem, the achievement of some goals is more difficult to measure than for others. For example, the SunWise program aims to educate school-aged children of the risks of overexposure to the sun. While the program aims to reduce the incidence of skin cancer, tracking individuals twenty to thirty years into the future to measure their health makes it difficult to evaluate the effectiveness of the program.

Even if a voluntary initiative targets a measurable environmental output, a third obstacle in gauging the effectiveness of a voluntary initiative is a lack of data on a measurable output. There are two potential sources of data. First, data may be available through a national pollution-reporting database, such as the Toxic Releases Inventory (TRI). The TRI, however, does not target greenhouse gases while several voluntary initiatives in the U.S. focus on environmental improvements in this area. For example, the Commuter Choice Leadership Initiative targets carbon dioxide, the AgStar and the Coalbed Methane Outreach programs target methane, and the SF6 Emission Reduction Partnership targets sulfur hexafluoride. None of these chemicals are part of a national pollution-reporting database, and hence, there are no available data on firm emissions.

In spite of a pollutant's exclusion from routinely collected data by the U.S. EPA, data may still be available if the voluntary initiative requires some type of auditing and reporting activities. While many programs do encourage firms to submit annual reports, there is concern about the validity of these data. Firms may intentionally misreport their data, either in an attempt to skew their performance or out of fear that U.S. EPA will use these data to regulate the firm more closely. This problem can be partially mitigated by requiring participants to submit their data to a neutral, third party, as the National Metal Finishing Strategic Goals Program does. Even so, firms may correctly report only the positive aspects of their activities. Most programs do not require firms to submit detailed auditing of their emissions. The lack of data on firms' environmental outputs, either from a national pollution database or from firms themselves, is a serious obstacle in measuring the environmental and economic effectiveness of voluntary initiatives.

Perhaps the most serious obstacle in evaluating the effectiveness of a voluntary initiative is forming a reasonable counterfactual baseline with which to make a comparison. In establishing the effectiveness of a voluntary initiative, most initiatives simply provide a before-and-after comparison of pollution levels. For example, the 33/50 program encouraged firms to reduce emissions of toxic chemicals relative to firm emissions in 1988. Since participating firms on average met these goals, the program was deemed to be a success (U.S. EPA, 1996). However, of the total reductions in emissions between 1988 and 1994, large reductions took place between 1988, the baseline year chosen for the program, and 1991, the year the program actually started and hence cannot be attributed to the program itself (GAO, 1994; Inform, 1995; Khanna and Damon, 1999). While a successful voluntary initiative may result in lower levels of pollution, a fall in pollution is not necessarily indicative of a successful voluntary initiative. In short, pollution may be declining for several reasons unrelated to the voluntary initiative. First, pollution may fall simply due to a general downturn in production. When the EPA contacted over 1,200 industrial facilities that had informed the agency of changes in generated waste between 1989 and 1990, they found "nearly 70 percent attributed some portion of their emission increases or decreases to production level changes" (GAO, 1994). Also, pollution levels may appear to fall if firms simply substitute production from a regulated substance to an unregulated substance.

A second factor unrelated to the effectiveness of a voluntary initiative that may result in lower pollution levels is technological innovation in an industry. If technological progress in an industry results in more efficient use of inputs and hence less pollution, it may appear that a voluntary initiative is more successful than it actually is. With the 33/50 program, EPA claimed to meet its goals for emission reductions a year prior to its target date. However, a number of researchers note that pollution levels were falling prior to the implementation of the program, possibly due to technological progress unrelated to the program (GAO, 1994; Inform, 1995; Khanna and Damon, 1999). A GAO report also concluded that "substantial reductions were reported for TRI chemicals not targeted by the 33/50 program, suggesting that production changes or other factors unrelated to commitments made under the program may be largely responsible for the companies' reported reductions" (GAO, 1994).

A third reason why observations of falling pollution levels do not necessarily imply a successful voluntary initiative is self-selection bias. In particular, many of the factors that influence a firms' decision to participate in a voluntary agreement also affect a firms' overall environmental performance. For example, naturally "green" or environmentally friendly firms are more likely to have lower levels of pollution. These firms are also more likely to join a voluntary initiative, either due to a genuine desire to improve their environmental performance or because their lower levels of pollution imply that these firms are closer to achieving the targets set by a voluntary initiative and hence face lower costs in joining. In the Green Lights program, for example, 593 out of the initial 2,308 participating firms were companies that sell, manufacture, and install lighting products (GAO, 1997b). Hence, these firms probably were already aware of possible opportunities for improving energy efficiency prior to joining the program, and any lighting improvements these firms made may have been undertaken regardless of firm participation in the program. Researchers should therefore be wary of crediting a voluntary initiative with reductions in pollution without first taking into account this self-selection bias.

Evaluating the effectiveness of voluntary initiatives in reaching an environmental objective is difficult for several reasons. One main obstacle, the lack of data on a measurable environmental output, can be overcome if voluntary initiatives encourage more defined and detailed goal setting and require more complete data collection and reporting. While voluntary initiatives should attempt to do more in these areas, one also needs to recognize the limitations in enacting these requirements. One of the main selling points of voluntary initiatives is their low cost and flexibility in empowering firms to identify and undertake abatement efforts on their own. As costly measurement and auditing processes are required of participating firms, voluntary initiatives may be less cost effective instruments. Even with more complete data on emissions of participants, another obstacle that continues to plague the evaluation of voluntary initiatives is a lack of data on non-participants. To form a reasonable counterfactual baseline for use in assessing the effectiveness of a voluntary agreement, a rigorous analysis needs to compare the efforts of participants against what non-participants in the industry are doing. This requires a more comprehensive and broader auditing and reporting program of emissions than is currently in place.

In spite of these difficulties, a few voluntary initiatives have been evaluated in a rigorous fashion, including the 33/50 program (Khanna and Damon, 1999), Responsible Care® (King and Lenox, 2000), and Climate Challenge (Welch, Mazur, and Bretschneider, 2000). These initiatives have been successfully evaluated because they do not suffer from the limitations discussed above. In particular, these initiatives are older and hence ready for evaluation, and the initiatives also have some type of measurable environmental output for which data are available or easily calculated, both before and after the start of the voluntary initiative. Because firms participating in these initiatives are generally large public companies, financial and other production data are also available to control for production changes and other industry effects. Given the availability of data on a measurable environmental output and on finances and production, researchers are able to estimate a two-stage model to take into account the self-selection issue raised above. A rigorous analysis involves a first stage participation equation that estimates the probability of participating in the voluntary initiative. Estimates from this regression then are used in a second stage estimation of the determinants of pollution levels to provide an unbiased estimate of the effect of the voluntary initiative on firm pollution (Khanna and Damon, 1999).

6. CONCLUSION

In the past decade, voluntary approaches have been increasingly used as a component of U.S environmental policy. Authorities mostly have relied on public voluntary programs designed at either the federal, state, or regional level. Other types of voluntary approaches, such as negotiated agreements and unilateral commitments, have been used much less frequently in the U.S. In general, voluntary approaches have been used to complement existing legislation (for example, implementing aspects of the Pollution Prevention Act of 1990) and in areas where enacting new legislation is difficult for political reasons (for example, in the climate change area).

Proponents of voluntary approaches argue that voluntary approaches effectively provide environmental protection, improve economic efficiency, result in administrative, monitoring, and enforcement cost savings, lead to environmental awareness and attitudinal changes, and encourage innovation. It has been difficult, however, to provide evidence substantiating these claims due to a lack of data and the inherent difficulty of identifying what would have happened in the absence of a voluntary approach. In addition, the literature notes that self-selection, free-riding, and attempts by industry to pre-empt regulation are inherent to the "voluntary" nature of these approaches.

The difficulties in proving the effectiveness of voluntary agreements does not imply that voluntary approaches should not be used. Rather, understanding the difficulties and potential pitfalls of using voluntary approaches will hopefully lead to better designed voluntary approaches in the future. Key design aspects should include more stringent goal setting and improved auditing and data collection. In addition, voluntary approaches are more likely to result in significant environmental improvements when backed by a serious legislative threat.

Also, while it is important to strive for accurate measurement of the effectiveness of voluntary approaches, the inability to measure the effectiveness of all programs or all aspects of a specific program should not prevent the use of voluntary approaches. In fact, voluntary approaches may have the greatest potential in areas where it is especially difficult to measure progress. For instance, voluntary approaches that reduce technological uncertainties or share information between affected parties may lead to increased environmental awareness and attitudinal changes, which may, in turn, result in the correction of market failures at the root of many environmental problems. Credible identification of the environmental quality of firms or their products may encourage consumers and investors to demand greener goods, which provides incentives for firms to improve their environmental quality in the marketplace. To the extent that voluntary approaches can harness market forces, they hold the potential to ameliorate environmental problems and to be effective policy instruments.

7. ACKNOWLEDGEMENTS

Keith Brouhle is an Assistant Professor at the University of Alberta. Charles Griffiths and Ann Wolverton are both economists at the U.S. EPA's National Center for Environmental Economics. The views expressed in this chapter are entirely those of the authors and do not necessarily represent the views of the U.S. Environmental Protection Agency. We are grateful to Cheryl Hawkins for providing U.S. EPA voluntary program information and for her many helpful suggestions.

8. NOTES

[1] We do not include as voluntary initiatives pure information provision programs (such as the Green Vehicle Guide) or individual firm efforts to capture cost savings by improving efficiency and reducing waste (such as 3M's Pollution Always Pays (3Ps) program or Dow Chemical's Waste Reduction Always Pays (WRAP) program). The reason that we exclude these activities from the discussion of voluntary initiatives is that they do not typically involve any specific agreement proscribing actions firms must take or goals firms must achieve as a basis of participation.

[2] We include in the list both programs that are currently in operation and programs that are not currently active. Programs known to be no longer in existence are marked with a *. Programs that have undergone name changes are marked with a †. In particular, programs that we identify that have undergone a name

change are as follows: the Sectors Strategies Program was the Industry Sector Performance Program, Best Workplaces for Commuters was the Commuter Choice Leadership Initiative, the Water Efficient Product Enhancements Program was the Water Alliances for Voluntary Efficiencies, and the Environmental Stewardship Program was Today's Environmental Partnership.

[3] Within this category, the degree of potential involvement and negotiation with affected industries in the development of a voluntary program can vary. In general, negotiation is minimal but usually involves consultation with affected industries. However, a few programs, such as Energy Star, have involved more direct negotiation with affected industries. While this higher degree of involvement may favour the placement of these programs in the negotiated agreements category, we keep these programs in the federal voluntary programs category because they stop short of joint design of the program. They also do not offer explicit regulatory relief (see section 3.4).

[4] While the Common Sense Initiative is no longer active, the early work done for this initiative forms the foundation for several recent voluntary programs. The Sector Strategies program, for example, includes two of the original six industries that were part of the Common Sense Initiative. The four other industries of the Common Sense Initiative also are covered by current voluntary programs.

9. REFERENCES

Alberini, A., & K. Segerson, K. (2002). Assessing Voluntary Programs to Improve Environmental Quality. *Environmental and Resource Economics, 22,*157-184.

Arora, S., &. Cason, T. (1995). An Experiment in Voluntary Environmental Regulation: Participation in EPA's 33/50 Program. *Journal of Environmental Economics and Management, 28(3),* 271-86.

Arora, S., & Gangopadhyay, S. (1995). Toward a Theoretical Model of Voluntary Overcompliance. *Journal of Economic Behavior and Organization, 28(3*), 289-309.

Bizer, K. (1999). Voluntary Agreements: Cost-Effective or Smokscreen for Failure? *Environmental Economics and Policy Studies*, 2(2), 147-165.

Blackman, A., & Mazurek, J. (2001). The Cost of Developing Site-Specific Environmental Regulations: Evidence from EPA's Project XL. *Environmental Management*, 27(1), 109-121.

Boyd, J. (1998). Searching for the Profit in Pollution Prevention: Case Studies in the Corporate Evaluation of Environmental Opportunities. Discussion Paper. Washington, DC: Resources for the Future, No. 98-30.

Brau, R., & Carraro, C. (1999). Voluntary Approaches, Market Structure, and Compeition. CAVA Working Paper, No 99/08/01.

Bui, L. (2003). Public Disclosure of Private Information as a Tool for Regulating the Environment: Firm Level Responses by Petroleum Refineries to the Toxic Releases Inventory. Working Paper, Boston University.

Chicago Climate Exchange (2004). *Chicago Climate Exchange: About CCX.* Retrieved Feb. 29, 2004 from www.chicagoclimatex.com/about

Croci, E. (2003). Voluntary Agreements for CO 2 Emissions Reduction, Evaluation & Perspectives. *Energy & Environment, 14(5),* 663-76.

Darnall, N., & Carmin, J. (2004). Greener and Cleaner? The Signaling Accuracy of U.S. Voluntary Environmental Programs, Working Paper, North Carolina State University.

DeCanio, S. (1994). Energy Efficiency and Managerial Performance: Improving Profitability while Reducing Greenhouse Gas Emissions. In D. Feldman (ed.), *Global Climate Change and Public Policy*. (pp. 86-101). Chicago: Nelson-Hall.

Delmas, M., & Mazurek, J. (2003). A Transaction Cost Perspective on Negotiated Agreements: The Case of the U.S. EPA XL Program. In A. Baranzini, and P. Thalmann (eds.), *Voluntary Approaches to Climate Protection. An Economic Assessment of Private-Public Partnerships*. Edward Elgar.

Delmas, M., & Terlaak, A. (2002a). The Institutional Factors of Environmental Voluntary Agreements. In Hoffman and Ventresca (eds.), *Organizations, Policy, and the Natural Environment: Institutional and Strategic Perspectives*. (pp. 346-366). Stanford: Stanford University Press.

Fierman, J. (1991). The Big Muddle in Green Marketing. *Fortune*, June 3, 91.

General Accounting Office, GAO (1994). *Toxic Substances: EPA Needs More Reliable Source Reduction Data and Progress Measures*. Washington, DC: GAO.

GAO (1997). *Challenges Facing EPA's Efforts to Reinvent Environmental Regulation*. Washington DC: GAO.

GAO (1997). *Information on the Results of Four of EPA's Voluntary Climate Change Programs*. Washington DC: GAO

Glanchant, M. (1996). The Cost Efficiency of Voluntary Agreements for Regulating Industrial Pollution: A Coasean Approach. In C. Carraro and F. Leveques (eds.), *Voluntary Approaches in Environmental Policy*. Netherlands: Kluwer.

Hartman, R. (1988). Self-Selection Bias in the Evolution of Voluntary Energy Conservation Programs. *The Review of Economics and Statistics, 70(3),* 448-458.

Horowitz, M. (2004). Electricity Intensity in the Commercial Sector: Market and Public Program Effects. *Energy Journal, 25(2),* 115-137.

Horowitz, M. (2001). Economic Indicators of Market Transformation: Energy Efficiency Lighting and EPA's Green Lights. *Energy Journal, 22(4),* 95-122.

Howard, J., Nash, J., & Ehrenfeld, J. (2000). Standard or Smokescreen? Implementation of a Voluntary Environmental Code. *California Management Review, 42(2),* 63-82

Inform (1995). Toxics Watch, 1995. New York.

Kappas, P. (1997). Industry Self-Regulation and Environmental Protection, Working Paper, University of California – Santa Barbara.

Khanna, M, &. Damon, L. (1999). EPA's Voluntary 33/50 Program: Impact on Toxic Releases and Economic Performance of Firms. *Journal of Environmental Economics and Management, 38(1),* 1-28.

King, A., & Lenox, M. (2000). Industry Self-Regulation Without Sanction: The Chemical Industries' Responsible Care Program. *Academy of Management Journal, 36(3),* 243-266.

Konar, S., & Cohen, M. (1997). Information as Regulation: The Effect of Community Right to Know Laws on Toxic Emissions. *Journal of Environmental Economics and Management, 32(1),* 109-24.

Konar, S., & Cohen, M. (2001). Does the Market Value Environmental Performance. *Review of Economics and Statistics, 83(2),* 281-89.

Lutz, S., Lyon, T., & Maxwell, J. (2000). Quality Leadership When Regulatory Standards Are Forthcoming. *Journal of Industrial Economics, 48(3),* 331-348.

Lyon, T., & Maxwell, J. (2001). Voluntary Approaches to Environmental Protection. In M. Franzini and A. Nicita (eds.), *Environmental Economics: Past, Present and Future.* Aldershot, Hampshire, UK: Ashgate Publishing.

Lyon, T., & Maxwell, J. (2003). Self-Regulation, Taxation, and Public Voluntary Environmental Agreements. *Journal of Public Economics, 87(7-8),* 1453-1486.

Maitland, I. (1985). The Limits of Business Self-Regulation. *California Management Review, 27(3*), 132-147.

Maxwell, J., & Lyon, T. (1999). An Institutional Analysis of US Voluntary Environmental Agreements, Working Paper. Indiana University, 99-27.

Maxwell, J., Lyon, T., & Hackett, S. (2000). Self-Regulation and Social Welfare: The Political Economy of Corporate Environmentalism. *Journal of Law and Economics, 43(2),* 583-618.

Mazurek, J. (1998). *The Use of Voluntary Agreements in the United States: An Initial Survey.* Paris: OECD, ENV/EPOC/GEE (98)27/Final.

Morgenstern, R. & Al-Jurf, S. (1999). Does the Provision of Free Technical Information Really Influence Firm Behaviour? *Technological Forecasting and Social Change*, *61*, 13-24.

Nadeau, L., Cantin, J., & Wells, R. (2003). Participation in Voluntary Programs, Corporate Reputation, and Intangible Value: Estimating the Value of Paticipating in the EPA's EnergyStar Program. Working Paper, June.

Nash, J. (2002). Industry Codes of Practice: Emergence and Evolution. In T. Dietz and P. Sterns (eds.), *New Tools for Environmental Protection: Education, Information, and Voluntary Measures.* Washington, DC: National Academy Press.

National Academy of Public Administration, NAPA (1997). *Resolving the Paradox of Environmental Protection.* Washington, DC: NAPA.

Office of Technology Assessment (1991). Changing by Degrees: Steps to Reduce Greenhouse Gases. Washington, DC: OTA, OTA-O-482.

Organisation for Economic Cooperation and Development, OECD (1999). *Voluntary Approaches for Environmental Policy: An Assessment.* Paris: OECD.

OECD (2003). *Voluntary Approaches for Environmental Policy: Effectiveness, Efficiency, and Usage in Policy Mixes.* Paris: OECD.

Ozog, M., & Waldman, D. (1994). Weighting Nonrandom Samples in Voluntary Energy Conservation Program Evaluation. *Energy Journal, 15(1),* 129-41.

Segerson, K., & Miceli, T. (1998). Voluntary Environmental Agreements: Good or Bad News for Environmental Protection? *Journal of Environmental Economics and Management, 36(2)*, 109-131.

United States Environmental Protection Agency, U.S. EPA (1996). *1994 Toxics Release Inventory, Public Data Release*. Washington, D.C: EPA.

U.S. EPA (2002). *Achievement Through Partnership: A Progress Report Through 2000*. Washington, DC: EPA, EPA-240R-02-001.

U.S. EPA (2004a). *Partners for the Environment: List of Programs*. Retrieved Feb. 11, 2004 from www.epa.gov/partners/programs/index.htm

U.S. EPA (2004b). *EPA's Energy Star: The Quality of Our Environment Is Everyone's Responsibility*. Retrieved Feb. 24, 2004 from www.energystar.gov

U.S. EPA (2004c). *EPA's WasteWise: Preserving Resources, Preventing Waste*. Retrieved Feb. 24, 2004. from www.epa.gov/wastewise/index.htm

U.S. EPA (2004d). *San Francisco Area Green Business Program*. Retrieved Feb. 24, 2004 from www.epa.gov/region9/cross_pr/innovations/gbp.html

U.S. EPA (2004e). *EPA's Project XL Program*. Retrieved Feb. 24, 2004 from www.epa.gov/ProjectXL

Videras, J., & Alberini, A. (2000). The Appeal of Voluntary Environmental Programs: Which Firms Participate and Why? *Contemporary Economic Policy, 18(4)*, 449-61.

Welch, E., Mazur A., & Bretschneider, S. (2000). Voluntary Behavior by Electric Utilities: Levels of Adoption and Contribution of the Climate Challenge Program to the Reduction of Carbon Dioxide. *Journal of Policy Analysis and Management, 19(3)*, 407-425.

NEGOTIATED REGULATION, IMPLEMENTATION AND COMPLIANCE IN THE UNITED STATES

N.A. ASHFORD, C.C. CALDART

Massachusetts Institute of Technology, Cambridge, MA, USA

Abstract. Interest in the use of so-called voluntary approaches to supplement or replace formal environmental regulation is on the rise, both in Europe and in the United States. These approaches fall into two general categories: (1) industry-initiated codes of good practice focusing on environmental management systems or performance goals, and (2) negotiation between government and individual firms (or industry sector trade associations) focusing on regulation or compliance. This paper addresses the latter. In the United States, the motivations for engaging in such negotiation are manifold and sometimes contradictory. They include desires (1) to facilitate the achievement of legislated environmental goals by introducing flexible and cost-effective implementation and compliance measures, (2) to negotiate levels of compliance (standards) fulfilling health-based legislative mandates, (3) to negotiate legal definitions of Best Available Technology and other technology-based requirements, and (4) to weaken environmental regulation. In the United States, administrative agencies have long been experimenting with "negotiated rulemaking as a means of *setting regulatory standards*, and the Administrative Procedure Act was amended in 1990 to encourage further use of this process. U.S. agencies have also made frequent use of negotiation as a means of defining *compliance* responsibilities for individual firms. In addition, the Environmental Protection Agency (EPA) has sometimes acted outside of the authority given to it by its enabling legislation in an attempt to negotiate environmental policy and implementation. Two recent examples are the "Common Sense Initiative," in which EPA attempted broad-based negotiation focuses on particular industry sectors, and "Project XL", in which the agency attempted to negotiate flexible implementation of environmental requirements with individual firms. Although both programs are now moribund, each provides useful lessons for future efforts at environmental negotiation. This paper describes and analyses negotiated agreements in the United States in the context of EPA efforts to ensure environmental protection. These agreements can be described according to the following taxonomy: (a) negotiated regulation (either preceding formal regulation or as a substitute for formal regulation); (b) negotiated implementation (negotiations with an individual firm to establish the timetable and/or the means for meeting a particular regulatory standard; and (c) negotiated compliance (negotiation in the context of an enforcement action in which the firm is out of compliance with an applicable standard and there is an opportunity for extra-statutory environmental gains, such as encouraging cleaner production through the leveraging of penalty reductions). The criteria for evaluation used in this paper include: environmental outcomes, effects on stimulating technological change, time for development (time to completion), ease of implementation (likelihood of court challenge), stakeholder influence (ability of large firms to dominate outcome, environmentalists-industry balance of power), and administrative features.

1. INTRODUCTION

Negotiation - as an alternative or an adjunct to the adversarial process - is increasingly touted as the wave of the future. Negotiation, it is argued, is a more efficient use of societal resources, because it is more likely to produce a result that all sides can accept. Moreover, negotiation is said to be more likely to produce creative solutions, because it

E. Croci (ed.), The Handbook of Environmental Voluntary Agreements, 135 - 159

forces the parties to focus on cooperation rather than confrontation. This article surveys the use of negotiation in formulating and implementing environmental policy in the United States, and assesses the potential of negotiation to (a) foster improved environmental outcomes and (b) stimulate technological change.

2. MODES OF NEGOTIATION

In a broad sense, there are three major instances in which negotiation is used to *make* or *effectuate* policy within the federal administrative system of the United States. First, there is *negotiated rulemaking,* wherein negotiation is used to help set regulatory standards. Originally an informal process, negotiated rulemaking has now been formalized through legislation. Second, there is *negotiated implementation,* where negotiation is used to determine how a regulatory standard, once set, is to be applied to a particular firm (or other member of the regulated community). Under United States environmental statutes, negotiated implementation often occurs when a permit is being issued or revised, as was the case with EPA's *Project XL* initiative. Such negotiation also occurs when the regulated firm seeks a waiver or variance from the regulatory standard at issue. Of particular interest here are the *innovation waivers* that have been made available by Congress in certain environmental statutes. When such a waiver is granted by EPA, the firm is given additional time to comply with the standard so that it may perfect a promising innovative compliance technology.

Third, there is *negotiated compliance,* where negotiation is used to determine the terms by which regulatory standards will be enforced against a particular firm (or other regulated entity) that is out of compliance with a particular regulatory standard. By its nature, of course, almost all enforcement involves some amount of negotiation between the enforcing agency (or, in the case of citizen enforcement suits, the enforcing citizen) and the alleged violator. Of interest here are those compliance negotiations that result in (a) compliance through the use of innovative technology, and/or (b) environmental gains *beyond* compliance. Since the early 1990's, EPA has pioneered the use of what it terms "Supplemental Environmental Projects" in an attempt to meet these goals within the compliance context.

In addition, there is what might be classified as a fourth type of policy-relevant negotiation – *regulatory reinvention* – that was begun (at least under that name) in the Clinton administration, and continues today in evolving forms. The most prominent early example was EPA's *Common Sense Initiative* (CSI), wherein the agency assembled groups of interested parties to focus on regulatory issues concerning a particular industry sector (e.g., automobile manufacturing), with an eye toward developing "cleaner, cheaper, smarter" ways of reducing or preventing pollution. In contrast, EPA's 'Project XL,' mentioned above, focused on negotiations with individual firms. Both programs have now been phased out, and the Bush Administration's National Environmental Performance Track program is now occupying center stage in regulatory reinvention. This program focuses on creating partnerships with individual firms in which the firms agree to exceed regulatory requirements, implement environmental management systems, work closely with their communities, and set three-year goals to continuously improve their environmental performance, in exchange for reduced priority status for inspections, reduced regulatory, administrative, and reporting requirements and positive public recognition[1]. The program is too new to evaluate for the purpose of this paper.

3. NEGOTIATED RULEMAKING

Since the mid-1970s, many commentators in the United States have advocated the use of negotiated rulemaking as a more efficient, sensible alternative to the traditional "notice and comment" procedure typically followed by federal agencies in the development of regulations. Occasionally in the 1970s, and more often in the 1980s, EPA, OSHA, and other federal agencies used the negotiation process as an aid to the development of certain regulations. In 1990, Congress formally endorsed negotiated rulemaking with the passage of the federal Negotiated Rulemaking Act, and both the Clinton Administration and the current administration have been among negotiated rulemaking's strong supporters.

3.1. The performance of negotiated rulemaking as a means of saving time and limiting judicial challenge

Those who advocate negotiated rulemaking – including Congress - tend to identify two primary benefits that are expected to flow from its use: reduced rulemaking time, and decreased litigation over the final rule. Presumably, face-to-face meetings among the interested parties will be able to avoid the various bureaucratic quagmires that can delay the drafting of a rule within an agency, and will, on average, produce a proposed rule more quickly. Further, since the interested parties have agreed on the wording of the proposed rule in advance, the notice and comment procedure presumably will be less contentious and time-consuming, and the incentive for anyone to file a judicial challenge to the final rule presumably will be slight.

In practice, however, it is not at all clear that negotiated rulemaking delivers on either of these promises. Of all the federal agencies in the United States, EPA has used negotiated rulemaking the most often. A study [Coglianese, 1997] of EPA negotiated rulemakings has concluded that: (a) on average, the promulgation of EPA rules through negotiated rulemaking took no less time than did the promulgation of a "control" group of similar EPA rules through traditional notice and comment rulemaking; and (b) 50% of EPA's twelve finalised negotiated rulemakings were the subject of legal challenge, compared with a litigation rate of 26% for all EPA rules issued during the period from 1987 through 1991. To date, then, it has not been established that negotiated rulemaking actually returns the primary benefits touted by its proponents.

3.2. The performance of negotiated rulemaking as a means of securing a "better" rule

Nonetheless, there may be other advantages of using negotiated rulemaking, at least in certain circumstances, depending on the goals one wishes to achieve. Significantly, because it facilitates face-to-face discussions among rulemaking "adversaries" that might not otherwise occur, negotiated rulemaking holds out the potential that, as differences are understood and addressed, creative solutions may be found to difficult issues in such a way that a substantively '*better*' rule emerges. Such a result might come, for example, through the identification of opportunities for innovative technological responses within the regulated community.

As an initial attempt at determining whether this potential is being realized, this article examines three negotiated rulemakings used by EPA to set air emission standards under the federal Clean Air Act. In addition to the limitations imposed by the small number of examples examined, the problem with an analysis of this nature is that any attempt to identify a "better" result is a qualitative exercise: depending on the context, it can mean quite different things to different people. For the purposes of this article, we have sought to evaluate the quality of the final rule produced by negotiated rulemaking according to whether it produced a rule that was more - or less - protective of environmental than might have been expected had negotiated rulemaking not been used. Further, we have given particular attention to the extent to which opportunities to promote technological change were - or were not - seized upon by the negotiating committee.

3.3. Negotiated rulemaking and clean air act emission standards

Of the twelve negotiated rulemakings completed by EPA through 1996, we have chosen to focus on three that resulted in the promulgation of air emission standards under the Clean Air Act: EPA's woodstoves rule, coke oven emissions rule, and wood furniture coatings rule. We have chosen these three because they share a common set of features: a full committee stayed with the negotiations to the end; the rule negotiated was the rule actually proposed by the agency; and the rule set an air emission standard designed to protect the environment and/or public health.

3.3.1. The woodstoves rule

One of EPA's early forays into negotiated rulemaking was the development of a national New Source Performance Standard for "residential wood combustion units" (woodstoves). EPA came to regulate woodstoves as a result of lawsuits brought against the agency by the Natural Resources Defense Council (NRDC) and the State of New York. That suit sought to force EPA to regulate polycyclic organic matter (POM) as a hazardous air pollutant under Section 112 of the Clean Air Act. As part of its settlement of the POM litigation, EPA agreed to explore the possibility of regulating woodstoves-one of the primary contributors of POM - as "stationary sources" of air pollution under Section 111 of the Act. Interestingly, such regulation was desired not only by environmental groups, but also by woodstove manufacturers, who hoped that the promulgation of a national standard by EPA would discourage states from setting their own (likely differing) standards.

Section 111 of the Clean Air Act requires that a New Source Performance Standard (NSPS) reflect the level of emission limitation achievable through the application of the "best system of emission reduction...[that] has been adequately demonstrated." To devise such a national emission standard, EPA convened an advisory committee consisting of representatives from industry, environmentalists, certain states, a consumer group, and the agency itself. Agreement on a single national standard was complicated, however, by the fact that there were two major categories of woodstoves on the market- those that utilised catalytic combusters and those that did not. It was clear that, at least in the short term, the stoves with catalytic combusters were capable of meeting a lower (more protective) emission standard than those without catalytic combusters. Because catalytic combusters require a higher degree of maintenance, however, there was some question as to whether they would continue to deliver this greater level of emission reduction over the long term. Rather than resolve this technical issue, the negotiating committee agreed rather early on to adopt the industry position on the matter, and to-propose two standards - one for stoves with catalytic combusters and the other for those without. Thus, the opportunity to diffuse what may well be a superior emission-reduction technology throughout the woodstove industry was lost (as was an opportunity for innovation through the development of new woodstove technology).

This does not necessarily mean, however, that the woodstove rule was a "failure" from an environmental/public health perspective. It is questionable whether Section 111 actually empowers EPA to regulate residential woodstoves as "stationary sources" of air pollution, especially since the rule governs the manufacturers and retailers who *sell* the stoves rather than the individual homeowners who operate them. Thus, it could be argued that the process of negotiated rulemaking - in which the various players were able to agree on a rule despite its legal infirmities – resulted in a giant step forward, in that it produced national emission standards which otherwise either might not have been promulgated, or might have been successfully challenged in court.

On the other hand, the Clean Air Act was not the only regulatory alternative available to address the woodstove issue. The federal Consumer Products Safety Act (CPSA), which governs the design and sale of products "for use in or around" the home or school, clearly *does* cover woodstoves sold for residential use, and clearly contemplates regulation of manufacturers and retailers. It is not clear, however, that regulation under the CPSA would necessarily have produced a stricter emission standard for stoves without catalytic combusters. The CPSA requires that the benefits of a consumer products safety standard be justified by its costs, and the members of the non-catalytic industry doubtless would have argued that a stricter standard would have driven them out of the market. Further, unlike EPA, the Consumer Products Safety Commission - a chronically underfunded agency that is often reluctant to take on new issues - had no particular incentive to regulate woodstoves.

3.3.2. The coke oven emissions rule

Coke ovens are used to convert coal to coke, which is then used to produce steel. Air emissions from coke ovens come largely from leaking oven doors and lids. In 1992, EPA estimated that some *3.5* million pounds of toxic chemicals, including benzene, phenol, toluene, and polyaromatic hydrocarbons, were emitted to the air annually from coke ovens operating in the US Based on this estimate, EPA put the cancer risk to exposed individuals at 1 in 100.

Many of the materials emitted by coke ovens are subject to regulation as hazardous air pollutants under Section 112 of the Clean Air Act, and the 1990 amendments to the Act specifically required that Section 112 standards for coke oven emissions be promulgated by December 31, 1992. In early 1992, after meeting with representatives of the steel industry, relevant labour unions, states, and environmental groups "to discuss available data to be used as the basis of [a Section 112 regulation]," EPA convened a negotiated rulemaking committee that drew from all of these constituencies. After several negotiating sessions, the committee agreed on a draft rule that was proposed by the agency in December 1992, and was published as a final rule in October 1993.

In general, Section 112 of the Clean Air Act takes a two-tiered approach to the regulation of hazardous air pollutants. EPA is first to set technology-based emission standards, on an industry-category by industry-category basis. These are commonly known as the "MACT" standards, because they are to be set with reference to the application of the maximum achievable control technology that the industry category can currently achieve. Eight years later, the agency is to set a more stringent, health-based standard if further emission reductions are deemed necessary to provide "an ample margin of safety to protect public health." A health-based standard for carcinogens *must* be set if the technology-based standard fails to "reduce lifetime excess cancer risks to the individual most exposed to [the] emissions...to less than one in one million." For coke oven emissions in particular, however, Section 112 offers an alternative whereby a source may delay compliance with the health-based standard until 2020 if it meets a different, more stringent technology-based standard in the interim. The committee followed this framework in drafting its proposed rule, and steel industry representatives said afterwards that, because they viewed any likely health-based standard as "essentially a shut-down standard," they expected all plants except those that planned to go out of business in the near future to choose this "extended compliance" option.

At the conclusion of the negotiated rulemaking process, participants from environmental groups, labour, industry, and state governments all expressed their satisfaction with the negotiated rule. An EPA representative stated his belief that the negotiated rule would result in more emission reductions than would have been obtained through the conventional rulemaking process, and remarked that the agency had never before "been able to grapple with the economic and technological issues" addressed by the rule. It is probably more accurate to say, however, that this is a rulemaking that was made considerably easier because Congress had taken it upon itself to specify the dates by which - and the minimum amounts by which - the steel industry would be asked to reduce emissions. Indeed, the chief contribution of *negotiation* to the rulemaking process appears to have been to afford the industry the opportunity to negotiate a standard that actually is *less* stringent than that which was mandated by Congress.

For coke oven facilities choosing the "extended compliance" option, EPA was required to promulgate two sets of technology-based emission limits by December 31, 1992, to become effective in November 1993 and January 1998, respectively. Emission limits for coke ovens had traditionally been expressed in terms of a maximum permissible percentage of leaking doors, lids, and offtakes, and Congress adopted this approach in Section 112. For the 1993 limits, Congress specified the precise percentages EPA was to require. For the 1998 limits, Congress directed the agency to set percentages "reflecting the lowest achievable emission rate" (colloquially known as "LAER"), and also specified a set of percentages representing the *least stringent* permissible 1998 standard that EPA could set, and a second set representing a more stringent *default* 1998 standard that was to take effect if the agency failed to promulgate the 1998 limits by December 31, 1992.

In writing the rule that was promulgated by EPA, the negotiated rulemaking committee began with the 1993 limits specified in the statute, and with the least stringent permissible 1998 limits specified in the statute, but converted them to "statistically equivalent" limits based on thirty days' average performance. Thus, while the statute specified a maximum percentage that was not to be exceeded, the negotiated rule specifies an average percentage that must be achieved over a thirty-day period. This allows a facility to *exceed* the percentage specified in the statute for certain periods, so long as it is sufficiently *below* that percentage for other periods to maintain the required thirty-day average.

This change was made because the steel industry expressed concern that a straightforward application of the standards specified by Congress would necessitate the closure of most of the existing coke oven facilities throughout the country, as they would be unable to meet the specified maximum limits on a continual basis. Union participants in the negotiations, who were interested both in preserving jobs and in reducing workplace emissions, apparently helped to persuade the environmental group participants that this concern was a valid one. In addition, the statistical conversion to thirty-day averages allowed EPA and the environmental group representatives to point to regulatory limits expressed as numbers that were actually *below* the numbers specified by Congress in the statute. For example, the statute requires 8% leaking doors in the 1993 limits, while the regulation specifies 7% leaking doors. Even though this difference is simply an artifact of the statistical conversion of the statutory number to a thirty-day average value, the appearance is of a more stringent standard.

From a health perspective, however, the regulation may well be less protective than the numbers specified in the statute. There is evidence that short-term exposure to a certain amount of carcinogenic materials is more harmful than exposure to the same amount of those materials, in smaller daily increments, spread out over a longer term. The increased damage done on the individual days of high exposure levels allowed under the thirty-day average approach, then, may not be offset by the reductions in damage experienced on those days when emissions are below the required average.

Moreover, it appears clear that the negotiated 1998 limits were *not* set according to the "lowest achievable emission rate" (LAER) as that term is defined in the Clean Air Act. LAER is defined, in relevant part, as "the most stringent emission limitation that is achieved in practice by [the] class or category of source," with no consideration of the cost of meeting that emission limitation. That is, a LAER limit is to be based on the emission levels being attained by the best-performing existing plant within the particular industry class or category. The best-performing coke oven facility in operation in the United States at the time was the Jewell Smokeless plant, in Vansant, Virginia, owned by Sun Coal. This facility employs a *nonrecovery* coke oven technology, while all of the other coke oven plants in the country employ the older, and dirtier, *by-product recovery* technology. A nonrecovery plant can achieve an emission limit of *0.0%* leaking doors, lids, and offtakes. Further, nonrecovery plants produce far less wastewater, and far less hazardous waste, than comparable by-product recovery plants, and also generate excess energy that can be utilised elsewhere in the facility. From an environmental perspective, then, the nonrecovery technology is undeniably superior.

Although there was some talk within the negotiated rulemaking committee of basing the LAER limits on the performance of the Jewell Smokeless plant, the committee decided instead to consider the performance of byproduct recovery plants only. The committee apparently focused on the performance of a USX (United States Steel) plant in Clariton, Pennsylvania, which the committee appears to have deemed the best-performing byproduct recovery facility. Yet, as noted, the committee set the 1998 limits simply by specifying percentages that were calculated to be the "statistical equivalent" of the least stringent permissible limits specified in the statute. If the committee took this approach because it believed that this was the best the industry could do, this appears to have been a significant error in assessment.

The negotiated 1998 limits (expressed as thirty-day averages) are 4.3% leaking doors for tall doors and foundry doors, and 3.8% leaking doors for all other doors. As LAER limits, these limits were required by statute to be representative of the very best performance within the industry. An EPA survey of by-product recovery plants done six months after these limits were promulgated in 1993, however, found that most plants were easily meeting the 1998 limits, and that some plants were averaging *1% to 2%* leaking doors. In other words, the best performance in the industry was considerably better than what the1998 limits allow. Subsequent EPA surveys of the industry revealed that the performance of many of the plants worsened somewhat thereafter, but was still comfortably in compliance with the legally-applicable 1993 limits. This suggests that the plants may have initially been testing their technology to ensure that they could meet the 1998 limits. In August 1997, with the 1998 limits due to become enforceable within a few months, most of the plants were again meeting the 1998 limits on a continuous basis, and roughly three out of every five of the plants had *maximum* (as opposed to thirty-day average) values of *less than 2%* leaking doors.

The Clean Air Act also specifies that, by January 2007, EPA is to review the 1998 LAER limits for coke oven facilities, and "revise [them], as necessary...to reflect the lowest achievable emission rate as defined...at the time," with such revised limits to become effective on January 1, 2010. Rather than waiting until later to set the revised LAER standard, so that it could assess technological improvements made in response to the 1993 and 1998 limits, EPA adopted the recommendation of the negotiated rulemaking committee to set the 2010 standard as part of the 1993 rule. Again based on performance data from the United States Steel plant in Clariton, the limits for 2010 are only slightly more stringent than their 1998 counterpoints, and are considerably less stringent than what the current data indicate the best-performing by-product recovery plants could meet. The statutory criteria for LAER, then, simply were not met.

EPA was also required to promulgate Section 112 emission limits for *new* coke oven sources. Once again, the negotiated rule appears to fall short of the statutory mark. The problem is one of scope as well as one of substance. Section 112 defines "new source" as "a stationary source the *construction or reconstruction* of which is commenced after the [EPA] first proposes regulations under this section establishing an emission standard applicable to such source." By the terms of the statute, then, a "new" coke oven source includes both the construction of a wholly new coke oven *plant* and the reconstruction of an existing plant to install a new coke oven *battery*. Under the terms of the regulation, however, a reconstructed coke oven plant becomes a "new" source only if the new coke oven batteries "increase the design capacity" of the facility. This removes an entire class of reconstructed facility from the ambit of the new source standard, and allows existing plants that do not expand their operations to replace coke oven batteries without making any improvements in technology.

Moreover, new source limits under Section 112 are to be "not less stringent than the emission control that is achieved in practice by the best controlled similar source," without regard to cost. As the Jewell Smokeless nonrecovery plant in Virginia was the best-performing coke oven plant in the United States, one would have expected it to have been the model for EPA's new source standards. Indeed, Congress specified that, in setting new source limits for coke oven facilities, the agency "shall evaluate...the Jewell design Thompson non-recovery coke oven batteries and other non-recovery coke oven technologies." Nonetheless, the negotiated rulemaking committee chose to set two new source standards, one for nonrecovery batteries and one for by-product recovery batteries. New sources choosing nonrecovery technology must meet a limit of 0.0% leaking doors, lids, and offtakes, while new sources choosing by-product recovery technology need only outperform the 2010 limits: 4.0% leaking doors for tall and foundry doors, 3.3% leaking doors for other doors, 0.4% leaking lids, and 2.5% leaking offtakes.

A final noteworthy feature of the negotiated rule is its requirement that compliance monitoring be done on a daily basis, by "certified observers" who are independent of the coke oven facility, but whose funding comes from the industry. Although there have been problems in securing the true "independence" of the observers, there seems to be little question that the rule has enhanced both the frequency and the accuracy of the compliance monitoring. By all accounts, these improvements to the monitoring routine are a direct result of the negotiated rulemaking process.

Overall, however, the rule fashioned by the negotiators was not designed to secure optimal environmental performance from coke oven facilities. The rule provides a framework wherein facilities are assured that, at least until the 2020 statutory target date for health-based limits, emission limits will be attainable through the use of inferior, pre-1993 technology. Indeed, an EPA official noted at the time that companies choosing the "extension track" would be assured that any improvements made to their plants when the rule went into effect in 1993 would be the last they were required to make for almost 30 years. Although this could change if the agency decides to tighten the 2010 limits before the 2007 deadline, the regulation clearly is not designed to encourage diffusion of the cleaner (nonrecovery) technology within the industry, much less to spur any further wholesale improvements in coke oven technology. Further, while EPA touted the negotiated rule as a triumph for "environmental justice"- (because coke oven plants tend to be located in heavily-industrialised, lower-income areas), the effect of the negotiated new source standards will be to discourage the use of the cleaner technology in those areas until at least 2020.

This is not to say that the result achieved by the negotiated rulemaking committee may not represent an appropriate balancing of environmental and economic concerns in its approach to a troubled industry. A major stumbling block to tying emission limits to the performance of nonrecovery technology, apparently, was the relatively high capital cost of replacing an existing byproduct recovery battery with a new nonrecovery battery. In addition, there was a concern about jobs. A nonrecovery facility typically employs fewer workers than a by-product recovery facility. Requiring improved performance at existing by-product recovery plants, however, actually created jobs. Negotiated rulemaking appears to have been an ideal vehicle for the discussion of these issues, and for the sharing of information that appears to have been necessary to convince the environmental group representatives to accept the less stringent emission limitations favoured by industry.

However, had the goal instead been to "push" the industry towards markedly better technology, and thus to risk some short-term dislocation within the industry, it is not at all clear that negotiation would have been the best approach. The fact that EPA so grossly underestimated the performance capability of even the existing by-product recovery technology suggests that the agency's limited resources were directed more at ensuring a "successful" negotiation than at ensuring that its technological and economic data base was a reliable one. Reportedly, the negotiated rulemaking process took an immense amount of agency resources. Had EPA instead used those resources to take a hard look at what the industry could do, now and in the future, it is likely that the agency could have crafted a rule that met the environmental goals of the Clean Air Act, and that created meaningful incentives for the use of better technology.

3.3.3. The wood furniture coatings rule

Another Section 112 regulation that was drafted, in large part, through negotiated rulemaking was the hazardous air pollutant emission standard for the wood furniture industry. After a series of public meetings with representatives from industry, environmental groups, and state government in late 1992 and early 1993, EPA convened a negotiated rulemaking committee to attempt to formulate a rule governing wood furniture (surface coatings) nation-wide. The committee held its first meeting in July 1993, and a proposed rule, largely drafted by the committee, was issued in December 1994. The timing of this promulgation likely was influenced by (if not wholly determined by) the fact that the Sierra Club, a private, non-profit environmental group, had sued EPA in 1993 to compel the issuance of several rules under Section 112, and that a consent decree entered in that case called for the promulgation of this proposed rule by November 21, 1994. The final rule - virtually unchanged from the proposed rule - was promulgated on December 7, 1995, although portions of the rule were challenged in court by the chemical industry.

Based on the committee's work, EPA determined that wood furniture manufacturers performed four basic operations in producing a finished product - finishing, gluing, cleaning and washoff - and the proposed rule contained standards for each. All but the gluing operation standards were drafted by the committee. The standards for the gluing operations were developed "outside of the regulatory negotiation process, because adhesive suppliers were not represented on the Committee." EPA estimated that more than 11,000 facilities were included within the wood furniture industrial source category, and that approximately 750 of these would be considered "major" (as defined by the rule), and thus subject to these regulations under Section 112.

As EPA noted in the preamble to the proposed regulation, "a regulatory negotiation process...often requires concessions from some parties in exchange for concessions from other parties." Considered as a whole, the wood furniture rule might well be viewed as a compromise of the stringency of emission levels in exchange for a clear focus on pollution prevention (as opposed to simply "end-of-pipe" emission control).

For example, Section 112(d) specifies that EPA "may distinguish among classes, types, and sizes of sources within a category or subcategory in establishing [technology-based] standards" for the emission of hazardous air pollutants. Rather than distinguish among the technological and economic capabilities of particular wood furniture industry segments, however, the committee proposed - and EPA accepted - an industry-wide standard. Accordingly, EPA dismissed the suggestion that it require the use of "finishing materials with a very low or zero HAP [hazardous air pollutant] content," on the basis that such materials "have not been demonstrated to be feasible *for all industry segments."* Had EPA divided the industry into subcategories for regulatory purposes, however, it appears that lower emissions of hazardous air pollutants could have been achieved in certain sectors through the required use of these finishing materials where such use *would* be feasible.

Further, in the part of the rule dealing with restrictions on certain work practices known to be associated with the release of hazardous air pollutants, the committee specified a list of solvents to be forbidden from use in cleaning or "washoff" activities. Agency technical personnel believed that the committee's list of the chemicals to be so restricted was too narrow and needed to be expanded. Here again, despite these technical concerns, EPA simply accepted the proposed rule as written by the negotiated rulemaking committee.

While the rule drafted by the committee is less stringent than it likely could have been, however, it *is* designed to encourage pollution prevention, and could ultimately result in changes in technology and practices that reduce emissions below the levels required by the rule. Further, the emphasis on pollution prevention has the advantage of providing protection both to the environment and to workers. Rather than focusing on the use of control technology to reduce emissions, the committee endeavoured to select a format that would "accommodate multiple compliance techniques for the various industry segments. For finishing operations, then, the committee chose to express the required emission limit in terms of kg (or pounds) of volatile hazardous air pollutants emitted per kg (or pounds) of solids contained in the finishing materials used. This method of expressing the limit was chosen, noted EPA, because sources are encouraged to reduce the quantity of HAP through reformulation methods."

Significant attention was paid to pollution prevention in the drafting of work practice rules as well. As noted supra, the use of certain solvents is banned in cleaning and washoff operations. In addition, the use of solvents in spray booth cleaning is prohibited except in limited circumstances, and sources are required to maintain a "solvent accounting system" to track the use of solvents in cleaning and washoff. As noted by the agency, "although it cannot be assumed that it will actually result in...reduction, the cleaning and washoff solvent accounting system may prompt facilities to eliminate inefficient uses of solvents."

The fact that this rule included a substantial emphasis on pollution prevention is not surprising. Both the decentralised industry profile (with thousands of small shops instead of a few large ones), and the relatively straightforward and uncomplicated opportunities for chemical substitution and use reduction, made this industry an ideal candidate for pollution prevention. Nonetheless, it does appear that the use of negotiated rulemaking facilitated the agency's focus on pollution prevention in the development of the rule. It seems likely that the active participation of industry representatives (who are in the best position to identify productive opportunities for pollution prevention) helped to both deepen and legitimise the committee's efforts to build pollution prevention into the rule.

Moreover, the committee negotiations produced an agreement, *outside of the parameters of the rule,* under which the industry will prepare a semi-annual "trends report," beginning in 1994, which is to contain "a brief discussion of technologies being used by the industry to reduce emissions, and a discussion of evolving technologies including new finishing materials, adhesives, and improved application equipment." This agreement reflects the belief - apparently shared by many committee members - that "new, lower emitting (both VOC [volatile organic compounds] and HAP) technologies...- are...on the threshold of demonstration." In addition, to help determine whether the rule actually results in the targeted reductions in hazardous air pollutant emissions, and to determine whether those emission reductions are being met through the substitution of other hazardous chemicals that are *not* regulated as hazardous air pollutants, the trends report is to include a chemical use and emission survey from a representative sample of the industry.

3.3.4. Evaluation

Table 1 summarises the results of these three negotiated rulemakings in terms of the substantive criteria suggested at the outset: environmental/public health protection and technological change.

Table 1. *Technological and environmental impact of three negotiated air emission standards*

	Diffusion	*Innovation*	*Short-term environment gain*	*Long-term environment gain*
Woodstoves	+/-	–	+	–
Coke ovens	+/-	–	+	–
Wood furniture	+(PP)	+(PP)	+	+

The first two columns focus on the particular rulemaking's potential to effect technological change within the regulated industry, where "diffusion" refers to the diffusion of a environmentally-superior *existing* technology within the industry, and "innovation" refers to the development of a *new* technology that either produces greater environmental gains than existing technology, or produces equal gains at a lower cost.

The second two columns refer to the rulemaking's potential to effect improvements in public health or the environment, where "short-term" gains are those that are achieved before new and better technology is developed, and "long-term" gains are those that are achieved when new and better technology is developed and fully implemented.

The woodstoves rulemaking did not seek to push the envelope of woodstove technology, and focused instead on the diffusion of existing control technology. It is assigned a "+/-" rating in the Diffusion column because it set a different emission standard for each of the two types of woodstove technology on the market, rather than seeking to devise a standard that would diffuse the superior technology throughout the industry. This resulted in short-term environmental gain, but did not create a strong, consistent signal designed to encourage the kind of innovation in woodstove technology that might produce greater environmental gain in the long-term.

The profile for the coke oven rule is quite similar. Rather than seeking to diffuse the cleaner existing (nonrecovery) technology, the coke oven rule focused on the use of readily-available control techniques to improve the performance of the dominant existing (byproduct recovery) technology, and has resulted in short-term environmental gain. Further, by setting a standard for new facilities that is not tied to the performance of the cleaner existing technology, and by setting a 2010 standard for existing facilities that many firms were meeting easily in 1993, the negotiated rule provides clear incentives for keeping the dirtier technology in operation longer, thus actually reducing long-term environmental gain.

The wood furniture coatings rule, in contrast, has both a focus on pollution prevention - denoted as "+(PP)" - and a focus on innovation. It can be expected to diffuse existing pollution prevention technologies and, especially given industry's agreement to prepare the semi-annual trends report, has a real potential to produce innovation (and, concomitantly, to produce long-term environmental gain).

4. NEGOTIATED IMPLEMENTATION

In contrast to its role when it is *enforcing* a regulatory standard (discussed in Section 5 below), an agency's role in *implementing* the standard (that is, when it addresses the question of the timing and the extent of the applicability of the standard to a particular firm) is a circumscribed one. Nonetheless, there are circumstances in which the agency may be able to use negotiation at this stage of the process to encourage innovation and/or incidental environmental or health and safety gains.

Over its history, EPA has made some use of negotiated implementation both within its explicit statutory mandates (with the use of innovation waivers made available under certain environmental statutes) and (with its Project XL program discussed later) outside of them.

4.1. Innovation waivers

Various United States environmental statutes have had provisions allowing EPA to issue *innovation waivers* to qualifying firms, thus allowing them additional time to develop innovative approaches to compliance. The Clean Air Act and Clean Water Act both contain provisions authorising EPA to grant innovation waivers in certain circumstances. Under these provisions, EPA is authorised to extend the deadline by which a firm must meet emission or effluent limitations, so long as the agency is persuaded that the firm is actively pursuing an innovative approach to compliance that shows real promise of coming to fruition. Innovation waivers are meant to focus squarely on the innovation of new technology, and are not designed to promote diffusion of an existing technology.

In concept, the innovation waiver makes a great deal of sense. Development of an innovative idea into an operational reality - which often requires several periods of trial and error - can take substantial time, during which a firm might otherwise find itself liable for penalties for violations of emission or effluent standards. The innovation waiver exempts the firm from such penalties during a designated trial period, and offers it the prospect of the cost savings that may be derived from the development of a superior technology. Although it may be unrealistic to expect EPA to use innovation waivers to promote radical process innovation, because of the long time generally needed to develop the innovation, the agency might well use such waivers to encourage both incremental process innovation and the acceleration of radical innovation already underway.

In practice, however, innovation waivers have been used sparingly by EPA, both because industry has been unsure of their application (and thus has been wary of risking non-compliance), and because the agency has not encouraged their use [Ashford et al., 1985; EPA, 1994]. Success will require EPA to give early, clear, and certain signals to the firm, thus minimising the risk of its technology being found unacceptable. Furthermore, good faith efforts resulting in significant, though not complete, achievement of the pollution reduction goal may need be rewarded by "fail-soft" enforcement strategies, such as a reduction of otherwise applicable penalties, if industry is to be persuaded to take a technological and legal risk that the innovation waiver often poses. In this context, one can make a case for "risk sharing" between government and industry in the interest of fostering innovative solutions.

4. 2. Extra-statutory efforts: Project XL

In an effort to add to those opportunities for flexibility that are specifically authorised by statute, such as innovation waivers, EPA endeavoured to incorporate flexibility into its regulatory implementation by agency fiat through its now defunct Excellence in Leadership Project, popularly known as Project XL. The Clinton White House announced this program, with considerable fanfare, in a 1995 policy statement, and EPA published a set of guidelines for approving Project XL proposals in 1996.

The basic idea of Project XL was to allow regulatory flexibility, in return for superior environmental performance, at selected facilities, on a facility-by-facility basis. As conceived, the cornerstone on which Project XL was to rest is negotiation among the regulators, the facility owners, and the affected community, resulting in a Final Project Agreement ("FPA") governing environmental performance at the facility. The underlying rationale for Project XL was the belief that, for appropriately selected (new and existing) facilities, such negotiations could produce a plan for limiting pollutant discharge from the facility that will both cost *less,* and reduce environmental and public health risks *more,* than would have been the case under existing regulations[2]. The program was far from a clear success, and no new applications were to be taken after January 2003. Few FPAs have been negotiated, and some of those that have are the subject of considerable debate and opposition.

A fundamental problem with Project XL was that it envisioned a kind of regulatory flexibility that has not been authorised by Congress. Because it was not authorised by statute, the regulatory plan set forth in the negotiated FPA did not supersede existing regulations. Thus, to the extent that the regulatory "flexibility" negotiated by the participants involved a failure to comply with certain regulations (even if also involves *outperforming* certain other regulations), the facility was operating in violation of the law. And, since relief from existing regulations is precisely what made this program attractive to the business community, most FPAs were expected to involve violations of applicable environmental regulations. Indeed, one source reported that an expression among EPA staff familiar with Project XL was that "if it ain't illegal, it ain't XL." This made Project XL an unsafe bet for the participating firm. For, even if EPA and the state give informal assurances that they will not take enforcement action that is inconsistent with the FPA, the agencies cannot guarantee that such enforcement action will not be taken under the "citizen suit" provision of the applicable federal statute.

In theory, the threat of a citizen enforcement suit was to be eradicated (or at least greatly minimised) by the inclusion of the affected community in the negotiation process. Yet this points to a second fundamental problem with XL: the difficulty of defining the relevant "community." Is it limited to those living near the plant, or does it include national and regional environmental groups with an interest in the issue? Does it include labour? Does it include those who speak on behalf of the protection of sensitive populations, or on behalf of disadvantaged neighbourhoods? These are high-stakes issues for two reasons.

First, any interested party who is excluded from the negotiation process is less likely to be satisfied with the result, and thus is more likely to challenge it, through a citizen enforcement suit, a public organising and publicity campaign, or both. Probably the best-known Project XL agreement to date, for example, pertains to Intel Corporation's newest semiconductor production site in Chandler, Arizona. The five-year project agreement, which covered operations at a 720-acre site, was negotiated among the company, federal and state regulators, and five Chandler residents. Although the participants apparently were satisfied with the FPA negotiated through this process, many non-participants were not. Two vociferous critics were the Silicon Valley Toxics Coalition, a California-based group that addresses pollution problems in the semiconductor industry, and the Natural Resources Defense Council, a national environmental group. These two groups, who were concerned about the national and industry-wide implications of this agreement as much as, if not more than, its local environmental impacts, mounted a high-profile campaign against the Intel agreement, and against Project XL itself. This level of opposition clearly indicates that the negotiating committee that devised the regulatory plan for the Intel facility was *not* representative of the "relevant" community.

Second, the composition of the negotiating committee is of obvious *substantive* importance as well. If important constituencies are left underrepresented, the agreement negotiated is much less likely to be the "right" result. The five community representatives who helped negotiate the Intel agreement were also members of a pre-existing Intel Community Advisory Panel, and were generally representative of a community sentiment that values the important role that Intel has played over the past sixteen years in helping transform Chandler from a small agrarian town into the third fastest-growing city in the United States. While this obviously is a legitimate perspective, it may well not be the one that places environmental and public health protection (much less the health concerns of particularly sensitive populations) at the forefront. Indeed, the tendency of local interests to sacrifice long-term environmental and public health interests in favour of short-term economic gain was one of the factors that drove Congress to begin setting *national* pollution standards in the 1970s.

One of the beliefs underlying Project XL was that sufficient public involvement and scrutiny at a site could greatly diminish the need for a national regulatory presence. This is unlikely to be the case, however, unless the "public" is broadly and fairly represented, and unless its "involvement" is truly meaningful. At the Intel site, it was not at all clear that the regulatory flexibility negotiated by Intel - such as relaxed permitting requirements for new product lines - was offset by "superior" environmental performance. While EPA concluded that the Intel plant would outperform certain regulatory requirements, there appears to have been no showing that the facility attained, much less outperformed, the current state of art for the semiconductor industry. For example, based on a comparison of projected toxic emissions from the new Intel facility to reported emissions from similarly-sized semiconductor facilities from 1992 through 1994, EPA was able to conclude only that "Intel is well within, if not exceeding, the standard for the industry."

Had groups such as the Silicon Valley Toxics Coalition and the Natural Resources Defense Council been involved as full-fledged negotiating participants at the Intel site, it is likely that any resultant FPA would have been substantively different from the one actually negotiated. It is questionable, however, whether Intel would have agreed to negotiate a FPA with such groups participating. Indeed, when these and other environmental groups requested that the Intel agreement be augmented with legally-enforceable pollution prevention requirements, Intel was not receptive. Both Intel and EPA countered that additional pollution prevention requirements requested by environmental groups would give external actors too much control over the XL process. Although this clearly does not represent the sentiments of all companies regarding all situations, the hesitancy that many firms would feel about sitting down as equal participants with environmental groups in site-specific negotiations is another factor that would tend to limit the success of an initiative such as Project XL. In addition, meaningful involvement of the public, even where it is acceptable to the company, likely would considerably extend the time necessary to develop the FPA.

EPA appears to have recognised that a site-specific negotiated solution is fraught with potential problems, and that -- like negotiated rulemaking -- it cannot be expected to be done successfully without a substantial commitment of time and resources. A Project XL success story makes the point. In 1997, the agency completed negotiations on what has been characterised as a "small, focused" FPA involving an OSi Specialities organo-silicone plant on the Ohio River. According to a company attorney who participated in the process, the negotiations were "enormously burdensome" for the agency. "Unless they can think of a more efficient way to do it," he opined, "I'd be surprised if the program survives." To some degree, of course, the amount of time and resources that the agency would devote to a Project XL negotiation is a function of the relative novelty of the XL concept within EPA, the level of mistrust of the XL process within the environmental community, and the pressure on the agency to "make good" on its promise to deliver increased regulatory flexibility without sacrificing environmental goals. Real negotiation of environmental policy, even if it is only the policy for a single facility, requires considerable effort.

5. NEGOTIATED COMPLIANCE

Roughly 90% of firms cited with noncriminal violations of federal environmental statutes in the United States resolve the matter through a negotiated settlement, rather than through an administrative hearing or court trial. The settlement of an enforcement action often offers an agency an excellent opportunity to promote pollution prevention, rather than conventional end-of-pipe control technology. The firm's attention has been commanded, and a need for creative (and less costly) approaches to compliance may well have become apparent. Outside of the enforcement process, an agency has little statutory or regulatory authority to require firms to implement pollution prevention; the regulated community can choose the means by which it will comply with federal requirements. But once an enforcement action is initiated, a window of opportunity for pollution prevention opens, because the means of achieving compliance likely will be subject to negotiation between the agency and the violator.

5.1. The environmental protection agency's supplemental environmental project (SEP) program

EPA has sought to capitalise on this opportunity by encouraging the use of *Supplemental Environmental Projects* (SEPs) to promote pollution prevention. SEPs are environmentally beneficial activities, which the violator agrees to perform and/or fund as part of its settlement with EPA, and which the violator is not otherwise legally required to perform. In the settlement process, EPA and company attorneys typically agree both on a penalty and on a set of activities designed to achieve and maintain compliance. In 1991, EPA adopted a SEP policy authorising agency enforcement personnel to reduce the amount of the penalty in exchange for the execution of a SEP. Encouraged by initial results from this approach, the agency has revised and expanded its SEP policy since that time.

The key to the SEP policy is the trade-off between penalties and SEPs. Current EPA penalty policy anticipates that, unless the SEP policy is invoked, the penalty assessed in any enforcement action will be the sum of (a) the amount of the *economic benefit* gained by the violator as a result of non-compliance (typically, the investment earnings from delayed capital expenditures, together with any avoided operation and maintenance costs), and (b) a *gravity* component (calculated according to agency guidelines) that is meant to reflect the relative seriousness of the violations. Under the present SEP policy, SEPs may be used to reduce this amount, so long as the final penalty paid is at least as large as what EPA characterises as the *minimum penalty:* the larger of (a) the economic benefit plus 10% of the gravity component or (b) 25% of the gravity component.

Currently, there are seven categories of acceptable SEPs: pollution prevention, public health, pollution reduction, environmental restoration and protection, assessments and audits, environmental compliance promotion, and emergency planning and preparedness. The key feature linking these various categories is the expectation that the project will result in some benefit to the environment or public health. Some SEPs, such as an off-site stream restoration project, offer direct, predictable public benefits while returning no direct benefit to the violator. Others, such as an agreement by the violator to conduct a comprehensive environmental audit of its facility, offer potential (and far less predictable) benefits both to the public and to the violator. In general, *pollution prevention* SEPs - which involve expenditures by the violator to implement technology or practices that reduce its generation of pollution - offer the greatest potential for the development of innovative production technologies and practices with widespread application.

So long as it does not reduce the penalty below the acceptable minimum, EPA will (depending on the assessed merits of the project) credit up to 80% of the after-tax cost of most approved SEPs (net of any savings - such as reduced operations costs - that the SEP may offer to the violator) against the amount of the penalty. In order to encourage certain types of projects, however, the agency revised its policy in 1995 to offer a credit of up to 100% for SEPs judged to be "of outstanding quality" according to a set of specified criteria[3]. Two of the six criteria specified in the most recent version of the SEP policy are: (a) the extent to which the project develops or implements *pollution prevention* techniques or practices; and (b) the extent to which the project develops or implements *innovative* technological approaches.

EPA reports that, from Fiscal Year 1992 through Fiscal Year 1994, it negotiated more than 700 SEPs, with an estimated total value (i.e., cost to violators) of over $190 million. Of these, approximately 14% were pollution prevention SEPs, with an estimated total value of approximately $57 million. EPA estimates that these pollution prevention SEPs will reduce the discharge of toxic chemicals and the production of hazardous waste by a total of some 65 million pounds.

A case study analysis of ten pollution prevention SEPs negotiated by EPA through Fiscal Year 1992 - selected because they reflect a range of technological responses - found that the technologies utilised included chemical substitution, process change, and closed-loop recycling [Becker and Ashford, 1994]. Representatives from all nine of the firms involved expressed support for the SEP policy. They indicated that they were glad to have had the option to implement a pollution prevention project in exchange for some penalty reduction, and noted their belief that the SEPs took some of the "sting" out of the enforcement process without eliminating the significant economic and psychological impacts of the enforcement action. Several company representatives also stated that the SEP process helped their firm to recognise other opportunities for environmentally beneficial improvements.

The technological changes undertaken by firms through pollution prevention projects can be categorised according to *the locus* of the change and according to the *degree of innovation* of the change. The majority of technological changes made by the SEP case study firms were diffusion-driven. A smaller number can be considered incremental innovations, and only one case can be considered a major innovation. There was a fairly even distribution of technological changes across the spectrum of primary, secondary; and ancillary processes[4]. If a random case-study selection process had been used, the sample would have been more heavily weighted toward diffusion-driven changes to ancillary production processes. The larger universe of EPA settlements containing pollution prevention consisted mainly of the adoption of off-the-shelf technologies. This suggests there are unexploited opportunities in enforcement for stimulating innovative technological change. Realisation of this potential likely would require changes in attitudes and knowledge levels, both within industry and within EPA. One move in this direction has been the agency's more recent willingness to allow up to two years for the completion of selected pollution prevention SEPs, as a longer-term time window is essential if more significant innovation is to take place.

6. REGULATORY REINVENTION: EPA'S "COMMON SENSE" INITIATIVE

Under the Clinton Administration, EPA determined that fundamental changes in approach would be necessary if significant additional progress in protecting the environment was to be made, and if the environmental challenges of the future were to be resolved satisfactorily. The agency referred to this as the need for "regulatory reinvention." In July 1994, EPA began its *Common Sense Initiative* (CSI), which it termed the "centrepiece" of its regulatory reinvention efforts. The primary goals of CSI were to find "cleaner, cheaper, smarter" ways of reducing pollution, and to formulate proposed changes in the existing regulatory structure to effectuate them. As with Project XL, negotiation among interested parties was the means by which EPA hoped to achieve the goals of the program. Unlike XL, however, the focus of the negotiations was *industry-wide.* To carry out CSI, the agency assembled six advisory committees, one for each of six industrial sectors: automobile manufacturing, computers and electronics, iron and steel, metal finishing, petroleum refining, and printing. Each advisory committee consisted of representatives from EPA, the relevant industry sector, state and local regulatory agencies, national and local environmental groups, labour, and community organisations. The work of these committees was overseen by a separate Council, the membership of which was drawn from the same sources. The Council was chaired by the EPA Administrator, and each of the six sector committees was chaired by an EPA official. The work of the Council and the committees was assisted by EPA staff.

This industry-sector structure was based on a fundamentally sound premise: that, for a variety of reasons, different industries often differ in their technological and economic potential for reducing pollution, and also in the way in which they respond to various types of regulatory signals. By bringing together people who are knowledgeable about the opportunities for reducing pollution within a particular industry, and who have a stake in how, when, and under what terms that reduction will occur, EPA hoped to harness the potential of each industry to a fuller extent than it had heretofore been able to do. The agency also hoped that, by creating an atmosphere in which innovation and flexibility were emphasised, the focus of the committees would be on pollution prevention rather than end-of-pipe pollution control. In December 1998, arguing that the CSI approach had been proven a success, EPA announced that CSI itself would be phased out, but that the lessons learned from the initiative would be expanded to other industry segments in a future action plan.

In fact, the results of the CSI experiment were mixed. On the one hand, as EPA points out, the initiative brought together six groups of people representing a diverse set of interests, and encouraged an ongoing dialogue on issues that are important to the future development of environmental policy. This is a valid point. If CSI succeeded at nothing more than promoting a better understanding of the issues among different stakeholders, and of each other, among those likely to participate in environmental policy-making and implementation affecting these industries, it arguably had a positive impact.

On the other hand, however, CSI has been criticised for its lack of substantive results. A series of reviews of CSI have raised this issue, including a 1997 report issued by the U.S. General Accounting Office ("GAO"), a research arm of Congress [GAO, 1997] (hereafter "GAO Report"). In general, GAO and other reviewers found that the CSI process moved considerably more slowly than most of the participants would have liked. The reasons for CSI's slow pace, GAO found, were multifold: the time necessary to collect and analyse data; the variations in the participants' understanding of the technical issues involved; the time taken by the participants "in reaching consensus on the approaches needed to address large, complex issues or policies;" the time taken by participants "discussing how they would carry out their work and developing their own operating standards;" and the difficulties experienced by some participants in making the necessary time commitment. None of this should be particularly surprising. Indeed, when one adds to this list the overall need to establish a degree of trust among the participants in each sector group sufficient to permit a meaningful discussion on substantive issues, it is not particularly difficult to understand why substantive progress was slow in coming.

Nonetheless, there appears to have been a growing feeling among participants that a failure to meaningfully step up the pace of substantive progress would mean the death-knell of the initiative. The automobile and petroleum refining industries ended their participation, and other participants indicated that they would leave unless EPA made changes - in response to the various reviews of the project - that make for a more efficient process. To address this issue, GAO had proposed that EPA

> ...provide an improved operating framework that (1) more clearly defines the Initiative's "cleaner, cheaper, smarter" environmental protection goal - including its expected results - and (2) specifies how the Council and its subcommittees and workgroups will accomplish their work, clarifying issues such as how and when consensus will be achieved, how the Initiative's goal should be interpreted and applied to individual projects, and to what extent representatives of all stakeholder groups should be included in activities at each level of the Initiative, including its projects and workgroups [GAO, 1997, note 13 at 7] [5].

EPA indicated at the time that it would introduce reforms of this nature, but GAO faulted the agency for not having done much of this at the outset. It is not at all clear, however, that this would have been the right approach. It is arguable that, had EPA attempted to dictate terms of this nature to the participants at the beginning of the process, rather than allowing the participants to first address these issues on their own, it would have engendered considerable resentment among some of the participants.

Moreover, the changes envisioned by GAO were unlikely to address the more deep-seated issues that slowed or prevented substantive results along the lines originally anticipated by EPA. It is likely that a major factor inhibiting real progress was the fact that, in contrast to negotiated rulemaking, the CSI negotiations did not proceed within a formal legal context, with a known and meaningful set of potential consequences. In negotiated rulemaking, the participants all know that, regardless of whether they reach agreement on a proposed rule, a rule is likely to be issued. The "stakes" for each participant thus are fairly clear: if we don't negotiate, the agency is going to go ahead and promulgate a regulation without us, and the result may be something we don't like. In the CSI negotiations, however, the consequences of inaction usually were both far less clear and far less dramatic. Indeed, in most cases the failure of a negotiating committee to agree on a particular "regulatory reinvention" proposal would have had no greater practical effect than simply the preservation of the status quo.

Accordingly, the chief factor likely to be motivating industry's participation in CSI-type of negotiations is the opportunity to push for regulatory alternatives that are less expensive (to industry) than the status quo. Industry's interest, then, is likely to be in "streamlining" - or eliminating - current regulation, and not in extending the scope of regulation into new areas. And, since the environmental representatives should not be expected to agree to a cheaper alternative if it does not also represent increased environmental benefit, progress in these type of negotiations may be slow in coming, especially in those industry sectors where few easy and obvious "win/win" (i.e., cheaper *and* cleaner) regulatory improvements present themselves.

Thus, it should not be surprising that the petroleum and automobile industries decided to abandon their participation in the CSI Initiative. Effective participation in negotiations of this nature takes a considerable commitment of resources. As noted by the American Petroleum Institute in a letter to EPA explaining the withdrawal of its member companies from the CSI negotiations, the companies "believe the refining industry's resources...can be more productively directed toward other approaches."

Another systemic problem one would expect to encounter in negotiations of this nature stems from the participants' unequal access to relevant data. If effective strategies to encourage pollution prevention are to be crafted by consensus, reliable technical information - especially information relating to the technological potential for pollution prevention - is likely to be important. Much of the relevant data, of course, will be in the hands of industry. Without a clear incentive to make these data available to the other participants, industry is likely to prefer to pick and choose what it will share, thus making meaningful negotiations all the more difficult. This reportedly has been a major issue, for example, in the computer and electronics work group. Firms reportedly have been reluctant to divulge information because "they feared that regulators would use data to extract further concessions," and because they believed that environmental groups would "use any information divulged during CSI meetings to mount lawsuits." This, in turn, contributed to a sense of mistrust among the environmental group participants.

This is not to say that cooperative approaches are not capable of producing any meaningful results of substance. There are cleaner/cheaper opportunities in a number of industries that may be able to be realised without the "push" of additional regulatory pressure, and cooperative approaches could bring some of these to light. The CSI metal finishing work group, for example, began a successful demonstration of a new technology for filtering chromium from air releases that should decrease chromium emissions while reducing costs by about 90%, and announced agreement on an emission reduction program that is reported to rely, in part, on pollution prevention strategies. And the CSI printing work group developed an education and outreach project designed "to achieve fundamental change" by incorporating the philosophy of pollution prevention into everyday work practices. In general, however, the bulk of the CSI negotiations reportedly did *not* focus on pollution prevention strategies, let alone innovation, thus falling well below EPA's original expectations.

In 1999, two years after the GAO report, EPA issued a report by an independent contractor evaluating some 40 CSI projects [Bruninga, 1999]. The report concluded that, although there had been a small number of sector-specific modifications, EPA had made little progress in addressing broad regulatory changes through CSI, and CSI successes were not being integrated into core EPA programs.

7. CONCLUSION

Negotiation should hardly be viewed as a panacea for the various difficulties that typically confront the policymaker. Used in the right context, however, negotiation can be a useful tool in the establishment, implementation, and enforcement of environmental and occupational safety and health policy. Negotiation can facilitate a better understanding of issues, concerns, facts, and positions among adversaries. It can also promote the sharing of relevant information, and can provide an opportunity for creative problem-solving. Whether negotiation will be better than other, generally more adversarial mechanisms as a means of fostering improved environmental, health, and safety outcomes, or of stimulating meaningful technological change, will depend on the situation in which it is used. In general, negotiation would appear to work best a means of securing these goals in situations in which the necessary regulatory signals for improvement and innovation are already in place.

This is one of the reasons that EPA's use of *negotiated compliance,* as embodied in its SEP policy, has been as successful as it has been. To the firm that is the target of the enforcement action, the "stakes" are clear: so long as it believes it faces higher costs (in the form of a larger fine and/or higher transaction costs) if it does not identify and execute a SEP that is acceptable to EPA, the firm has a meaningful incentive to participate in good faith in the SEP process. And, because the agency has structured the program to allow maximum credit for pollution prevention projects, pollution prevention can become the focus, *and the goal,* of the negotiations. The pollution prevention results of the SEP program have been relatively modest - mostly diffusion and, sometimes, incremental innovation - but this is in keeping with the relatively modest nature of the financial incentives typically involved, and with the relatively short time period within which the SEP typically must be identified and completed. Especially because negotiation is the traditional means of resolving enforcement disputes, even outside of the SEP process, negotiation appears to work well here.

One would also expect negotiation to work well in those *negotiated implementation* situations that have a clear, formal focus on technological change, such as the innovation waiver opportunities created by certain environmental statutes. The chief signal to innovate - the new regulatory standard - is already in place (or clearly on the horizon) before negotiation over the waiver or variance begins, and the statutes typically provide an extended period of time for the firm to develop and test the proposed innovation. Thus, so long as the new standard is stringent enough to command the firm's attention, firms should have a meaningful incentive to negotiate time to pursue an innovative compliance alternative.

The fact that EPA's innovation waiver program has thus far not lived up to expectations appears largely due to a failure of leadership and administration. This, in turn, may have contributed to what appears to be a reticence by Congress to include innovation waiver provisions in its revisions to existing statutes. If EPA could develop and promote its innovation waiver program the way it has the SEP program, the innovation waiver might become a much more important means of securing environmentally beneficial technological change.

In contrast to negotiated compliance and negotiated implementation, *negotiated rulemaking* is a situation in which the chief regulatory signal for improvement and innovation is *not* already established, at least not in full. Rather, one of the functions of negotiation in this context is to *establish,* either in part or in full, the stringency of the regulatory standard. If the goal is innovation, this may well be problematic. If the nature of the regulated industry is such that it will require a dramatic impetus - such as the promulgation of an unexpectedly stringent standard, or the fear that such a standard will be promulgated -before it will be motivated to innovate, negotiated rulemaking may well be inadvisable. Since negotiated rulemaking seeks consensus among the participants, and since such an industry is unlikely to agree to a standard that it views as having a "dramatic" impact, negotiated rulemaking is unlikely to produce a standard of this nature. In such situations, negotiated rulemaking's focus on consensus can effectively *remove* the potential to spur innovation [Goulding and Murphy, 1998]. In situations in which the desired technological change is likely to come more easily, negotiated rulemaking should be expected to have a better chance of success. Here, the advantages of negotiation, such as information-sharing and creative problem-solving, may work to encourage productive technological change. The key to the willingness of industry representatives to explore the technological options in good faith is likely to be tied to what they perceive the likely "default" standard to be. If they believe that, in the absence of a negotiated rule, the agency will promulgate a stringent rule on its own, their willingness to focus on creative technological solutions is likely to be higher. The agency can facilitate this process by making clear at the outset that promoting technological change will be a focus of the regulation. If technologically literate stakeholders, such as trade unions or sophisticated non-profit groups, are involved, the dominance of industry's technical expertise may be minimised, and outcomes that advance the state of the technology may emerge.

Another important difference between negotiated rulemaking and negotiations over SEPs and innovation waivers, however, is that the scope of the negotiations in negotiated rulemaking is (at least) industry-wide, rather than firm-specific. Interest in the negotiations thus is much stronger, and the number of participants who must be involved, if the negotiations are to succeed, is an order of magnitude higher. Accordingly, management of the negotiation process becomes a formidable task, and the agency must have the resources to be able to keep pace. There is always the risk that the process itself, and not the ultimate results of the process, will assume centre stage, and that a focus on technological change will give way to a focus on achieving consensus.

Many of these same concerns are germane when negotiation is used in an *extra-statutory* sense, as was the case with EPA's Project XL and Common Sense Initiative, in an attempt to change regulatory policy. If the focus is industry-wide, the resource demands will be large. Further, where there is no meaningful incentive for industry negotiators to move away from the status quo - that is, where there is no impending "default" standard or requirement that they perceive as onerous - they may well be interested only in those regulatory changes that save them money.

In the last analysis, it must be recognised that negotiation is a process that facilitates *market solutions* to questions regarding the appropriate ends or means of compliance. That is, the relative bargaining power of the stakeholders largely determines the outcome, unless it is checked at the end of the process by a government agency with a strong sense of trusteeship for the congressional policy it is charged with implementing. Agencies who see themselves as *mediators* of the negotiation, or who otherwise relinquish their statutory role as trustees, help to promote a market-like result through the operation of the consensus process. In this case, negotiation is unlikely to produce impressive environmental gains linked to technological change. When this happens, the relative success of the negotiations likely will depend on whether some *other* factor - such as a court ruling or a scientific study - can produce the kind of incentives that are likely to promote technological change. If a superior result is to be achieved, it likely will require the participation of agencies with both the means and the will to take a firm position in support of the environment, and in support of the development of new technologies.

8. AKNOWLEDGEMENTS

The authors gratefully acknowledge the Fondazione Eni Enrico Mattei for its support of the earlier research underlying this now expanded and updated article. (See Caldart CC, Ashford NA. Negotiation As a Means of Developing and Implementing Environmental and Occupational Health and Safety Policy. *Harvard Env. Law Rev.* 1999; 23(1):141-202).

9. NOTES

[1] Approximately 350 firms have joined the program from a diverse cross-section of the economy. In contrast to Project XL, regulatory flexibility seems to relate to discretionary activities of agency inspection and reporting policies, rather than extensive exclusion of individual firms from mandatory regulatory provisions. See http://www.epa.gov/performancetrack

[2] Negotiation between the agency and the facility owner (sometimes also involving environmental groups and/or local community groups) is commonplace in the permitting process. Project XL negotiations were different, however, in that they purported to *replace* current standards with an alternative approach, while traditional permit negotiations generally are over the proper way to *apply* current standards to the facility in question. Thus, XL purported to be the negotiation of environmental *policy,* albeit on a facility-by-facility basis.

[3] Five criteria were specified in the 1995 policy: benefits to the public or environment at large; pollution prevention; innovativeness; environmental justice; and multimedia impacts. In 1998, a sixth criterion – community input – was added.

[4] Becker and Ashford, 1995 at 224A. The distinction between primary, secondary, and ancillary manufacturing and production processes is an important one for innovation. An example in the context of casting and plating metal screws makes the point. The primary process is the casting of the screw. The secondary process is electroplating. The ancillary process is cleaning or degreasing the screw using organic solvents. If the environmental problems facing the firm is created by the latter activity, it might be relatively easy for the firm to search for and find an alternative, non-polluting cleaning process, and no innovation would be required. If the electroplating is the process that needs to be modified, at least a new process might have to be brought into the firm - usually by the diffusion of alternative plating technology - but the firm would be uncomfortable about changing a proven method and taking a chance on altering the appearance of its product, even if it is a separate operation. The most resistance could be expected by demands on the primary process. Here innovation might be necessary, and the firm would not be likely to invest in developing an entirely new casting process merely to reduce a penalty.

[5] In addition, several environmental justice groups, as well as representatives from the State of Michigan withdrew from the CSI negotiations.

10. REFERENCES

Ashford N., Ayers C., Stone R. (1985), Using regulation to change the market for innovation, *Harvard Env Law Rev,* 9(419):443-62.

Becker M., Ashford N. (1995), Exploiting opportunities for pollution prevention in EPA enforcement agreements, *Env Sci Tech,*29(5):220A-6A.

Bruninga S. (1999), "CSI Successes Not Being Integrated Into Core EPA Programs, Stakeholders Say" *Environment Reporter* 29(50) Friday, April 23.

Coglianese C. (1997), Assessing consensus: the promise and performance of negotiated rulemaking, *Duke LJ,* 46.

EPA, Office of Water (1994), *Providing waivers from NPDES permit compliance for industrial pollution prevention technology: the industrial pollution prevention project (IP3). Analysis of Sections 301(K) and 307(E) of the Clean Water Act*, Washington D.C., USA

Goulding A., Murphy J. (1998), *Regulatory realities: the implementation and impact of industrial environmental regulation,* Earthscan Publications Ltd, London, 120.

United States General Accounting Office (1997), *Regulatory reinvention: EPA's common sense initiative needs an improved operating framework and progress measures*, July.

CHAPTER 4

DESIGN, NEGOTIATION AND IMPLEMENTATION OF ENVIRONMENTAL VOLUNTARY AGREEMENTS: NATIONAL AND SECTOR APPROACHES

EFFICIENCY STANDARDS VERSUS NEGOTIATED AGREEMENTS IN THE EUROPEAN ELECTRICAL APPLIANCE SECTOR

P. MENANTEAU

Laboratoire d'Economie de la Production et de l'Intégration Internationale Département Energie et Politiques de l'Environnement (LEPII-EPE) Centre National de la Recherche Scientifique – Université P. Mendes France, Grenoble, France

Abstract: Energy labelling and minimum efficiency performance standards have proved to be very effective in stimulating energy efficiency improvement in the domestic appliance sector. But standards suffer from long and often difficult implementation periods because of the resistance of the industrial sector. As a consequence, the question has been raised as to whether similar results could not be obtained more easily and at lower cost with voluntary agreements, which offer flexibility margins in the achievement of commitments. This paper analyses the specific advantages of voluntary agreements compared with efficiency standards in the domestic appliance sector. We conclude that voluntary agreements may be an effective instrument in this respect but in certain conditions. The alternative of regulatory measures must remain a credible, realistic threat if voluntary agreements are to have a really significant impact on performance improvement.

1. INTRODUCTION

Because of the increase in the use of lighting and the rising ownership of household appliances and electronic equipment, domestic electricity consumption for specific uses (household electrical appliances and lighting) has increased dramatically over the last twenty years in industrialized countries. According to the International Energy Agency, household appliances are the second greatest source of electricity consumption in the OECD countries and the third source of greenhouse gas emissions (IEA, 2003). Consumption will probably continue to grow at a steady rate despite the expected saturation in ownership level of certain appliances. The IEA has projected that electricity consumption by domestic appliances will continue to increase by 25 % between 2000 and 2020 despite the energy management policies already introduced [the increase would have been 60 % without any kind of energy policies (Ibid.)].

The possibility of controlling growth in electricity consumption, especially in the residential and tertiary sectors, is one of the conditions for reducing greenhouse gas emissions in virtually all the industrialised countries. The technological opportunities for improving the energy efficiency of electrical appliances are numerous (IPCC, 2001), but because of market risks these opportunities have not been sufficiently explored by manufacturers. Public policies introduced in this field

E. Croci (ed.), The Handbook of Environmental Voluntary Agreements, 163 - 178

are designed to accelerate the penetration of more energy efficient technologies and to inject more momentum into the process of technological change.

The strategy adopted by the European Union to accelerate the diffusion of energy efficient technologies has been to associate a consumer information device - a labelling programme - with a regulatory device in the form of minimum energy performance standards (MEPS). Energy labelling of domestic cold appliances thus became compulsory in all the European member States in 1995 and subsequently for other domestic appliances, in particular, washing machines, dishwashers, clothes dryers, and residential lighting equipment. This was followed in 1999 by MEPS designed to eliminate the least efficient products from the market.

This combined approach was effective in transforming the household appliance market, with labelling acting as an incentive to innovate and thereby complementing the regulatory approach which is generally not very effective in stimulating innovation. However, regulations raise a lot of opposition among manufacturers and consequently often require considerable time for implementation. The more flexible solution of voluntary agreements thus emerges as a possible alternative that is easier to implement while remaining just as effective.

The aim of this paper is to examine whether negotiated agreements can be as effective in stimulating the diffusion of more energy-efficient technologies in the household appliance sector as the combined action of labelling and performance standards. First, we illustrate the effectiveness of this combination of information and regulatory measures by referring to the example of cold domestic appliances, and we show that the synergy of the two instruments makes it possible to go beyond the usual limits of the regulatory approach in stimulating technical progress. We then examine the advantages of negotiated agreements from the point of view of the manufacturers and the public authorities and the reasons why the European Commission is increasingly interested today in adopting this type of approach to improve energy efficiency in the household appliance sector. We provide details of the negotiated agreements on washing machines that became applicable at the same time as the regulations concerning cold appliance performance standards, thus making it possible to observe the similarities and differences between these two instruments. Our discussion concludes that negotiated agreements can be truly effective only if there is a constant, credible threat of regulations.

2. THE SYNERGIC EFFECTS OF LABELLING AND EFFICIENCY STANDARDS

The European policy concerning transformation of the domestic appliance market was implemented essentially through two complementary measures: labelling programs to improve consumer awareness and MEPS. This combination may seem paradoxical in that performance standards are intended to set regulatory efficiency levels when the price signal is ineffective in promoting energy efficiency, whereas the primary purpose of labelling is in fact to provide a market signal to stimulate the purchase of efficient appliances by better informing consumers. Experience in the European Union over the last few years has proved, however, that these two instruments can co-exist very effectively and have very interesting characteristics where stimulating technological change is concerned.

2.1. Impact of energy labelling on purchasing behaviour

Lack of information for consumers is generally considered to be one of the main barriers to improving energy efficiency through the adoption of more energy efficient technologies. Comparison labels and endorsement labels are two ways of solving this problem by providing information on the energy efficiency of appliances, thereby encouraging consumers to compare products and choose the most efficient.

Comparison labels enable consumers to compare the energy efficiency of all the products in a particular category (refrigerator/freezers, clothes dryers, washing machines, etc.). The European Label – EUR- or EnergyGuide – USA are examples of such labels. Endorsement labels simply identify appliances which are particularly energy efficient. An example is the Energy Star program in the USA. The first type generally applies to all the products on the market, whereas endorsement labelling is a voluntary scheme in which manufacturers may participate.

Following the example of the USA and Canada, Europe introduced a framework for energy labelling in 1992 (comparison labels). The program became effective in 1995 for domestic cold appliances and has gradually been extended to cover other household appliances. This measure has undoubtedly contributed to transforming the domestic appliance market even if its impact is difficult to distinguish from the general trend in improved energy efficiency resulting from improved knowledge (cf. supra). An analysis of sales from 1994 to 1999 shows a clear trend towards greater overall energy efficiency in the domestic cold appliance market in Europe, with a significant shift in sales towards more efficient appliances (classes A, B and C) at the expense of the less efficient classes (E, F and G) (Figure 1). By the end of the 1990s, there was a 30% improvement in average energy efficiency compared with the beginning of the decade (CEC, 2000).

Energy labelling thus led to a transformation of the cold appliance market which is the result of a change in consumer preferences and changes in the marketing strategies of manufacturers and retailers (ECU, 1998). The influence of labelling on the innovation strategies of manufacturers can be interpreted as follows: anticipating changes in consumer preferences or future regulations, manufacturers discontinued certain models that had become difficult to sell (expensive and not energy efficient), improved - sometimes marginally - the appliances destined to remain on the market and gradually introduced new more efficient products. In 1998 at the Confortec electrical appliance show, all the manufacturers had introduced new more efficient models into their product ranges, and some had focussed on energy efficiency by presenting mainly class A and B appliances.

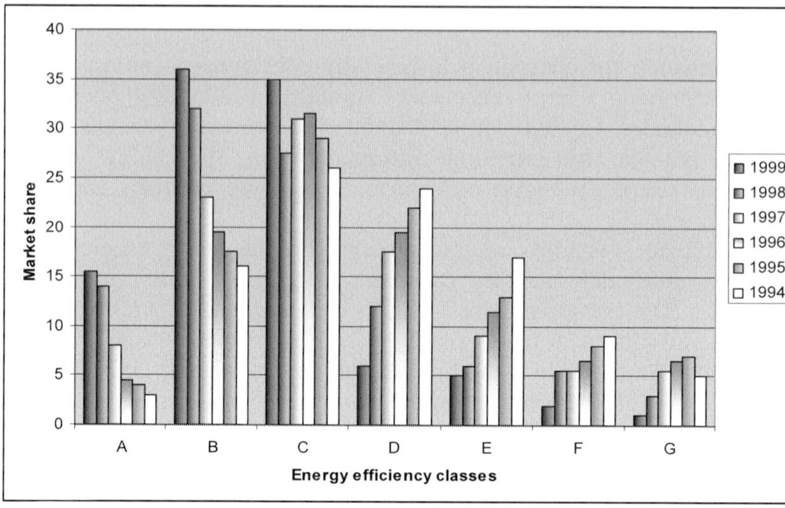

Source: CEC, 2000

Figure 1. *Cold domestic appliance sales in Europe*

Labelling is thus a powerful instrument for differentiating products, and one which can promote innovation among manufacturers who wish to use this device to improve their competitive position or to gain an edge in new market niches.

But labelling programs clearly have their limits. Their success depends to a great extent on the differences in efficiency between appliances and the related financial stakes for purchasers. Furthermore, labelling does not prevent the least efficient models from remaining on the market nor consumers from buying them. For this reason, programs imposing MEPS for household appliances generally accompany labelling programs.

2.2 Labelling and efficiency standards: a necessary complementarity

By definition, efficiency standards are based on a regulatory process which affects all the manufacturers in a particular country or economic region. The aim of such standards is to complement labelling schemes or to replace them in cases where the energy price signal is not strong enough to encourage consumers to purchase the more efficient appliances, which may happen even though comprehensive information is supplied.

In Europe, domestic cold appliances were the first to be subjected to performance standards. According to the Directive 96/57/CE of the European Parliament:

> Member States shall take all necessary measures to ensure that refrigeration appliances covered by this Directive can be placed on the Community market only if the electricity consumption of the appliance in question is less than or equal to the maximum allowable electricity consumption value for its category as calculated according to the procedures defined in Annex I.

The standard was chosen so as to obtain an improvement of 15% in the average energy efficiency of new appliances. From September 1999, appliances not belonging to efficiency classes A, B or C could no longer be sold, which meant that 40 % of the appliances on sale in 1996 had to be withdrawn from the market.

The effectiveness of the regulations is evident from an examination of the new products introduced on the market:

> All D, E, F and G appliances have been removed from the market, with the exception of chest freezers, for which E-class appliances can still be sold (Appliance Efficiency, 2000).

Are we to understand that labelling has just been a preparatory step to the introduction of legislation on energy efficiency, the latter being ultimately the most effective instrument for transforming the market? Does labelling still have an impact or can such schemes be discontinued to leave efficiency standards to do the job alone?

When efficiency standards are introduced, manufacturers are encouraged to improve their products so as to comply with legislation, but it does not necessarily encourage them to develop new highly efficient products if they are not required to do so by the consumers. To promote innovation, very stringent energy efficiency levels must be imposed, so that manufacturers will be compelled to innovate, or provision must be made for a gradual tightening up of regulations taking into account the improvements already made. But without additional incentives, the energy efficiency of appliances would remain overall much the same, since (most) manufacturers would simply ensure that their products were positioned just beyond the regulatory performance level. Manufacturers can in that case oppose the introduction of new more stringent standards by arguing that the new targets are not realistic from a technological or economic point of view.

The advantage of labelling programs is not simply that they facilitate the introduction of standards by defining efficiency classes that can be used to determine the authorised efficiency levels. Labels also have a very important role in encouraging differentiation and are thus an incentive to technological progress. With labelling, manufacturers have the possibility of differentiating their appliances from standard products, something they can achieve through innovation. This will gradually have an impact on all the appliances on the market and ultimately lead to higher efficiency standards.

By stimulating the arrival of new more efficient products on the market, labelling schemes thus condition the effectiveness of regulations. Such schemes must be constantly reviewed if they are to remain a way of differentiating between products. If efficiency classes are not redefined regularly, the combined result of labelling and standards will be that most appliances will be positioned in the highest efficiency classes and it will be impossible to identify new appliances that are even more energy efficient. A labelling scheme which can evolve and which operates in conjunction with MEPS that are periodically revised thus seems to be a particularly effective method and one that appears well suited to the transformation of the household appliance market.

2.3. The limits of regulatory measures

The regulatory approach nevertheless has certain limits. In particular, it is reputed to provide little incentive for technical change. Faced with the introduction of performance levels, manufacturers generally react by proposing products that meet the minimum requirements but they are not encouraged to go beyond these. Furthermore, the apparent simplicity of the regulatory approach should not hide the real difficulties involved in implementation. The great majority of manufacturers are opposed to the introduction of MEPS which they consider to be a limitation on their room to manoeuvre in terms of technological innovation. For them, such standards represent an additional constraint that results in higher production costs. Coupled with costs related to new environmental regulations (elimination of CFCs for refrigerators), this increase in production costs will be reflected in the selling price and could have a negative effect on the household appliance market. Finally, manufacturers are not a priori convinced that greater energy efficiency is necessarily desired by consumers, especially if it means sacrificing certain features to which they have become attached (for example, American style two-door refrigerators which consume considerably more electricity).

Without the cooperation of manufacturers, defining MEPS generally becomes a long and complex process (cf. infra). Thus the initial proposal to introduce performance standards in Canada in 1984 met with strong opposition from Canadian household appliance manufacturers' associations, which succeeded in blocking the initiative until 1988. Their position with respect to standards radically changed, however, following adoption of federal standards by the United States in 1987, a decision that engendered certain economic risks for Canadian industry (Varone, 1998).

On the other hand, the introduction of a single federal regulation in the United States was carried out relatively quickly because of manufacturers' concerns about the increasing number of specific performance standards in different American states. The constraint of a federal standard was then considered to be preferable to the risk of development of a national market that was totally heterogeneous from the point of view of technical requirements. If it had not been for this very specific context, federal regulations on energy efficiency may not have seen the light of day, or at least not as rapidly.

Similarly, the process leading to the adoption in Europe of a regulatory measure on the energy performances of domestic cold appliances was very long. According to European Community law, member States cannot introduce national legislation that might limit the free movement of goods and services within the European Union. The Commission thus quite logically opposed the decision of certain member States, at the beginning of 1990s, to introduce MEPS for household appliances, but agreed, on the other hand, to prepare common regulations for the Union within a relatively short time. The proposed directive was presented to Parliament and the Council of Ministers in December 1994, which was more than 2 years after The Netherlands had informed the Commission of their draft regulation on the energy efficiency of refrigerators[1]. This considerable time lapse may be explained by the difficulty of negotiations with manufacturers who contested the principle of regulations but also with certain member States which considered that such regulations would have very different economic consequences from one country to the next (Bertoldi, 1999).

While regulatory measures appear well suited to improving energy efficiency in the household appliance and office automation sectors, it cannot be denied that they present implementation problems on account of opposition from manufacturers. Adopting an approach of consultation and negotiation between public authorities and private actors with a view to defining objectives for energy efficiency improvement may, in these conditions, prove to be just as effective and even quicker to put into practice.

3. VOLUNTARY AGREEMENTS: AN ALTERNATIVE TO REGULATIONS?

Since the early 1990s, voluntary agreements have been considered an instrument of environmental policy in their own right (S. Baecke and alii, 1999). They are no longer limited to certain sectors such as waste management, electricity generation, or the high energy-consuming industries but have been gradually extended to other sectors, including fairly recently to the home appliance market.

Recently the European Commission negotiated agreements with manufacturers of televisions and video cassette recorders, as well as with washing machine manufacturers, with the aim of improving the energy efficiency of these appliances (CCE, 2000). Reflecting the position of certain member States and a large majority of manufacturers, the Commission is showing a growing interest in such negotiated agreements, which are increasingly seen as an alternative to what are felt to be overly restrictive regulations.

3.1. More effective than regulations from a theoretical point of view

In theory, voluntary commitments have a number of features which, in economic terms, make them more effective than regulatory measures.

When it comes to defining efficient environmental objectives in economic terms, public authorities are penalised by their poor knowledge of existing technical options and the cost of implementing them. There is information asymmetry between manufacturers, who are very well informed about technologies and costs, and the regulating authority. The consequences have two aspects:

- it is in the interest of firms subjected to the regulations to overestimate pollution abatement costs to encourage the regulating authority to define less restrictive overall objectives.
- it is impossible to impose differentiated objectives to take into account the particular situation of each firm. This means that marginal pollution abatement costs differ for each firm, which for the economist characterises an inefficient solution.

Voluntary agreements have a theoretical advantage in this respect in that distribution of the objectives among the different firms is left to the firms themselves. Pollution abatement objectives can thus be allocated among the firms according to their particular technical possibilities and implementation costs. Cost minimisation is reached if the allocation leads to the equalisation of private marginal abatement costs; the distribution of objectives is then optimal (Glachant, 1999). In reality, burden sharing is not necessarily optimal, but the principle of negotiating individual commitments introduces an element of flexibility not found in the regulatory approach.

Furthermore, the inter-firm negotiations and cooperation that are necessitated by voluntary commitments in any given sector can contribute to a collective learning process that is beneficial to each individual firm:

> When using voluntary agreements, intense collective learning improves information of the firms and allows them to implement their private pollution abatement objectives at lower cost (M. Glachant, 1999).

This works for a relatively homogeneous business sector and in a context of general uncertainty:

> All the firms are in the same situation of uncertainty concerning available techniques and related costs, and are more encouraged to co-operate with one another to make up for the lack of information (C. Defeuilley, 2000).

Another advantage of voluntary agreements compared with regulations is that the negotiation framework gives firms the chance to participate directly in defining the objectives and the target dates for implementation. It is true that public authorities may also consult industry when drawing up regulations, but it is the regulator that has the final word. In the case of voluntary agreements, the objectives are defined jointly by manufacturers and the public authority.

Voluntary agreements also have a number of advantages for public authorities:

- similar environmental objectives can be reached in a shorter time and at a lower cost than in the case of regulations because of the voluntary nature of manufacturers' participation.
- where there is asymmetry of information (general case), negotiating with companies can provide the regulating body with the opportunity to obtain information about technologies and implementation costs.
- finally, since commitments are partly self-monitored by participating companies, public administration and monitoring costs are reduced.

But voluntary agreements have different limits. When a few firms make a voluntary commitment, the others may feel that they do not have to make any significant effort to reduce pollution and the overall impact on pollution abatement may be limited. Voluntary agreements must therefore involve a large majority of manufacturers in the market or define an overall objective for energy efficiency improvement if there is to be any effective benefit to the environment.

Moreover, the real environmental impact of a voluntary agreement cannot be measured simply in terms of achieving objectives. The objectives themselves must correspond to a real effort on the part of firms and not simply be part of a general trend in energy efficiency improvement. Since such agreements are by definition the fruit of negotiations where each party does not have the same information, the regulating authority does not know the real effort that will be required from the firms involved. The objective agreed upon may correspond to the general trend in energy efficiency improvement (business as usual scenario) and require no additional effort from the manufacturers. In this case the agreement would have no environmental impact in itself.

Finally, voluntary agreements can be upsetting for individual markets. Where a limited number of companies are signatory, non-participating firms benefit from short-term advantages (continued use of high-pollution, low-cost technologies, limited R&D investments, etc.) compared with those who agree to contribute to a joint effort to respect pollution abatement commitments (free-riding behaviour). On the other hand, an agreement limited to a few firms with a certain technological lead could give them a strong market position and end up creating unfair competition.

Table 1. *Effectiveness of regulations and voluntary agreements*

	Regulations	*Labelling and regulations*	*Negotiated agreements*
Participation of firms	-	+/-	++
Implementation time	-	-	++
Administrative costs	-	-	+
Incentive to innovate	-	+	+/-
Environmental impact	+	++	+/-

3.2. The example of negotiated agreements for washing machines

In 1996, the European Committee of Manufacturers of Domestic Equipment (CECED) drew up a proposal for a voluntary agreement to improve the energy efficiency of washing machines sold in Europe. This proposal led to the first negotiated agreement with the European Commission on energy efficiency in the domestic appliance sector (CECED, 1997). It was followed by several other proposals concerning televisions and video cassette recorders, dishwashers, electric water heaters and refrigerators, while during the same period no new regulations governing MEPS were introduced. For the Commission, negotiated agreements modelled on the one described below are now a credible alternative to regulatory measures.

3.2.1. Content of agreement

The proposal was discussed in depth by manufacturers and the European Commission and an agreement was finally concluded in December 1998 whereby manufacturers agreed to:

- improve the energy efficiency of washing machines sold in the European Union: the objective was a 20% improvement over the period 1994-2000, corresponding to a reduction in the energy consumption of a standard wash cycle from 0.30 kWh/kg in 1994 to 0.24 kWh/kg in 2000.
- gradually phase out production and importation of the least efficient models in two successive stages: in the case of standard washing machines[2] elimination of models in energy label classes E, F and G from 31 December 1997, followed by those in class D from 31 December 1999,
- inform consumers about the conditions of use of washing machines and their impact on energy consumption (choice of programmes), conduct research programmes on low-temperature washing techniques, and increase cooperation with detergent manufacturers.

Furthermore, the CECED agreed to monitor improvements in the energy efficiency of washing machines and report regularly to the European Commission. This provided a way of checking that manufacturers' commitments were genuine and that the objectives were reached.

The Commission's approval confirmed the acceptance of the proposal by the public authorities and the implicit agreement not to introduce regulations imposing MEPS for this type of equipment for the duration of the agreement (Bertoldi and Bowie, 1997).

3.2.2. A transparent procedure for defining objectives

The procedure for defining the objectives adopted in the context of these agreements is particularly interesting. Indeed, it might be quite legitimate to question the reality of the additional efforts made by manufacturers in the framework of certain agreements, especially when the objectives negotiated correspond to spontaneous technological progress. In the present case, the objectives accepted by manufacturers were defined on the basis of a preliminary analysis conducted by the public authorities for the purpose of drawing up regulations (GEA, 1995). This technical-economic analysis was used to estimate an "optimum" energy efficiency level among the different technological options available (least life cycle cost analysis), taking into account overall cost and return on investment of each option in relation to a reference situation. The energy efficiency improvement recommended on the basis of this study, considered to be technically possible and economically viable, was 25% compared with 1994. The proposal from manufacturers made explicit reference to this study, suggesting a target of a 20% improvement over a 5-year period (1994 –1999). [3]

Unlike certain voluntary agreements where the objectives are defined more or less unilaterally by industry and where improvements are hard to distinguish from the general trends in technological progress and/or spontaneous development of the markets, the energy efficiency improvement targets here were based on an independent analysis. They took into account the state of technology and prospects for improvement, as well as the economic consequences for consumers. As F. Moretti, Chairman of the CECED working group on energy efficiency, points out:

> As it is based on the conclusion of a SAVE Study, it is guaranteed that the total saving target is well aligned with public and political expectation, but the method how to reach the goal is essentially left to the manufacturers. (Moretti, 2003)

It is therefore reasonable to assume that these goals were in the end similar to those that the European Commission might have introduced in the context of MEPS that would have been based on the same preliminary studies[4].

3.2.3. Monitoring of commitments and ways of imposing sanctions

For the Commission, negotiated agreements on the energy efficiency of household appliances must comply with a certain number of conditions if they are to be a viable alternative to regulatory measures (Bertoldi and Bowie, 1997):

- the manufacturers signing the commitment must account for at least 80% of the appliances sold on the European market

- the quantitative targets must correspond to a significant improvement in the energy efficiency of the appliances over a reasonable timescale
- finally, the system must include an independent monitoring procedure for verifying that improvements are coherent with the stated objectives.

In accordance with these conditions, the agreement in question involved most of the manufacturers in the sector since the CECED represents over 90% of the market in a sector where imports from outside the European Union are very limited.

A procedure for monitoring and reporting was also set up. It was based on a data base containing information on the energy performance of washing machines marketed in Europe and monitoring by an independent observer of the appliances sold. A report was submitted to the Commission once a year setting out the improvement achieved.

The agreement also included measures to dissuade free-riding. Thus, each signatory was bound to respect the terms of the agreement concerning the import or marketing of inefficient models and to help achieve the average sales-weighted energy efficiency improvement. In the case of non-compliance, the CECED would first ask the manufacturer to comply with the agreement, and if this failed the manufacturer concerned would be deemed no longer party to the negotiated agreement. Such decisions are made public in the press, which provided a strong incentive for manufacturers to respect their commitments and avoid tarnishing their image.

3.2.4. Results in line with commitments

The report submitted to the Commission by the CECED at the end of 1999 indicated that manufacturers had complied with the commitments they had made. The average sales-weighted energy consumption of washing machines at this time was 0.228 kWh/kg, representing a 24% improvement in energy efficiency compared with the reference situation in 1994. The initial objective of 20% for the period 1994-2000 was achieved and even surpassed before the target date (end of 2000). Similarly, in accordance with their commitments, manufacturers had withdrawn the least energy-efficient models (classes D, E, F and G) from the market; the few models remaining in class D had disappeared by the end of 1999. Finally, the report submitted to the Commission by the CECED also mentioned additional action taken by manufacturers to help control the energy consumed by washing machines (improvement in energy efficiency of low-temperature cycles, cooperation with detergent manufacturers, consumer information).

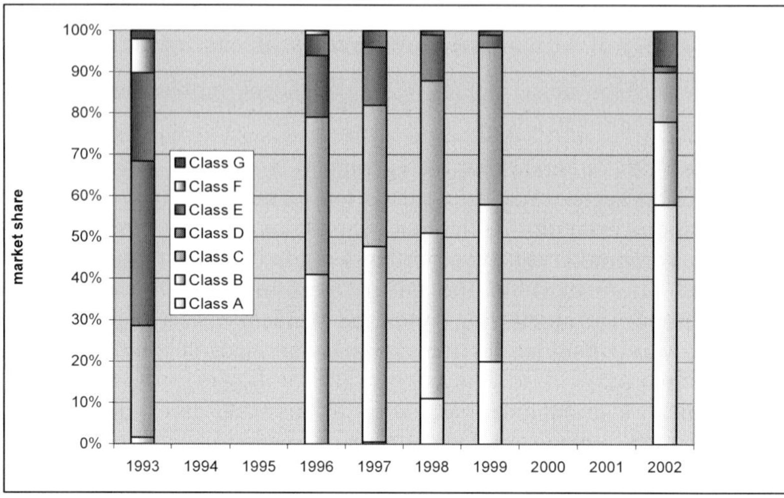

Source: Moretti, 2003

Figure 2. *Evolution of sales of washing machines according to energy classes*

For manufacturers, the results obtained demonstrate the effectiveness of the negotiated approach since the objectives defined in consultation with the public authority were reached before the planned date (CECED, 2000). Given the success of the first voluntary commitment, the manufacturers prepared a second proposal, submitted to the European Commission at the end of 2002, which adopted the same structure as the first:

- an improvement in the average sales-weighted average energy efficiency of 12% compared with the situation in 1999 (a 33% improvement compared with 1994) corresponding to an objective of 0.20 kWh/kg for the year 2008;
- an end to the import and sale of appliances in energy class D by the end of 2003;
- support of manufacturers for additional measures to achieve energy savings (labelling, financial incentives, etc.) as well as different commitments similar to those in the previous agreement concerning information for consumers

The first results published at the end of 2003 seem to indicate that manufacturers have continued their efforts to improve the energy efficiency of washing machines. Sales weighted energy efficiency thus reached 0.208 kWh/kg at the end of 2002, the target of 0.20 kWh/kg being fixed for 2008. Similarly, the objective of totally eliminating class D models, irrespective of their characteristics, was well underway since the number of class D models sold represented no more than 1% of sales.

3.2.5. A specific characteristic: flexibility

The commitments made by manufacturers within the framework of these negotiated agreements are not very different from the constraints imposed by the regulatory measures if we consider in particular the MEPS imposed on domestic cold appliances. In the present case, the manufacturers are also committed to a precise calendar for the withdrawal of those appliances with the highest energy consumption.

The fundamental difference in relation to regulatory measures concerns the flexibility allowed by the agreements. The objectives indicated above and recalled in Figure 2 concern the most popular models, but special provisions are made to take into account the significant differences existing between the national markets within the European Union: small washing machines and those with low spin speeds, which are in widespread use in the countries of southern Europe, benefit from an additional period of time in which to adapt. Under the second voluntary commitment, the complete elimination of all class D models is only programmed for the end of 2003.

At the time the first agreement was concluded, between 10 and 11% of machines sold in the European Union did not meet the new requirements, but for certain manufacturers this proportion was in excess of 30% of sales (CEC, 2000). Moreover, if all the washing machines sold in Europe had had to achieve the same improvement in energy efficiency, average prices would have increased by 1 to 2% in Northern Europe but by as much as 8 to 14% in Southern Europe and the United Kingdom where the proportion of machines in the low efficiency classes is highest (CEC, 2000).

For manufacturers, this approach allowing a rapid improvement in energy efficiency in northern European markets, where consumers are more sensitive to the energy efficiency criteria, and a more gradual change in those of southern Europe, thus better takes into account the characteristics of the market and ultimately proves to be less costly than non-differentiated regulatory measures.

Table 2. *Flexibility in the agreements for washing machines*

	Target date	*A*	*B*	*C*	*D*	*E*	*F*	*G*
General								
Load > 3 kg & Spin speed > 600 rpm	1998					**X**	**X**	**X**
	2000				**X**			
Exceptions								
Load < 3 kg	1998						**X**	**X**
	2000					**X**		
	2004				**X**			
Load > 3 kg & Spin speed < 600 rpm	1998					**X**	**X**	**X**
	2004				**X**			

Source: CECED, 2000, 2002.

4. CONCLUSION: VAS ARE AN EFFECTIVE INSTRUMENT FOR MARKET TRANSFORMATION UNDER CERTAIN CONDITIONS

Regulatory measures that impose MEPS for all available appliances have proved their effectiveness. In association with energy labelling, which encourages consumers to differentiate products, they can be effective in stimulating technological progress. However, the regulatory approach assumes a strong commitment on the part of public authorities, and preparation times may be long because of opposition from manufacturers. Negotiated agreements, by virtue of their greater flexibility, which makes them easier to implement, can be an interesting alternative to regulations.

These agreements have a great deal in common with the MEPS introduced for cold appliances. They anticipate the gradual removal of the least efficient models from the market. A first analysis would therefore suggest that they have no particular interest for manufacturers compared with regulations. However, they include an important dimension of flexibility which the regulatory approach does not have. With negotiated agreements, the manufacturers have the possibility of stepping up the introduction of new more efficient models on the more dynamic markets and delaying the discontinuation of less efficient models on other markets, rather than having to simultaneously improve the efficiency of all their appliances, which would be much more costly.

Similarly, negotiated agreements offer several advantages from the point of view of public decision-makers, namely more rapid implementation, a cooperative approach that enables access to non-public information, lower preparation and monitoring costs on account of the involvement of manufacturers, and so on. The essential question, however, concerns their environmental efficiency. To ensure improvements from an environmental point of view, manufacturers must make commitments that impose a genuine additional effort that goes beyond the general trend in energy efficiency.

In the case of the agreements for washing machines, it was possible to impose fairly ambitious objectives because of the threat of regulatory measures. Indeed, the possibility of regulatory measures in the household appliance sector became very real following the introduction of MEPS for domestic cold appliances. Since the bargaining power was then in the hands of the public authority (in this case the European Commission), it was able to impose constraining targets (similar to those that would have been obtained by energy efficiency regulations). In exchange, manufacturers have obtained a certain flexibility regarding implementation periods and methods.

For negotiated agreements to be effective at the environmental level, it is essential that the possibility of regulatory measures remain a realistic threat. During the negotiating process, the level of constraint imposed, and thus the type of incentive to be offered to firms, depends on the respective powers of the companies and the public authority. A very restrictive requirement (realistic threat of regulations) may result in ambitious objectives that force firms to make real additional efforts. On the other hand, if the threat of regulations is not really credible, the public authority's negotiating power is limited and companies have considerable room for manoeuvre, with the consequent risk of accepting commitments which are not very different from general market trends.

The credibility of the regulatory threat depends directly on the information the public authority has regarding the firms' room for manoeuvre, the technological opportunities available and the implementation costs. Considerable preparatory work is thus essential so that the regulating body can obtain a maximum of information and negotiate ambitious targets. This means that paradoxically negotiated agreements do not necessarily involve shorter implementation times or lower preparation costs than the regulatory approach if their aim is to achieve the same level of environmental efficiency.

5. NOTES

[1] The Directive was adopted in September 1996 and took effect in the different member States in September 1999, in other words 5 years after a first proposal was presented to Parliament.
[2] These rules apply to the most popular models, that is those with a wash capacity of over 3 kg and a spin speed of over 600 rpm. Special measures were planned for models of smaller capacity or with slower spin speeds (cf infra).
[3] In the second negotiated agreement on the energy efficiency of washing machines (CECED, 2002), reference was similarly made to the SAVE II study sponsored by the European Commission which set energy efficiency improvement objectives for washing machines (Novem, 2001). However, in this case manufacturers did not take up the objective proposed in the study, which was to reduce energy consumption to 0.20 kWh/kg by 2003, suggesting instead that they reach this target only by 2008.
[4] For the purpose of comparison, the energy efficiency improvement sought by the introduction of performance standards for domestic cold appliances was of the order of 15% for the period 1996-99 (Bertoldi, 1999).

6. REFERENCES

Appliance Efficiency, "Confortec 2000: the market moves on", issue 1, vol.4.

Baecke, S., De Clercq, M., and Matthijs, F. (1999), *The nature of Vas: empirical evidence and patterns – Literature survey*, CAVA Working Paper, n°99.

Bertoldi, P. (1999), «*Energy efficiency equipment within SAVE: Activities, strategies, success and barriers*», The SAVE Conference, For an Energy Efficient Millennium, Graz.

Bertoldi P., and Bowie R. (1997), *"European union efforts to promote more efficient appliances"*, European Commission, Directorate General for Energy, ECEEE, Summer Study.

CCE (Commission des Communautés Européennes) (2000), *Plan d'action visant à renforcer l'efficacité énergétique dans la Communauté Européenne*, Communication de la Commission au Conseil, au Parlement Européen, au Comité Economique et Social et au Comité des Régions.

CEC (Commission of the European Communities) (2000), COLD II, *The revision of energy labelling and minimum energy efficiency standards for domestic refrigeration appliances*, contract DG-TREN SAVE.

CECED (1997), *Voluntary commitment on reducing energy consumption of domestic washing machines*, Final, September.

CECED (2000) *Voluntary Commitment on reducing energy consumption of domestic washing machines*, 3rd annual report to the CEC, Aug.

CECED (2002), *Second voluntary commitment on reducing energy consumption of domestic washing machines* (2002-2008), CECED, October.

Defeuilley, C. (2000) , *Contrat et politique publique d'environnement. Enseignements tirés d'Eco-Emballages*, Rapport final, Ministère de l'Aménagement du Territoire et de l'Environnement, Programme inter - institutionnel de recherches et d'études en économie de l'environnement, Dec..

Glachant, M. (1999), The cost efficiency of voluntary agreements for regulating industrial pollution: a Coasean approach, In *Voluntary Approaches in Environmental Policy*, C., Carraro and F. Lévêque (Ed.), Kluwer Academic Publishers

Group for Efficient Appliances (1995), "*Washing machines, dryers and dishwashers*" Final report for the European Commission. Directorate General for Energy.

International Energy Agency (IEA) (2003), *Cool appliances: Policy strategies for energy-efficient homes*, Paris.

IPCC, Climate Change (2001), *Contribution of Working Group III to the Third Assessment Report of the Intergovernmental Panel on Climate Change*, Cambridge University Press

Moretti (2003), *"Industry commitment: the case of wet appliance"*, 3rd International Conference on Energy Efficiency in Domestic Appliances and Lighting (EEDAL 03), Turin.

Novem (2001), *"Revision of energy labelling & targets washing machines (clothes)"*, Final Report, Study for the Directorate-General TREN of the Commission of the European Communities, SAVE study, March.

Varone, F. (1998), *Le choix des instruments des politiques publiques*, Ed. P. Haupt, Berne.

IMPLEMENTING THE DUTY OF ACCEPTANCE IN FLEMISH WASTE POLICY

The role of environmental voluntary agreements

M. DE CLERCQ, R. BRACKE

Centre for Environmental Economics and Management, Ghent University, Belgium

Abstract. In this paper, we focus on the implementation process of environmental voluntary agreements based on a Flemish case study concerning the introduction of the duty of acceptance in Flemish waste policy. At the moment, the duty applies for paper, batteries, vehicles, tyres and electrical and electronic equipment. Producers are obliged to set up free of charge take-back collection systems for the disposal of their products in the post-consumption phase. As such, market failure is corrected by internalising social costs into private ones, respecting the polluter pays principle. For the practical execution of the basic rules laid down in the legal framework, environmental voluntary agreements are concluded with sector associations. The use of environmental voluntary agreements allows producers to fulfil their individual obligations resulting from the duty of acceptance in a collaborative sector-based approach. This entails several operational advantages and allows setting up efficient collection and disposal networks taking into account the specific characteristics of each waste product rather than implementing a uniform system. The paper investigates how the agreements have been negotiated and implemented paying particular attention to the relationship between the underlying environmental legislation and the voluntary character of the agreements. The agreements clearly have a support function aiming to implement existing legislation in the most efficient way. Next to the conclusion, the management of these agreements is discussed. It appears that the policy process entails much more than just fixing overall collection and recycling targets, but should be regarded as an on-going process with evaluations and consultations with all stakeholders. In this regard, the importance of alternative instruments in case of non-compliance to the agreement and back-up policies to create market opportunities for recycled materials is stressed.

1. INTRODUCTION

Household waste generation is one of the few environmental problems in which, despite intense policy attention, results are still lacking. The highly wanted disconnection between the amount of household waste generated and economic growth seems to be unattainable up until now. One of the most promising new approaches to tackle this challenge is extended producer responsibility. Researchers and policy makers such as the European Commission and the OECD have repeatedly promoted the principle of extended producer responsibility. The 6th Environmental Action Programme refers to producer responsibility as one of the elements to be addressed in the context of the waste recycling strategy and in the recently adopted Communication towards a thematic strategy on the prevention and recycling of waste (COM(2003) 301), the Commission emphasizes that producer responsibility continuously must be used in Community and national legislation to

E. Croci (ed.),The Handbook of Environmental Voluntary Agreements, 179 - 202.

promote the recycling of end-of-life products (EC, 2003). In fact, policy programs based on this principle have already been put in place at national and community level. The Directive on end-of-life vehicles and the Directive on waste electrical and electronic equipment (WEEE) are community-level examples. During the nineties, many countries (e.g. Switzerland, Germany, the Netherlands, France) have implemented waste management policies based on extended producer responsibility, especially for packaging waste.

Following this principle, producers are responsible for the environmental impacts of their products throughout the entire life cycle, including the collection and final disposal. Producers have to set up systems to take back and recycle the products they have put on the market with the aim of recovering raw materials, closing material loops and reducing the amount of waste for final disposal. The organisational and financial responsibility is shifted from government to producers. As such, market failure is corrected by internalising social costs into private ones, respecting the polluter pays principle. It is expected that the financial responsibility will lead to an integration of waste management concerns into product development and production process decisions. This way, economic incentives are provided for innovation towards environmentally friendly products: the material choice, the repair-friendliness, the durability, the modular construction, the use of non-hazardous substances as well as the recyclability might improve (EEB, 2001). The strength of extended producer responsibility lies in its ability to simultaneously put in operation life cycle thinking, the waste minimisation hierarchy and the polluter pays principle (OECD, 1997).

Extended producer responsibility comprises a variety of different policy instruments as, for example, legal take-back requirements, public waste funds or privately organised waste management institutions like the German Green Dot Program for packaging waste (Runkel, 2003). In order to implement this principle in Flemish waste policy, the Flemish Parliament introduced the duty of acceptance in the Waste Decree in 1994. The decree enables policy makers to oblige producers to accept and process their products in the post-consumption phase. At the moment, the duty of acceptance applies for waste paper, used batteries, end-of life vehicles, waste tyres and WEEE. In the coming years, more waste streams will be submitted (amongst others waste wood, waste carpets, waste mattresses, waste from medical appliances, waste lamps…). This shows that the Flemish government attributes an important role to the duty of acceptance in its future waste management policy.

In order to fulfil the collection and recycling obligations the duty of acceptance entails, environmental voluntary agreements are concluded between government and associations of producers making a collective solution at sector level possible. The agreements set out detailed requirements for the collection and management of the different waste streams. This sector-based policy approach allows setting up efficient collection and disposal networks taking into account the specific characteristics of each waste product rather than implementing a uniform system.

The paper is structured following the policy process. Section two discusses the introduction of the duty of acceptance. In the following section, the negotiations leading to the environmental voluntary agreements are analysed. Section four deals with the implementation process. The evaluation and follow-up policy is the subject of the last section. The paper ends with some concluding remarks.

2. THE INTRODUCTION OF THE DUTY OF ACCEPTANCE

2.1. Waste Generation in the Flemish Region

Comprised of the Dutch speaking part of the country, the Flemish Region has the largest population share (59%) of Belgium counting almost 6 million inhabitants. In addition, the population density of 442 inhabitants/km^2 is quite high and with a Gross Domestic Product of € 25.048 per inhabitant in 2002, Flanders is one of the wealthiest European regions. Both figures indicate that landfilling capacity is limited which, together with the high public environmental concern, puts waste management high on the policy agenda.

In 2002, over 3 million tons of household waste was gathered. On average, each inhabitant is responsible for 556 kg of waste. Fortunately, as a European benchmarking study showed, the degree of separate collection of household waste is among the highest of the Members of the EU (Suikerbuijk, 1999). In 2002, 69,5% of the waste was separately collected. Since 1995, the total amount of household waste going to final disposal has dropped continuously. Moreover, by charging higher levies on landfilling than on incineration, the authorities have succeeded in making incineration more attractive and as a result only 4,4% of household waste was landfilled in 2002 (VMM, 2003).

Despite these positive trends, the short-term target on the total amount of household waste was exceeded by 8% in 2002 (VMM, 2003). The long-term target for 2007 is a total amount of household waste of 490 per inhabitant, of which 340 should be collected separately. Therefore, the amount of waste generation for final disposal should drop by 19 kg and the amount of separately collected waste should drop by 47 kg per inhabitant. Additional efforts are thus unavoidable. However, the cost of separate collection weighs strongly upon the governments budget and the marginal cost of enhancing separate collection rises sharply. This forms a possible barrier for further improvements. The duty of acceptance is one of the most far-reaching instruments policy makers have introduced in the Flemish waste policy to tackle this challenge.

2.2. Waste Policy in the Flemish Region

Belgium is a federal state composed of three regions: Flanders, Wallonia and Brussels. The regions are mainly competent for territory related matters, e.g. economy, housing, land use planning and infrastructure. Thus, environmental policy is mainly dealt with by the regional governments. The national government is left with only limited powers in this field: product standards, nuclear waste and the negotiation and the implementation of the international engagements of the country. In this respect it is important to notice that there is no hierarchy of legal systems in

Belgium. Hence the national government cannot impose its will on regional governments in environmental matters.

At the regional level, each region has its own environmental law and its own regulations concerning the collection, recycling and final disposal of waste. When the regions have different legislations, this can lead to serious environmental or economic consequences. The need for a harmonised policy in the three regions is strong, but sometimes hard to realise in practice due to a different level of ambition and different policy agendas.

The Waste Decree of 2 July 1981 on the Prevention and Management of Waste marks the start of Flemish waste policy. The decree is concretized in Waste Plans. The first Waste Plan (1986-1990) primarily concerned waste disposal. Its objective was the closedown and sanitation of landfilling areas and the optimal use of incineration capacity. In the following Waste Plans, the emphasis shifted to prevention and recycling.

The first Environmental Policy Plan for the period 1990-1995 already referred to the possibility of using environmental voluntary agreements for achieving these waste management goals. An agreement with the packaging waste sector was the first objective. Based on these experiences, other agreements could follow. However, as well the packaging waste story as the first consultations with other sector associations proved that it was quite impossible to reach the agreements the Flemish government had in mind. On a voluntary base, the contribution of the private sector was directed at avoiding far reaching regulations and additional costs (Bracke, 2003). In reaction to this, policy makers decided to first establish a legal basis. This should enhance the negotiation position of the government representatives to be able to conclude ambitious agreements with the sectors. As such, when in 1994 the Waste Decree was renewed because some international obligations had to be incorporated in the Flemish waste policy and new instruments were needed to support the aims of recycling and prevention, the duty of acceptance was introduced.

2.3. The Duty of Acceptance

The introduction of the duty of acceptance was realised in two steps. First the general principle was laid down in the Waste Decree in 1994. Next, waste products are placed under this regime and collection and treatment targets are stipulated.

The Waste Decree of 1994 enables the Flemish government to put a duty of acceptance upon certain waste products. The duty applies for end-sellers, wholesalers, producers and importers. End-sellers are required to take back free of charge end-of-life products of which consumers want to discard if a new product of the same kind is purchased. They can pass on the collected waste products to their wholesaler. Finally, producers and importers must accept all end-of-life products collected by retailers and wholesalers for appropriate treatment. The regulation clearly assumes a take-back system based on reverse logistics through the distribution chain. The general principle is that a costumer can return his old product to the retailer if a new product of the same kind is bought (1/1). However, the possibility to organise other collection systems through negotiating environmental voluntary agreements is provided for. The crucial point is that producers and importers have to set up take-back systems for waste products with a view to useful application and environmentally safe disposal. The Flemish Government is qualified to assign waste products and to define additional rules concerning the way the duty of acceptance has to be fulfilled. This was done for the first time with the approval of the Flemish Regulation concerning Waste Prevention and Management of 17 December 1997.

Table 1 shows the waste streams that became subject to the duty and the environmental targets concerning the collection and treatment thereof as stipulated in the Flemish Regulation. The selection of which waste products were submitted to the duty of acceptance was done on an ad-hoc base (Rekenhof, 2002). Nevertheless some arguments are obvious. Batteries and accumulators, waste tyres, end-of-life vehicles and WEEE are of particular importance in terms of the growth rate of the waste generation, the hazardous character and the increasing number of materials used. Moreover, the selection of end-of-life vehicles and WEEE fits within the framework of the respective European Directives. Paper waste is targeted because of the high volume of this waste stream and the resulting financial burden on the waste management budget. A decisional framework that enables to decide rationally for which waste streams the duty of acceptance might be an efficient and effective policy instrument is however lacking.

***Table 1**. Waste products subject to the duty of acceptance*

Waste product	*Environmental targets*	*1/1*	*1/0*
Paper Waste	Recycling rate of 60% in 1998; 80% in 1999-2000; and 85% from 2001 on		1/06/1998
Used accumulators and batteries	From 2000 on: a collection rate of 75% for batteries and 95% for accumulators and a recycling rate of 50%		1/06/1998
Waste tyres	From 2000 on: a collection rate of 100% and a valorisation rate of 90% of which at least 25% retreading	1/07/1999	1/07/2004
End-of-life vehicles	From 2005 on 85% of the weight has to be re-used or usefully applied; 80% of the weight has to be re-used or recycled	1/07/1999	1/07/2004
Waste electrical and electronic equipment (WEEE)	From 2001 on: a recycling rate of 95% for metals and 20% for synthetic material and useful application of all other synthetic material	1/07/1999	1/07/2004

The last two columns of table one indicate the obligation towards the retailers. 1/1 indicates that retailers only have to accept a waste product if a new product of the same kind is bought, whereas under the 1/0 system they have to accept all waste products even if no new product is bought. The difference between the 1/1 and the 1/0 obligation entails important implications for both producers and policy makers.

Producers on the one hand fear that, when in 2004 the 1/0 rule is adopted, this could lead to a huge import of waste products from neighbouring countries that must be accepted free of charge. As Flanders is a small and open economy completely integrated in the European Union and most end-of-life products have a negative value, this could have serious financial impacts because the costs cannot be recovered by a disposal contribution that has to be paid when purchasing the product (see 4.2.). For policy makers on the other hand, the 1/0 regulation is interesting because this way, producers can be made responsible for waste that is not collected through the distribution chain but that is collected by the cities and municipalities. In practice, policy makers threat with the 1/0 to make producers finance the collection efforts made by the local authorities (SERV, 2003).

Notice that the duty of acceptance covers both orphaned and historic waste. The former consists of end-of-life products for which no responsible producer can be identified anymore. The latter comes from products that were put on the market before the regulation came into force. In the first years, this covers an important quantity as most products targeted are long lasting consumption goods.

The stipulations set out in the Flemish Regulation, as summarized in table 1, should be seen as minimal requirements. Once these are established, the Flemish government enters in dialogue with the sector associations with the aim of reaching a consensus for an environmental voluntary agreement.

3. NEGOTIATING ENVIRONMENTAL VOLUNTARY AGREEMENTS

3.1. The Use of Environmental Voluntary Agreements

At the beginning, Flemish environmental policy was mainly characterised by command and control regulations. However, since the beginning of the nineties, more emphasis was put on co-operation and consultation with target groups. In Belgium, the first environmental voluntary agreement was concluded in 1988. Since then, about thirty agreements were concluded at the Federal or Regional level. Chronologically, two groups can be distinguished with 1994 as hinge point. In 1994 a decree on environmental voluntary agreements was approved.

The decree describes environmental voluntary agreements (called environmental policy agreements) as ‘any agreement between the Flemish region on the one hand and one or more representative organisations of companies on the other hand, with the objective to prevent environmental pollution, to limit or to eliminate its consequences or to promote an efficient environmental management’. Agreements are legally binding for all parties and the commitments have the force of law. The Region is authorized to convert the content of the agreement partly or entirely into regulations. The agreement cannot replace legislation, nor can it deviate from it in a lenient way. Neither does it reduce the competence of other public authorities than the Region. On the other hand, the Region cannot implement any regulation that imposes stricter requirements than those agreed to in the agreement. This rule can only be deviated from in case of urgent necessity or to respect international or European obligations. The decree provides for a participation and counselling procedure, during which all sections of society have the opportunity to formulate their remarks. An agreement can be closed for no longer then five years.

For the agreements that were concluded before 1994, the label 'voluntary' is appropriate. Due to the unclear juridical statute, the enforceability of these agreements was limited. Next to this lack of enforceability, the lack of transparency and public involvement, poor reporting and the voluntary character leading to non-ambitious targets were criticised often (SERV, 1997). The legal framework set out in the decree was instated with the aim to create an atmosphere wherein the use of environmental voluntary agreements would be more successful, because in many policy plans, the closing of an agreement was announced but never realised (Wille, 2000).

Up to this date, only eight agreements have been closed under this decree. Next to the six agreements related to the duty of acceptance, there is one agreement concerning the collection of old medicines and one concerning the storage of petrol for private use. The decree seems to have two consequences. First, the goals of the agreements seem to be narrowed down to waste management objectives. The agreements concluded before 1994 have more divergent objectives. Second, the decree has reduced the amount of agreements concluded. This contrasts to the initial goal of the decree. Moreover, it took about four years before the first agreement based on the decree was concluded and since the decree two agreements have been concluded outside this legal framework.

The more stringent character of the agreements concluded under the decree seems to have the consequence that this instrument is only used if industry is forced to by law and is no longer concluded as a voluntary engagement of industry anymore. By the introduction of the duty of acceptance in waste management, a legal basis was created, explaining the use of this instrument in this policy field.

3.2. The Role of Environmental Voluntary Agreements

Theoretically, producers/importers can opt for an individual or a collective solution. In the former, a producer/importer needs to submit to OVAM (the Public Waste Agency for Flanders) a waste management plan specifying how he will comply with the requirements concerning the duty of acceptance laid down in the Flemish Regulation. OVAM sees to it that the arrangements set out in the waste management plans correspond to the general lines reflected in the environmental voluntary agreement.

This option is presented as an alternative for firms who do not want to participate in the environmental voluntary agreement but must be interpreted as an instrument to stimulate participation by making free-riding less attractive. In fact, the duty of acceptance is intended to be executed in a collective way (OVAM, 2004). In practice, the use of individual waste management plans is exceptional because a sector-based collective execution by means of a voluntary agreement entails several advantages.

Firstly, in collection and recycling networks organised on a collective basis, economies of scale can be exploited bringing down the operational costs. Secondly, a collective system is by far the best way of dealing with historical waste and in particular with orphaned waste (OECD, 2001). Thirdly, an environmental voluntary agreement can provide an opportunity for all stakeholders to be involved in a process of consensus building. Finally, the government's task of control and enforcement is reduced. If no agreements would have been signed, the authorities would be faced with a huge amount of quite diverging waste management plans from individual companies. In fact, the numerous amounts of producers/importers and the complexity of the sector due to e.g. parallel import by retailers make a sector-based agreement a necessity.

Up till now, six negotiated agreements have been signed (see table 2). For waste paper two agreements have been concluded, one with the informative press and one with the advertising sector. The publishers of the printed advertisement sector deposit 0,37 eurocent into a fund for each kilo of paper they have put on the market in the Flemish Region. These contributions are then transferred to the municipalities depending on the amount of waste paper that was separately collected. For the publishers of the informative press, the financial responsibility is fulfilled very weak by granting advertising space to the Flemish Region for a total value of € 3,22 million, VAT excluded, on a yearly basis. These agreements are unique as they only put in place the financial responsibility. The adoption of the agreements did not end the legal responsibility of the municipalities concerning the collection of scrap paper. Both agreements are target of fierce criticism (OVAM, 2000). First of all, municipalities claim that the cost of collecting waste paper amounts to 5 eurocent per kilo so the polluter pays principle is not completely applicated. Secondly, many question the link between the achievement of waste management targets and the grant of advertising space. This requires very little effort as the publishers can just add an additional page in their publication and the authorities have to finance the design of the advertisements (Wille, 2000).

***Table 2**. Negotiated agreements (NA) and waste management organisations (WMO)*

Waste product	*Signing of NA*	*WMO*	*Date of set up WMO*
Paper Waste ▪ Press ▪ Advertisement	 17/04/1998 17/04/1998		
Used batteries	19/06/1997	Bebat	21/08/1995
Waste tyres	14/01/2000	Recytyre	9/02/1998
End-of-life vehicles	19/01/1999	Febelauto	15/06/1999
Waste electrical and electronic equipment	26/01/2001	Recupel	1/06/2001

The other agreements oblige the formation of a waste management organisation. A waste management organisation is a non-profit organisation set up by a few large producers or by a sector association in order to fulfil the duty of acceptance for its members. Also producers/importers that are not a member of the producers' association can join the waste management organisation in order to broaden the collective approach. The sector associations who are organised on a national scale stress the importance of expanding the agreements to the entire country, as they want to implement a collection and disposal network on national level. Next to the benefit of operational efficiency, diverging regulations could lead to serious economic impacts resulting from cross-regional waste disposal or consumption without paying the disposal contribution (infra 4.2). The Flemish government recognises the need for harmonisation and co-operates with the other regions in order to reach similar agreements in the Brussels and the Walloon Region. This necessity of reaching converging agreements in each region enhances the complexity and the time needed to close agreements in the Flemish Region.

3.3. Actors Involved in the Policy Process

All agreements are signed by a number of associations: the waste paper (press) agreement by 4, the waste paper (advertisement) agreement by 6, the end-of-life vehicles agreement by 8, the battery agreement by 2, the waste tyres agreement by 8 and the WEEE agreement by 12. These associations are almost exclusively associations of producers/importers. However, a lot more stakeholders can be identified. We will categorise them in three groups:

- Group I: the first group consists out of actors appointed as responsible by the legislation on the duty of acceptance as laid down in the Flemish Decree on the Prevention and Management of Waste. This group consists of the producers/importers, the suppliers and retailers;
- Group II: actors in this group have to assist the first group to make a successful execution possible. This group consists of consumers; public or private waste collectors, sorters, handlers, processors and second-hand shops; waste management organisations and the federal, regional and local government;
- Group III: this group collects all other actors who might have an interest in an agreement like: the SERV (Flanders' social and economic council), the MiNa-raad (Flanders' official advice council on environmental issues), trade unions, environmental pressure groups, …

In only two agreements, actors that do not belong to the first group have signed. In the agreement on end-of-life vehicles, Coberec (confederation of Belgian recovery companies) and FEVAR (Federation of car parts and recycling companies) have signed. The agreement on WEEE is signed by the KVK (Federation of second-hand shops).

Should all actors participate in the negotiations and sign the environmental voluntary agreement? Full participation enhances the policy acceptance and thus creates a social basis for a voluntary execution. On the other hand, the diversity of interests might be so big, making the conclusion of an agreement impossible. Especially waste management agreements tend to involve more partners and more different sectors (EEA, 1997). Besides, the decree on environmental voluntary agreements provides for a public enquiry procedure. This might reduce the need to involve all of the potential partners in the negotiation process or to sign the agreement. Moreover, signing an environmental voluntary agreement brings along commitments. So it might be the better option for actors from the second group not to sign the agreements. On the other side, if they do not participate, the chance that their interests are not taken into account enhances.

But, even if the actors of the second group would be willing to cooperate, the producers are in the position to veto this. Taking the responsibility of the producers/importers as a starting-point, they are the first to be involved in the negotiations. Mostly, they prefer being the only contracting party as this allows them to implement a collection and treatment system according to their own aspirations. They opt for a specific financial and logistic system. This implies they can decide which actors they want to involve in their system. As a result, other actors involved are set aside and the producers are able to implement a tailer-made solution. Especially, the public and private collection and disposal sector claim they should be more actively involved in the negotiation process as the duty of acceptance has the most impact on their sector (VVSG, 2002).

In this regard it is important to notice that, except for the waste paper case, the sector of producers/importers is dominated by a limited number of multinational organisations. Although in though competition, they have converging interests associated to the duty of acceptance like avoiding a deposit-refund system and insisting on a nation wide and collective solution. Moreover, the companies are organised in very powerful associations. This contrasts with the waste management sector that is traditionally characterised by a large number of family-run companies who do not possess the powerful sector associations the producers have. In fact, actors like dismantlers, shredders and recycling companies prefer legislative action as they would be more protected and because they feel their point of view is not duly taken into account in an environmental voluntary agreement (Onida and Paqout, 2000). In this regard, it is interesting to notice that the agreements contain quantified collection and recycling targets for which the burden is shifted to the recovery sector, but only vague targets related to ecological product design for which the producers would be responsible.

In order to meet the demands of the actors from group II and III, a workgroup on the duty of acceptance was installed by OVAM. This workgroup consists of a number of organisations that did not sign the agreements but who are indirectly involved. The aim is to search for solutions for observed bottlenecks within actors of the second group. Participant of these meetings however state that the motivation is low because they feel that they lack the power to induce significant changes. As a result, the frequency of the meeting has continuously diminished (Rekenhof, 2002).

3.4. The Negotiation Time

The process of concluding an environmental voluntary agreement can be split into an informal and a formal phase. The informal phase starts with the first consultation between government and sector representatives and ends with the publication of the draft agreement in the Belgian Official Journal. This marks the start of the formal phase, prescribed in the Flemish decree on environmental voluntary agreements, that ends when the final version is published in the Belgian Official Journal. The time needed to conclude the informal phase is hard to determine because of the lack of transparency of this phase and especially because of the difficulty to determine the starting-point (Rekenhof, 2002).

Nevertheless, the beginning of the informal phase must be situated for all agreements in the beginning of the nineties. In fact, the informal phase consisted out of two periods of intense negotiations interrupted by a break in the contacts. Already in 1990 a study group waste tyres published a report on the subject and the first negotiations were held in 1995. In 1993 a workgroup with representatives of government and industrial representatives was set up by OVAM with the aim of analysing the situation on end-of-life vehicles and to come up with policy proposals. The first concrete result of the growing policy attention on WEEE was a report made by OVAM published in 1994.

The first attempts are thus situated before the introduction of the duty of acceptance in the Flemish Decree on Waste Management and Prevention. However, as well the packaging waste story as the first consultations proved that it was quite impossible to reach the agreements the Flemish government had in mind. On a voluntary base, the contribution of the private sector was directed at avoiding far-reaching regulations and additional costs. As such, it was decided to first lay down a legal basis. This should enhance the negotiation position of the government representatives to be able to conclude ambitious agreements with the sectors. Most actors agree that the introduction of the duty of acceptance, or the threat to do so has accelerated the negotiation procedure (Rekenhof, 2002). The informal phase thus accelerated again after the approval of the Flemish Regulation in December 1997. But even then, for most agreements it took another couple of years before a consensus was reached. For waste tyres and WEEE this resulted in situations wherein the duty of acceptance already started without being executed by an environmental voluntary agreement. For waste tyres the agreement was signed six months to late. This implies that the draft agreement was reached in time. Moreover, the waste management organisation was already set up. For WEEE on the other hand, the agreement was signed a year and a half to late. In this period, the government conducted a tolerating policy, awaiting a draft agreement to be concluded. The government is trapped into a difficult position, as it has to fulfil a dual role: cooperating to reach an agreement and sanctioning non-compliance (OVAM, 2004). If the government would emphasis sanctioning, the change that an agreement would be concluded would be reduced significantly.

For all agreements, the formal phase was less time-consuming than the informal one. On average, it took about 7 months to go through the official procedure. The decree on environmental voluntary agreements outlines that after the publication of the draft, anyone can comment and as well the SERV and the MiNa-council have to give a non-binding advice. Remarkably, the advices given hardly had any impact. Only the agreement on end-of-life vehicles and the agreement on WEEE received a positive advice with only minor remarks. The agreement on waste paper (advertisement) received a conditional positive advice from the SERV. All other advices were negative. It might be surprising that nevertheless all agreements were closed and no agreement has been adapted to the remarks made. The explanation is the fact that the possibility for remarks comes too late in the policy process. After the lengthy and sometimes difficult negotiations that have lead to the draft agreements, the motivation to adapt the draft agreement to the remarks is limited. In practice, this problem is partly tackled by inviting the SERV and MiNa-council to the negotiations. The official advice given on the draft should be more considered as their official opinion towards the agreement. The government as well as the sectors know these opinions, but did not agree on putting them into the agreement.

4. IMPLEMENTING THE AGREEMENTS

4.1. Organising Collection and Recycling Networks

The waste management organisations have the task of organising the collection and treatment of the different waste streams. The first task is to supply adequate collection equipment to the collection points. Table 3 shows the different collection points for the different waste streams. Next, the collected waste is gathered and transported to specialized recovery operators. Furthermore, the waste management organisations administer the monitoring mechanisms for reporting to the authorities and set up commercial advertising campaigns in order to achieve a high level of consumer participation.

***Table 3**. Collection system of the various waste products*

Waste product	*Collection System*
Waste Paper	▪ House-to-house collection organised by the municipalities ▪ Municipal collection sites (300)
Used batteries	▪ Approximately 20.000 collection boxes in market stores, schools, jewelleries, photo stores, public places…
Waste tyres	▪ Municipal collection sites (300) ▪ End sellers (tyre dealers, car body workshops, garages) (11.000)
End-of-life vehicles	▪ Automobile dealers (2.103) ▪ Recognised car dismantling centres (16)
Waste electrical and electronic equipment	▪ End sellers (2.178) ▪ Municipal collection sites (300) ▪ Re-use centres (40)

Table 3 clearly shows that the way the collection is organised differs for the various waste streams. As a result, the 1/1 principle (meaning that a retailer has to accept waste products if a new product of the same kind is bought) using reverse logistics through the distribution chain only gathers a limited amount of the total amount collected. This shows the aim of organising efficient collection networks, taken into account the specific characteristics of each waste stream, rather then implementing a uniform collection system through the sales chain (Wille, 2000). Notice that the framework legislation in the Waste Decree was left rather vague in order to be able to implement efficient, tailor-made environmental voluntary agreements with each sector.

Producers and retailers complain about the obligation to accept waste products. The operational efficiency of the collection is hampered as this results in a high number of collection points that only gather a limited fraction of total waste generated, leading to high collection costs. Retailers claim they are not waste collectors and that they are confronted with additional costs from the storage of accepted waste.

The waste management organisations co-operate with collectors and recyclers from both the private and public sector to fulfil their waste management tasks. This co-operation is organised in a number of ways. For the collection of end-of-life tyres, batteries, paper and electronics, agreements are concluded with local authorities who administer municipal collection sites. Transport and recycling activities are out-sourced to specialised companies on the basis of competitive tenders for used batteries and WEEE. Within the agreements concerning waste tyres and end-of life vehicles, the waste management organisations have developed a certificate to which all operators (e.g. waste tyre collectors, scrap dealers, dismantlers, shredders) have to comply if they want to participate in the network.

One important disadvantage of waste collection systems based on the free of charge tack-back principle is that only end-of-life products with a negative value are gathered. This implies that the collection networks do not have a full coverage of the waste stream. This especially holds for car wrecks, waste tyres that are suitable for retreading and certain fractions of WEEE. People who want to discard of these wastes look for alternative collectors. Dismantling companies, scrap dealers and shredders who recover spare parts and the metal content and exporters of waste to developing countries are active in these alternative networks.

As such, dual collection and treatment networks arise: on the one hand the networks that are lead by the producers through their waste management organisation and on the other hand the alternative networks. In the negotiated agreements, the producers engaged themselves to more strict environmental standards for the management of waste fractions, entailing higher cost and a competitive disadvantage compared to the alternative network. Moreover, it is a well-known fact that many of the companies of the alternative network do not operate within the prescribed environmental standards due to a lack of control from the authorities on their activities. The alternative networks developed naturally as a result of free market forces and already existed before the introduction of the duty of acceptance. Their activities were profitable because the collectors paid them when disposing of waste or because they recovered and sold spare parts and metals. Only a limited number of these companies have been selected by the waste management organisations and have become part of the network under control of the producers.

As such, in spite of the fact that the duty of acceptance is directed towards the producers, the biggest economic impact is found in the collection and recovery sector. It is here that changes are necessary in order to achieve the targets to which the producers engaged themselves in the agreements. The companies in this sector have to take the investment risks. Traditionally, this sector is characterised by a fine-meshed network of many small actors who are specialised in a limited number of activities. Due to the introduction of the duty of acceptance, the evolution characterised by horizontal as well as vertical integration towards a sector wherein only a few large companies are active, is stimulated. First, the stricter standards for treating end-of-life products demand new investments in infrastructure. A lot of small family-run firms lack the necessary financial means to comply with these standards. Secondly, the waste management organisations try to limit the number of partners needed for the collection and treatment activities as this simplifies the monitoring and control of the network. Finally, the potential amount of waste collected is rather limited, leaving room for only a limited number of players.

As a result, large contracts are driving out small waste management firms that are not able to qualify. For example, Recupel stresses that potential partners who want to treat WEEE must possess an environmental management system like EMAS or ISO 14001, which is quite hard for small and medium sized companies. Of the increasing market dominance by large waste management companies, the expected results would include excessive waste management charges that would be borne by consumers. There is a concern that there may be collusive self-contracting that may lead to highly 'inbred' and monopolistic systems. Moreover, where dominance is achieved in one market, it may then be used to try leverage dominance in a different market. Also the OECD stresses its concern for these potential problems in the context of implementing extended producer responsibility (OECD, 1997). From an environmental viewpoint however, this shift is beneficial and might have been the governments hidden agenda.

A nice example of the existence of alternative networks and rising market concentration in the producers' networks is the case of the end-of-life vehicle agreement. A central element here is the creation of a nation-wide network of recognised dismantling companies. Traditionally, a large number of small actors like scrap dealers, second car dealers, junkyards, dismantling companies, and shredders are active in this business. Due to a lack of enforcement of environmental law, the environmental performance of these actors is questionable. The number of scrap dealers and dismantlers in Belgium is estimated over 2.000 (Schenk, 1998). Up till now, only 16 companies were able to invest in the necessary infrastructure and became recognised. In 2002, 4 of them gathered 80% of all cars collected by the producers' network. However, only 60.000 wrecks were gathered by the producers' network whereas the total number of end-of-life vehicles is estimated at about 140.000 (Febelauto, 2003). A high percentage of all car wrecks is still being treated by actors who do not possess the necessary certificate and thus not guarantee a high level of environmental performance. Moreover, the implementation of the agreement has led to a huge export of car wrecks from Flanders to neighbouring regions and countries where the implementation of similar agreements goes more slowly.

4.2. Financing the collection and recycling activities

The activities of the waste management organisations are financed by a collection and recycling contribution, payable by consumers on every new product. This way, the waste management costs are shifted from taxpayers (by paying lump sum municipal waste taxes) to polluters. The contribution has to be paid when purchasing a new product and not at the moment of disposal, in order to prevent that consumers would resort to illegal dumping. For tyres and electrical and electronic equipment, the contribution is separately invoiced. For paper and batteries, the contribution is quite small and therefore integrated in the product price. Producers and retailers prefer the separate mentioning as it helps explaining the price increases to their customers and it arises that the simultaneous rise in prices by all producers could be considered as a collusion practice by competition authorities. It also enhances the publics' environmental awareness.

Table 4 shows the collection and recycling contributions for the various waste streams. For electrical and electronic equipment, just a few examples are shown. Here, the size of the contribution depends on the type of product bought and reflects the recycling cost. In this regard it is interesting to notice that the primary aim of the contribution is to finance the costs associate with collection and disposal and not to implement a pigouvian tax. Producers themselves can choose the level of the contribution but are subject to approval from the authorities. The contributions are regularly reviewed. For cars, no contribution has to be paid as the producers claim that recycling car wrecks is a profitable business.

***Table 4**. Collection and recycling contribution in 2002 (VAT included)*

Waste product	*Collection and recycling contribution*
Paper Waste	0.37 eurocent/kg
Used batteries	12 eurocent
Waste tyres	€ 1 – 1.5
Waste electrical and electronic equipment	
▪ Refrigerators	€ 20
▪ Washing machines	€ 10
▪ Televisions	€ 11
▪ Tuners	€ 4
▪ Video-players	€ 6
▪ Water boilers/mixers/razors	€ 1
▪ PC	€ 3

The financial responsibility imbedded in the duty of acceptance is fulfilled in a collective way whereby each producer contributes to the general collection and recycling cost according to his market share. This financial responsibility according to current market shares also holds for historic and orphaned waste. This is thus a collective financial responsibility model whereby all producers are jointly responsible for the total costs linked to the management of the waste products. Compared to the individual financial responsibility principle, the direct link between the producer and his real waste management cost is missing and therefore innovating producers will not be rewarded (EEB, 1999). Innovation in eco-design by one producer will probably get diluted and lost in the collective system. In order to keep the incentive for innovation in eco-design, there is a need for differentiated fees that are not yet in place. In this regard it is interesting to notice that when discussing the directives on end-of-life vehicles and on WEEE, both the European Commission and the Parliament advocate an individual financial responsibility model with the purpose of guaranteeing maximum effectiveness of the producer responsibility. It should however be noted that the Parliament, the Commission neither the Council was able to prescribe an individual acceptance duty in the WEEE Directive. Producers are therefore still free to opt for a collective disposal scheme (Vedder, 2002).

5. EVALUATION AND FOLLOW-UP POLICY

5.1. Evaluation of the environmental voluntary agreements

In the agreements a yearly evaluation is provided to allow the government to adjust the modalities of execution if required. Each year, the waste management organisations have to report to OVAM and to the general public on the execution of the agreements. The agreements stipulate which data should be provided. These data allow the government to check the achievement of the targets or the progress made towards achieving them. These guidelines are minimal requirements and in practice much more information is gathered and reported. Next to reporting about the activities of the past year, an operational plan for the year ahead is added. Based on these reports, OVAM informs the Flemish Parliament. The results of the agreements are presented in table 5, the corresponding targets are mentioned between brackets.

***Table 5**. Collection and treatment results*

Waste management organisation	*Collection*	*Treatment*
BEBAT	2003: 68% *(75%)*	Recycling: 70% *(50%)* Energy recovery: 11%
Recytyre	2002: 63% *(100%)*	Retreading: 5% *(25%)* Re-use: 7% Granulating: 22% Energy recovery: 66% *(65%)*
Febelauto	2002: 59,166 car wrecks	Re-use: (6%) *(re-use and recycling 80%)* Recycling: 70% Energy recovery: 1% *(5%)*
Recupel	2002: 35,875 ton or 3.6 kg/inhabitant	Ferrous metals: recycling 99% *(95%)* Non-ferrous metals: recycling 95% (95%) Synthetic material: 41% recycling, 28% energy recovery *(80%)* and 31% landfilling *(0%)*

When looking at the results of the agreements from the point of view of compliance, the image is not very nice. For the agreements on paper waste, there exists no adequate monitoring mechanism to check the recycling targets. The reporting was incomplete and the financial payments from the publishers to the municipalities were only executed with long delays (OVAM, 2000). For none of the other agreements the environmental targets are completely fulfilled. Within the agreement concerning waste tyres, only 5% of all collected tyres are retreaded whereas the goal was 25%. However, the sector argues that a much higher percentage is actually retreaded, but that these tyres are usually not collected by their network because of the positive value of waste tyres that are suitable for retreading. The collection rate of batteries is with over 60% quite high, but not enough to reach the original targets (Ameels, 2002). The industry claims that the growing percentage of rechargeable batteries with a longer life span (e.g. portable computers, mobile phones) sold, explains this. Within the agreement concerning WEEE, still 30% of all synthetic material is being landfilled (Recupel, 2003). As the recycling goals for end-of-life vehicles are set for 2005, the goal achievement of this agreement cannot be judged yet. For 2002, a re-use and recycling rate of 76% was reported.

OVAM acknowledges that some targets are difficult to achieve. Moreover, sometimes the non-achievement of targets is not due to a lack of producers' efforts. Therefore, the annual evaluations of the agreements by OVAM are not limited to a simple target check. In this starting phase, OVAM takes a broader view and most attention is paid to the efforts made by the waste management organisations for creating an efficient and effective collection and disposal network and adequate monitoring mechanisms. Moreover, in order to prevent negative evaluations that could lead to use of the unilateral cancellation clause provided in the decree on environmental voluntary agreements, OVAM is actively involved in the execution phase. For example, government representatives are invited as observers in the board of directors of the waste management organisations. When important decisions have to be made, the waste management organisations know the governments' opinion on the issues involved.

The reporting clearly shows how the environmental voluntary agreements have shifted management tasks from the regional administrations to the private sector. The collection and recycling networks set up by the producers rely on self-control and reporting. The data provided are as much as possible certified by an independent party. The governments' task is limited to control for example by sample surveys.

Also the governments' responsibility of enforcement towards free-riders is shared by the waste management organisations. They report to the government producers who are not a member. The government then will try to convince the free-riders to participate in the collective solution by threatening them with the obligation to draw up an individual waste management plan if producers refuse to participate. As the waste management organisations are set up by and have close contacts with the sector of producers, they have a better knowledge of possible free-riders than the government. The fear of being criticised because of free-riding could play an important role in this regard. In competitive markets where there is a high degree of closeness between the final market and consumers (e.g. automobiles, electronics, tyres, batteries), the potential loss for companies not participating or complying could be rather high due to decreasing sales as a result of high consumer preoccupation with environmental behaviour. Knowing that waste management organisations present their member lists to the authorities and help them searching free-riders, this strategy made the amount of free-riders negligible. As such, the control and enforcement task of the government is directed at controlling operators in the alternative networks.

Already in its first year of operation, 24 persons were working permanently for Bebat. Each producer/importer pays a contribution of 10 eurocent per battery to Bebat. This way, the total income of Bebat amounts about 8 million Euros each year of which 15% is spent for administrative purposes and 3-4% on auditing recycling companies (Ameels, 2002). In 2002, Recupel employed 25 people. According to our own estimates, Recupel received about 50 million Euros in 2002 of which about 10% is spent on administration, monitoring and auditing. This indicates that former public tasks are shared by the private waste management organisations.

For Recytyre and Febelauto less information is available. These organisations employ fewer employees, as they do not coordinate a centralised financial contribution scheme. Febelauto employs about five full time equivalents, Recytyre two. Consultations with various actors, the establishment of a collection and recycling network and creating an adequate monitoring scheme are their main tasks.

5.2. Alternative and Back-up Policies

Although the creation of a legal framework on the duty of acceptance was a necessary precondition for getting the negotiations started, one shortcoming of the regulation is the lack of an adequate sanction mechanism. Within the legal framework, it is possible to make a report of an offence of an individual producer. This however is only effective to motivate individual producers or importers to participate in an existing negotiated agreement. In order to force a sector to negotiate and sign an agreement and to guarantee the compliance of the agreement, this sanction possibility is not suitable. The threat legally suing all producers and importers lacks credibility. What is needed is a stimulating or sanctioning mechanism that has an effect on the whole sector. As a stimulating mechanism, a labelling scheme could be a possibility. An example of a sanction mechanism is a product tax.

An interesting case in this regard is the battery case. In Belgium, the federal government imposed a tax on all sold batteries (16 July 1993), which would come into effect in January 1994. This tax was an implementation of the general ecotax law. This law determined a whole range of products that had to be taxed with the primary goal of discouraging the use of these products in favour of less polluting substitutes. The battery industry strongly opposed this law and started negotiations with the ecotax commission proposing a voluntary collection scheme. Eventually, the ecotax law for batteries was changed on 7 March 1996, stating that batteries were exempted from the ecotax if a voluntary collection and recycling scheme was set up and certain collection percentages were met. If these objectives were not met, the ecotax on batteries would be levied. The battery industry set up Bebat (21/08/1995) as a waste management organisation to co-ordinate the collection and recycling of batteries. The collection percentages reached are quite high, comparing them with the collection percentages before and with present collection percentages abroad. One of the most important determinants explaining these good results is the existence of the ecotax as an alternative instrument with more severe consequences (Ameels, 2002).

Independently of the federal ecotax, the Flemish government submitted used batteries to the duty of acceptance from June 1998 on and started negotiations with the battery industry to reach an agreement. Remarkable is the fact that the battery industry insisted to keep the federal ecotax in force in order to be able to motivate all individual battery producers to keep financing Bebat, which existed already a couple of years (SERV, 2003).

Next to a complementary stimulating or sanctioning mechanism, supporting back-up policies are required. One logic consequence of enhancing recycling efforts is the fact that more recycled materials are produced. The problem is often insufficient demand for these materials. This explains the hesitation of firms towards investing in recycling activities. This is worsened by the fact that the more recycled materials there are, the lower becomes the price of these materials. What is needed are back-up policies that make recycling attractive, e.g. by making recycled materials more attractive compared to raw materials and by creating new outlets for recycled materials. Only if recycling is perceived to be economically the best option, a big step towards closing material loops can be made. If governments cannot fulfil this precondition, it will stay very hard to steer waste management practices in this direction, certainly by the use of environmental voluntary agreements (OECD, 2001).

5.3. Learning and Preparing New Agreements

Sometimes it is stated that the most positive aspect of the agreements is the fact that much is learned from these first experiences with the duty of acceptance. Especially the relations between the government and the producers' associations have improved. At the beginning, Flemish environmental policy was mainly characterised by command and control regulations. This was certainly the case when the first agreements were negotiated. As more waste streams became subject to the duty of acceptance, some learning has occurred. Within the first two agreements (on waste paper), negotiations were held in a mistrustful atmosphere leading to very difficult and lengthy negotiations (Wille, 2000). As a result, the consensus reached was not in line with the objectives the government had when starting the negotiation process. After the first evaluation in 2001, the Flemish parliament claimed that both agreements had to be renegotiated. On the other hand, the most recent agreement (on WEEE) is promoted as an example for other negotiated agreements to be closed in future and as an example for dealing with WEEE in Europe and can in no way be compared with the disappointing agreements on paper waste concluded a couple of years before.

The attitude of the sector associations entering a negotiation process has clearly changed. At the beginning, the main aim was to prevent new regulations coming in force whereas more recently, producers more actively participate with the aim to reach an effective agreement with feasible targets. As sectors notice that several sectors become subject to the duty of acceptance, they realise that it becomes inevitable and they actively participate in the negotiations in order to be able to implement a system according to their own aspirations. The authorities on the other hand have learned not to focus solely on recycling and re-use targets but to take a broader view at the problems involved. In this regard it can be stated that the general policy-style evolved strongly, putting more emphasis on consultation and co-operation with target groups.

In order to be able to apply the knowledge build up in these first experiences in a concrete manner, a working group was gathered under the supervision of OVAM during the period 2002-2003. Thee aim of the project was to formulate a global vision (code of good practice) on environmental voluntary agreements concluded to execute the duty of acceptance with the aim to (OVAM, 2004):

- reach a certain degree of uniformity in future environmental voluntary agreements. Agreements should be concluded following a number of fixed objective principles;
- streamline the negotiations for agreements to be closed in future. This way lengthy discussions should be avoided and agreements should be concluded faster;
- be a reference to solve problems that could occur during the execution of an agreement.

To reach these objectives, all actors involved in the policy process should support the vision. As a result, representatives from all groups (see 3.3.) were involved in the project: the waste management organisations (Recupel, Bebat, Recytyre and Febelauto), the producers/importers (FEE, Agoria and VEV), the distribution (Fedis, Nelectra and Federauto), the waste operators (Febem, Coberec, VMH and KVK), the local authorities (VVSG, Interafval) and the advisory bodies (SERV and MiNa-council). After several meetings, a draft version was drawn up. In the following meetings, the workgroup aimed to reach a final version. Pretty soon however, OVAM had the feeling the discussion was not really productive anymore and the aim to reach a consensus was abandoned. Based on the draft version, OVAM unilaterally wrote the reference document. Despite the fact that a consensus proved to be impossible, it is believed that this document will facilitate future agreements. Amongst others, the reference document provides for:

- a definition of a producer/importer and a solution for the problem of e-commerce;
- an assessment of the degree of individual versus collective responsibility towards the achievement of the targets;
- a description of the way these agreements should support prevention of waste;
- guidelines for selecting partners for logistic or treatment operations;
- a description of the way the system should be financed. E.g. orphaned and historic waste should be financed according the current market shares;
- a list of all tasks provided by third parties that should be financed by the waste management organisation;
- a description of the monitoring and reporting obligations.

Next to establishing adequate guidelines for negotiating and concluding environmental voluntary agreements, the selection of waste streams strongly influences the implementation process. An important conclusion of a survey carried out in 2002 was that the selection of which products are to be submitted to the duty of acceptance was done on an ad-hoc base (Rekenhof, 2002). A decisional framework that enables to decide rationally for which waste streams the duty of acceptance might be an efficient and effective policy instrument is lacking.

The first effort in this regard was a study ordered by OVAM. The assignment was to create such a decisional framework and to investigate for which waste streams, the duty of acceptance might be a promising instrument (WES, 2001). The result however was disappointing. Representatives of OVAM noticed the study did not really enhance their know-how.

In fact, the idea that an adequate decisional framework can be designed is now abandoned. The emphasis now is on gathering as much information as possible on the waste streams that are to be submitted to the duty of acceptance. Instead of starting the negotiations as soon as possible, OVAM first tries to enhance its knowledge of the waste stream in question. As well the sector structure (production, import, distribution) as best available techniques (BAT) for managing the waste are taken into account. A study with this purpose was carries out by VITO (Flemish Institute for Technological Research) in cooperation with ERM (Environmental Resources Management, an environmental consultant agency) in 2003. This should support OVAM for negotiating the agreements to be closed in future. The first steps have already been taken. On December the 5th of 2003, the Flemish Parliament approved the revision of the Flemish Regulation on Waste Prevention and Management. The following waste streams are be submitted to the duty of acceptance: waste oil and fats, waste photo-chemicals, waste farm films, medical electrical and electronic appliances, waste wood, waste carpets and waste mattresses. The future will tell wether the knowledge build-up will lead to more efficient negotiations and effective environmental voluntary agreements.

6. CONCLUSION

Flanders is one of the first European regions where policy makers have experimented quite extensively with the instrument of extended producer responsibility. In order to implement this principle in Flemish waste management, the duty of acceptance was introduced in the waste legislation. Negotiated agreements are concluded with target groups stipulating the practical execution of the basic rules laid down in the legal framework. The environmental voluntary agreements have a support function directed at implementing policy targets determined outside the agreements.

This sector-based policy approach allows setting up efficient collection and disposal networks, taking into account the specific characteristics of each end-of-life product, rather than implementing a uniform regulation. Moreover, the agreements create an opportunity to collectively fulfil the individual take-back obligations, thus entailing several advantages like operational efficiency of the collection and coverage of orphan waste. The role of the waste management organisations is crucial in this regard. A lot of management tasks are passed on from the region administrations to these waste management organisations. As the collection and disposal networks set up by the producers rely on self-control and reporting, the government's task of control and enforcement was reduced to the control on the alternative networks. As a result, government can spent more resources to tailor-made policy making.

The analysis pointed to the impact of the agreements on the recovery sector. The monopolistic position of waste management organisations creates a trend towards market concentration with potential monopolistic consequences. Active government involvement is necessary to avoid collusive self-contracting that may lead to highly inbred and monopolistic systems due to market dominance by one or a limited number of large waste management companies. A high number of small sized companies are currently struggling to survive because they have no opportunity to participate in the collection and disposal networks of the producers. This proves that the producers, who consist mainly of large multinational companies organised in powerful associations have succeeded in shifting most the burden associated with the duty of acceptance to the small actors in the less organised recovery sector. The sector structure explained how producers have succeeded in limiting their responsibility to set up a waste management organisation. The associated costs are passed on to their costumers by invoicing a collection and recycling contribution.

Although the compliance of the agreements seems disappointing at first face, the lessons learned from the experiences might be more important than the actual compliance. Especially since the government attributes an important role to the duty of acceptance in its future waste policy. The relations between the government and the producers' associations have improved. Instead of trying to avoid additional regulation, producers now actively participate in order to design an effective agreement with feasible targets. Government on the other hand has learned not to focus solely on recycling and re-use targets but to take a more broad view at the problem involved. In this regard it can be stated that the general policy style evolved strongly, putting more emphasis on consultation and co-operation with target groups. As a result, the most recent negotiated agreement on WEEE is generally considered to be an example for other agreements to be closed in future and for dealing with WEEE in Europe and can in no way be compared with the disappointing agreements on paper waste concluded a couple of years before. Convinced that this innovative policy approach contributes to the overall goal of sustainable development, the government has already planned to start up additional sector-based policy programs based on the duty of acceptance in the coming years.

7. REFERENCES

Ameels B. & De Clercq M. (2002), The Belgian agreement upon the collection and recycling of batteries. In M. De Clercq, *Negotiating environmental agreements in Europe: critical factors for success* (pp. 113-130). Cheltenham: Edward Elgar.

BEBAT (2003), *Annual Report 2002*. Brussel: Bebat.

Bracke R. (2003), Milieubeleidsovereenkomsten ter uitvoering van de aanvaardingsplicht. In Vlaamse Milieumaatschappij, *Milieu- en natuurrapport Vlaanderen: beleidsevaluatie* (pp. 125-162). Mechelen: VMM.

Rekenhof. (2002), *Gebruik van convenants in de Vlaamse Gemeenschap*. Brussel: Rekenhof.

De Clercq M. (2002), *Negotiating environmental agreements in Europe: critical factors for success*. Cheltenham: Edward Elgar.

European Commission (2003), *Communication from the commission: towards a thematic strategy on the prevention and recycling of waste*. COM(2003) 301.

European Environment Agency (1997), *Environmental agreements: Environmental effectiveness*. Copenhagen: EEA.

Febelauto (2003), *Annual report 2002*. Brussel: Febelauto.

Sociaal Economische Raad voor Vlaanderen (2003), *Advies over drie ontwerp milieubeleidsovereenkomsten: afvalbatterijen, afvalloodstartbatterijen en afvalbanden*. Brussel: SERV.

Vlaamse Milieumaatschappij (2003), *MIRA-T: Milieu- en natuurrapport Vlaanderen: Thema's*. Mechelen: VMM.

Lymberidi E (2001), *Towards waste-free electrical and electronic equipment*. Brussels: European Environmental Bureau.

OECD (1997), *Extended and shared producer responsibility: phase 2 framework report*. ENV/EPOC/PCC(97)20/REV2. Paris: OECD.

OECD (2001), *Extended producer responsibility: a guidance manual for governments*. Paris: OECD.

Omida M., & Paquot A. (2000), *Voluntary agreements in the field of waste management: assessment of practical experiences and related issues*. CAVA working paper No 2000/2/10. Waste Management Unit, DG Environment, European Commission.

Openbare Afvalstoffenmaatschappij voor het Vlaamse Gewest (2004), *Aanvaardingsplicht: referentiedocument voor milieubeleidsovereenkomsten*. Mechelen: OVAM.

Openbare Afvalstoffenmaatschappij voor het Vlaamse Gewest (2002), *Inventory of household waste 2001*. Mechelen: OVAM.

Openbare Afvalstoffenmaatschappij voor het Vlaamse Gewest (2000), *Evaluatie van de milieubeleidsovereenkomsten oud papier*. Mechelen: OVAM.

Recupel (2003), *Annual Report 2002*. Brussel: Recupel.

Recytyre (2002), *Annual Report 2001*. Brussel: Recytyre.

Runkel M. (2003), Product durability and extended producer responsibility in solid waste management. *Environmental and Resource Economics*, 24, 161-182.

Schenk M. (1998), *Altautomobilrecycling: technisch-ökonomische zusammenhänge und wirtschaftspolitische implikationen*. Wiesbaden: Duetscher Universitäts Verslag.

Sodhi M.S. & Reimer B. (2001), Models for recycling electronics end-of-life products. *OR Spektrum*, 23, 97-115.

Spengler T. & Ploog M. & Schröter M. (2003),. Integrated planning of acquisition, disassembly and bulk recycling: a case study on electronic scrap recovery. *OR Spektrum*, 25, 413-442.

Suck A. (2000), *Changes and limits of environmental agreements in waste management policy in Germany*. CAVA working paper No 2000/2/17. Ghent: Centre for Environmental Economics and Management.

Suikerbuijk M. (1999), *Benchmarking of Flemish waste management policy compared with neighbouring waste markets*. Mechelen: OVAM.

Vedder H. (2002), *Competition law, environmental policy and producer responsibility: experiences in the Netherlands from a European perspective*. Groningen: European Law Publishing.

Vereniging van Vlaamse Steden en Gemeenten (2002), Producentenverantwoordelijkheid en aanvaardingsplichten in het huishoudelijk afvalstoffenbeleid. *De Gemeente*. Brussel: VVSG.

Wille D. (2000), *The integration of voluntary approaches into existing legal systems: an important instrument in the Flemish waste management*. CAVA working paper No 2000/2/11. Mechelen: OVAM.

ENVIRONMENTAL VOLUNTARY AGREEMENTS IN PORTUGAL

Characterisation and implementation

M.F.M. CABUGUEIRA

Universidade Portucalense, Economics Department, Porto, Portugal

Abstract. Environmental Voluntary Agreements (VAs) were implemented in Portugal for the first time at the end of the 80s, but it was during the 90s that the VAs became an important environmental policy instrument with the creation of the "Environmental Adaptation Contracts".
At the present moment the Portuguese Environmental Authorities are implementing a third generation Environmental Contracts — Contracts for the Continuous Improvement of Environmental Performance — with the objective of giving support to the companies that are willing to adopt "over-compliance" measures.
In this paper we will review the Portuguese experience with VAs: we will start with an historical review of its implementation and a characterization of the different experiences; we will then, move to a closer analyses of the "Environmental Adaptation Contracts", presenting a characterization of the contracts, making a comparison with the European experience with negotiated agreements and we will conclude with some comments on its application.
Throughout the text we will use the Portuguese experience as a reference to make a theoretical review of the VAs characterization and to emphasise the factors that motivate public and private participation on this "voluntary" environmental regulation process.

1. INTRODUCTION

Facing continuous difficulties in the implementation of the environmental regulations, the Portuguese Environmental Authorities turned to Voluntary Approaches for the first time in 1988/1990. (Figure 1).

This first experience included a group of four environmental agreements signed between the Portuguese Environmental Agency (at that time the DGA, *Direcção Geral do Ambiente*[1]) and the Industrial Associations representing: the pulp paper industry (1988), leather industry (1989), glass packaging industry (1990) and the cardboard packaging for liquids industry (1990). All four agreements aimed at reducing the environmental impact of these industries in a period till 1992 (that was later extended to 1995).

E. Croci (ed.), The Handbook of Environmental Voluntary Agreements, 203 - 219

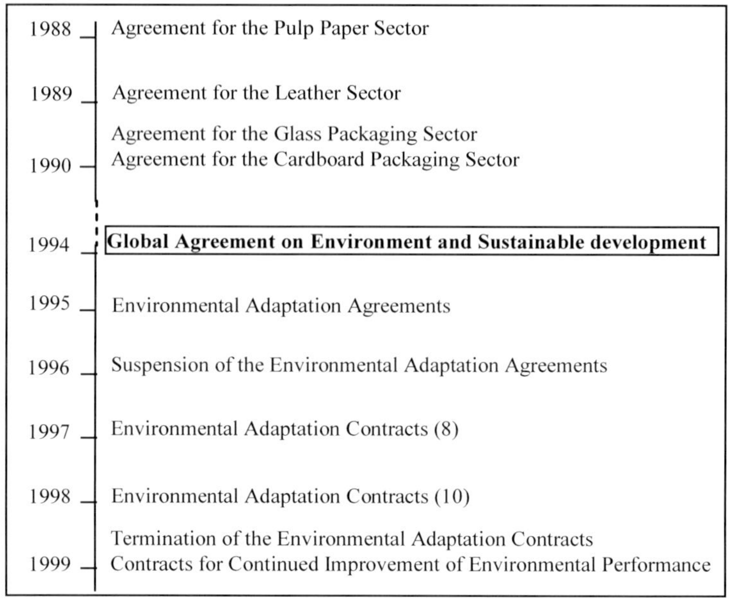

Figure 1. *Voluntary Approaches in Portugal Time Line*

Following a positive evaluation of the results achieved with the pulp paper agreement, a "Global Protocol on Environment and Sustainable Development" was signed, in 1994, creating a framework for the implementation of further Voluntary Approaches.

This "Global Protocol on Environment and Sustainable Development" was signed by the Ministry of the Environment and Natural Resources, the Ministry of Industry and Energy and the Ministry of Agriculture, the Portuguese Agricultural Confederation and the Portuguese Industrial Confederation.

In that document, it was clearly stated that the "environment preservation" was an important part of the public and private economic policies towards global competitiveness, and it was recognized that the environmental protection efforts should follow the principles of "precaution", "co-responsibility" and "co-operation".

Both public and private entities acknowledged the necessity to make considerable investments in "environmental protection". With particular relevance, the public authority recognised that, under the principle of co-ordination, it had an important role to play in creating incentive schemes to support company's efforts in this field.

Voluntary Agreements were underlined as preferential environmental policy instruments, capable of incorporating these principles and necessities.

Supported by the Global Protocol, a second generation of Voluntary Approaches were implemented in 1995 with the creation of the Environmental Adaptation

Agreements (EAAs). Even though this was an important step to confirm the voluntary instrument as part of the Portuguese environmental policy, the lack of legal support to the enforcement of the EAAs condemned them to an unavoidable failure.

It was necessary to change the legal status of the environmental agreements and to place it under the framework of the "public administrative law" to re-establish the credibility of the Voluntary Approach. This was accomplished with the implementation of the Environmental Adaptation Contracts (EACs), in 1996.

The EACs were successfully implemented till the end of 1999 , by the time the Portuguese Environmental Authority, showing increased maturity in the use of these flexible environmental policy instruments, started implementing a third generation Voluntary Approach: the Contracts for Continuous Improvement of the Environmental Performance (CCIEP). These contracts are legally binding, as the EACs, but have the ambitious objective of pushing the companies to over-comply with its environmental obligations.

1.1. The environmental adaptation agreements

The EAAs were introduced as a complementary instrument to the public environmental regulation. They were created with two different purposes: on one hand, they responded to the incapacity of the regulator to safeguard the full application of the "command and control" regulation implemented under the pressure of the EU directives; on the other hand, its flexibility allowed for an intervention on the environment while protecting the economic situation and the competitiveness of the Portuguese companies.

Using the Global Protocol has a background framework, 8 EAAs were implemented: one regional agreement between the Portuguese Environmental Agency, the Portuguese Industrial Agency and the companies that adhere to a wastewater-treatment system in the region of Águeda; and seven industrial agreements, signed by the following industrial sectors: metallurgical and metal-mechanic; vegetal oils; chemicals; decorative stones; dairy products; tomatoes; hog raising.

As the name implies — "Environmental Adaptation Agreements" to the public regulation — these agreements were signed with the specific intention of programming a gradual adjustment of the participating companies to the environment regulation that was applicable at the time.[2]

In terms of environmental objectives, all EAAs aimed at improvements in water pollution, air pollution and waste management, with the exception of the regional agreement, which had as only concern the water management problem in that particular region.

In comparison to the "direct public regulation" requirements, the EEAs integrated some important innovative features.

As a starting point, the participating industrial sectors were obliged to produce a report on the "environmental situation of the specific sector". Using that report, a specific "environmental adaptation plan" should be established for each sector.

Even though such "plans" had some degree of generality throughout all the EEAs, it is important to recognize the existence of "custom-made" features, especially in what concerned the implementation schedule, conferring flexibility to the instrument.

These two initial requirements had two relevant impacts: first they forced the creation of environmental information, information that came to respond to a lack of knowledge, shared by regulator and companies, on the environmental impact of each sector ; second, the associated, obligation to employ technical expertise assistance in the elaboration and implementation of the "environmental adaptation plan", meet the technical insufficiencies that were visible in each sector.

Besides these obligations to create environmental information and a implementation plan, the environmental agreements also determined that, under the principles of co-responsibility and co-ordination: (1st) the moratorium for compliance to the environmental legislation was only granted to those companies accepting and fulfilling with the agreement requirements throughout the adaptation schedule; (2nd) under the period of the agreement the application of the environmental legislation was more flexible, and any changes in that legislation should be negotiated with the participating companies; (3rd) the environmental effort of the companies should be taken into account whenever the companies applied to public incentive schemes.

1.2. The environmental adaptation contracts

Facing a disappointing result in the implementation of the EAAs, the environmental authority suspended, in 1996, its application and proposed the creation of the Environmental Adaptation Contracts (EACs).

The explicit change in the name of the environmental agreements, from "agreements" to "contracts", wasn't a simple cosmetic change. It reflected, an important modification in the legal framework of the agreements that changed from being non-legally binding to become legally binding. This transformation placed them under the administrative law that rules contracts between public and private entities.

The substitution of the EAAs for EACs was accepted by several industrial sectors giving way to the signature of eight EACs during 1997 and ten at the beginning of 1998 (even though many other negotiations began but fail to be completed).

In 1997 four contracts were signed replacing EAA from 1995 — for the industrial sectors of vegetable oils, chemicals, decorative stones, dairy products — together with four new contracts for the olive oil[3], textiles, paper and lumber sectors.

In 1998, the tomato sector EAC replaced a 1995 EAA, and nine new contracts were created for the industrial sectors of footwear, plant protection, cork bark, nautical industry, ink, glue and similar products, printing, ceramics, electric and electronic, rubber and tires recapping.

Even though the public regulator maintained the EAAs environmental objectives (with an implementation timeframe till the 31st December 1999) and most of its implementation characteristics (specifically: the requirements in relation to the creation of the "environmental report" and "environmental adaptation plan"), three important innovations were introduced, giving a remarkable push to the effectiveness of the contracts. First, the change in the legal framework: the evolution from "agreements" to "contracts" created a legal bind for the participating companies and placed the instrument inside of the "public administrative law" framework giving particular empowerment to the public signatory.

Second, the increase commitment from the companies: the industrial scope of the agreements was maintained but a requirement for an individual commitment from each individual company was introduced strengthening the private responsibility for the success of the policy Even though the negotiations over the implementation schedule (and in some contracts the quantitative environmental objectives) were still conducted by the Industrial Associations, each participating company was now obliged to sign an individualised contract committing itself to the fulfilment of each step of the adaptation plan.

Third, the clear specification of the "environmental objectives" and the "implementation schedules", pressed for a better control, increasing the effectiveness of the instrument.

1.3. Contracts for continued improvement of the environmental performance

Showing increased commitment to the Voluntary Approaches as environmental policy instruments, the Portuguese environmental authority substituted, at the end of 1999, the EAC for a third regeneration agreement: the Contract for Continuous Improvement of Environmental Performance (CCIEP).

These new agreements are intended to support the companies' effort to over-comply with the environmental regulation through the implementation of an environmental management system in accordance to the EMAS regulation[4].

In these agreements, companies commit themselves to implement EMAS till 2004/5 in accordance to a pre-established schedule, being subject to a continuous control by an external committee. This external committee is constituted by: representatives from the environmental and industrial public agencies and the industrial associations.

As a compensation, the public authorities agreed to take this environmental effort and the degree of compliance to the established schedule into consideration, whenever the companies apply for public funding.

Until now, two contracts were signed: one, for the cement industry (January 1999 to December 2004) with the participation of 3 companies representing 6 different industrial units; and a second for the glass industry including 5 companies with 6 industrial units.

Two significant steps forward were given with the evolution from the adaptation Contracts to the Contracts of Continuous Improvement: first we enter in the realm of over-compliance, i.e., we are now facing environmental objectives that go beyond the "satisfying" standards publicly specified; second there is an evolution in terms of objectives, we move from a "target based voluntary approach" to a "performance based voluntary approach" (Storey, 1996).

In the progress from the EAC to the CCIEP, we move from a "centralised setting of private objectives", where the regulator pressure over the "voluntary nature" of the private action is more evident (Glachant, 1994), to a truly "voluntary action in environmental preservation" both on the intentions and the objectives. The companies, when agreeing to implement the EMAS are committing, not only, to fulfil certain environmental objectives, privately established, but also, to embody a continuous practice on environmental preservation management.

The voluntary recognition of these environmental objectives, which are not obligatory by law, place interesting questions in terms of what are the private motivations. In the literature the over-compliance action is normally related to commercial advantages and improved image (Arora and Cason, 1995) or an anticipation to public regulation, either as a response to a regulation threat (Segerson, 1997) or as a strategic positioning as "first mover". In the Portuguese case, the anticipation of a deeper public intervention in the regulation of those two economic activities seemed to be the most plausible explanation for the companies' over-compliance action.

2. THE PORTUGUESE ENVIONMENTAL ADAPTATION CONTRACTS

The Portuguese EACs were implemented during the period between 1997 (with the first contracts being signed in the 18th of March 1997, with the textile sector) and 1999.

The importance on the EACs in the development of the Portuguese environmental regulation and its' peculiarity in the overall EU experiences on negotiated agreements, justifies a more careful analysis of its characteristics.

Using the information presented by the DGA in the report on the implementation of the EACs (DGA, 2000), and aiming to analyse the most important aspects related to the negotiated agreements that are normally identified in the literature, an inclusive view of the EACs is presented following three steps: classification of the EACs (2.1.), characterization of the EACs (2.2.) and some notes on the implementation of the EACs (2.3.)

2.1. Classification of the Portuguese Environmental Adaptation Contracts

It is now normal to consider that the voluntary approaches are implemented through three types of instruments: unilateral commitments, public voluntary schemes and negotiates agreements (Croci and Pesaro, 1998 and Carraro e Lévêque, 1999).

In the case of the EACs, due to the process that supports its creation and implementation, they might be considered as "two level agreements" once they have characteristics of both negotiated agreements and public voluntary schemes (Croci, 2003).

While being negotiated, between the environmental and industrial authorities and the industrial associations, the EACs assume the form of negotiated agreements.

This characteristic allows for the flexibility in the implementation schedule and in the specification of the environmental objectives.

After the definition of the agreement structure, the companies were then invited to participate. The acceptance procedure implied an individual commitment to the obligations of the agreement. This commitment was formalised through the signing of an individual contract. This is the voluntary "participation program" characteristic of the EAC.

An important observation is the fact that, even though the agreements are adopted as substitutes of the existing public regulation and there is the contractualisation of the companies' commitment towards the environmental adaptation objective and its schedule, the voluntary nature of the instrument is preserved. Both options, either to participate or not to participate in the program constitutes a decision that the companies can take voluntarily.

2.2. Characterization of the Portuguese Environmental Adaptation Contracts

Making a Comparison with the European implementation of voluntary approaches at the time of the EACs[5], we can see, in Figure 2, that Portugal was the fourth country in term of number of negotiated agreements signed (following Netherlands with 107, Germany with 93, Austria with 20).

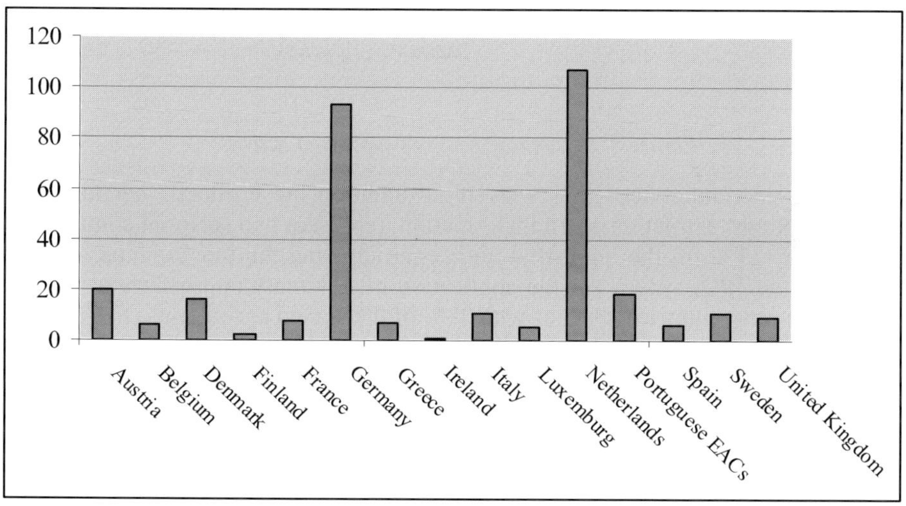

Figure 2. *Negotiated Agreements in the EU (EEA, 1997, Portuguese information added)*

The EACs covered all the spectrum of industrial transformation activities. They were used as environmental policy instruments to intervene in the sectors with high environmental impact, like the chemical sector, but also in the economically most fragile and significant sectors, like the textile, footwear and the cork bark sector and some transformation activities related to the agricultural sector (the transformation of tomato and dairy products).

The distribution by industrial activities is presented in Figure 3. If we take the most significant Portuguese industrial sectors - cork bark, footwear and textile - we have four EACs covering 42% of the industrial units signing the contracts.

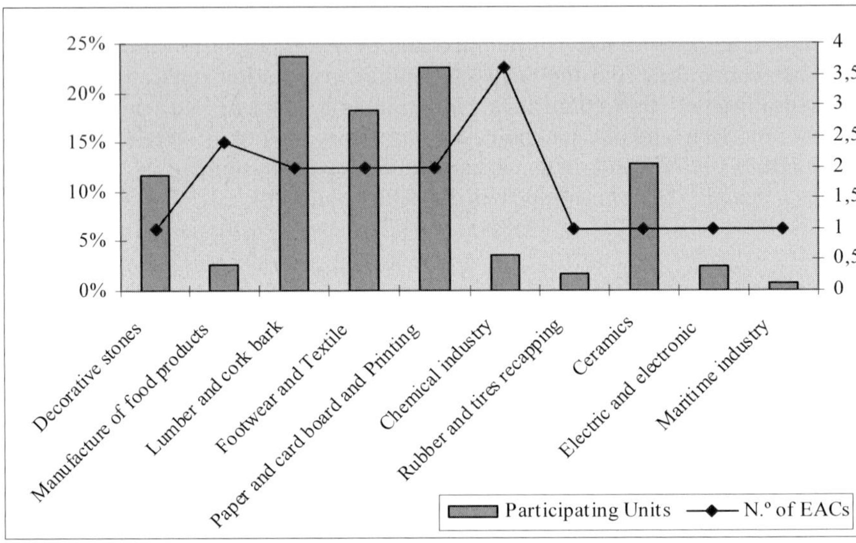

Figure 3. *Portuguese EACs by industrial activity*[6]

Other two economic activities with high tradition in Portugal, ceramics and decorative stones extraction and transformation, represent two sectorial contracts but account for 25% of the industrial units participating in the voluntary action. According the 1996 survey for the application of voluntary approaches in the EU, this was the most significant sector with 28% of the signed agreements (EEA, 1997).

The two sectors with chemical based effluents, paper and printing and the chemical industry (which includes the chemical sector, plant protection, ink, glue and similar products and 60% from the vegetable oils sector) represent 27% of the participating industrial units.

The food processing activities, which were the second most significant sector in EU terms with 12% of the negotiated agreements in 1996, only represent 3% of the industrial units signing the EACs.

To better characterise the Portuguese EACs we will now analyse the four distinguishing aspects of the negotiated agreements considered throughout the literature: the participating parties; the EACs role in environmental public regulation; the EACs environmental objective and its' legal statute.

2.2.1. Participants in the EACs

As an important characteristic of the Portuguese EACs, it must be emphasized the high involvement of the governmental agencies and the private Industrial Associations.

In all EACs, the Environmental Ministry was represented by the Environmental Agency (DGA).

Co-signing most of the EACs with the DGA on behalf of the Government, was the public Industrial Agency (DGI, *Direcção Geral da Industria*) and, for the contract for the decorative stones sector, the Geological and Mining Institute (IGM, *Instituto Geológico e Mineiro*), in representation of the Industrial Ministry.

For the two agricultural related contracts, the EACs for the dairy products sector and the olive oil sector, the co-signatory with the DGA was the agricultural institute (GPPAA, *Gabinete de Planeamento e Política Agro Alimentar*) in representation of the Agricultural and Fishing Ministry.

Only three EACs, for the tomato, textiles, paper and lumber sectors, had as sole representative of the public regulator the DGA.

For the private party, all contracts were negotiated and signed by sectorial Industrial Associations.

Even though the Industrial Associations assumed an important role, the participation in the contracts was not an exclusive option for those companies that were associated, the participation was open to all the companies that developed industrial activities in each sector.

In relation to the individual companies, they were required to make an individualized commitment. Each participating company was obliged, under the EACs, to individually accept the objectives and the implementation schedule of the contract. The formal participation of the companies, through the signing of a participation contract, made them individually liable for its execution.[7]

2.2.2. The EACs and the public regulation

In the structure of the Portuguese public regulation, the voluntary approaches have been introduced as instruments to support the technical and financial co-operation between the central administration and other public and private agents (*Decreto Lei 384/87*) and/or as an instrument to extend the fixed limits of compliance to the legislation (*Lei 11/87, artigo 35*).

As we established earlier, the EACs were implemented as a complementary instruments to the traditional command and control environmental regulation. More precisely, they were created as specific regulations for each sector, supporting and specifying the application of the general regulation that covered each environmental theme.

The EACs were set as instruments of gradual adjustment of the industry to the environmental legal requirements. The existence of an adaptation period to the new environmental norms was justified for two reasons: because of the administrative and technical difficulties of the regulator to fully implement and control the direct regulation; and, also, as a response to the economical and technological difficulties, faced by the companies, to implement the technical and managerial changes required to fulfil the established environmental rules in a short-term period.

2.2.3. The EACs environmental objectives

The objectives of the EACs focus, exclusively, on the production process of the companies covering all the environmental themes: air pollution, water pollution, waste management, and noise. The broad range, in terms of environmental protection objectives is clearly shown in Table 1, where it is specified the environmental themes for each contract.

Table 1. *Environmental themes of the EACs*

EAC	Water Pollution	Air Pollution	Waste Management	Noise
Vegetable oils	✓	✓	✓	✓
Dairy products	✓	✓	✓	
Tomato	✓	✓	✓	
Decorative stones	✓	✓	✓	✓
Cork bark	✓	✓	✓	✓
Lumber	✓	✓	✓	✓
Textiles	✓			
Footwear	✓	✓	✓	
Ceramics	✓	✓	✓	✓
Paper and card board	✓	✓	✓	✓
Electric and electronic	✓	✓	✓	✓
Rubber and tires recapping	✓	✓	✓	✓
Printing and paper transformation	✓	✓	✓	✓
Plant protection	✓	✓	✓	✓
Ink, Glue and similar products	✓	✓	✓	✓
Chemicals	✓	✓	✓	✓
Nautical industry	✓	✓	✓	✓

Most of the contracts explicitly establish, in their text, that the environmental objective was to achieve compliance with all the existing environmental regulation applicable to the particular industrial sector.

In same cases, some of the environmental regulation was not relevant for the specific activity and for that reason it wasn't considered. For instance, noise impact regulation was not considered in the dairy products and the tomato sector, since the two activities complied with the acceptable limits.

Two contracts are particularly interesting in what concerns the established environmental objectives: the textile EAC and the EAC for the chemical sector.

The textile EAC, was created with an exclusive preoccupation over water management. To be more precise, the aim of the contract was to support the implementation of an adaptation plan to the industrial water discharge norms, for the companies from the textile sector that were outside a particular water discharge scheme (SIDVA – *Integrated system for the depollution of the Ave river).* The main preoccupation of this contract was one of equity between companies. The contract intents to maintain the equilibrium between the companies that are part of the SIDVA water pollution reduction schemes and those that are outside of its geographical range, guarantying an equitable treatment in terms of environmental obligations inside of the sector.

The EAC for the chemical sector, even though maintaining the broad objective of adaptation to the environmental norms, establishes a very precise schedule, with intermediate environmental objectives for each year of implementation. This is the most explicit example of the result achieved throughout the negotiations in relation to the intention to adjust the "adaptation condition" for each sector.

2.2.4. The EACs legal statute

To conclude the characterization of the EACs, it is important to confirm its voluntary nature and to verify its legal status.

In spite of the fact that the EACs have the same legal origin as the EAAs and that both of them were considered to be "specific regulation" for each sector, we can establish that the EAAs were non-legally binding voluntary agreements while the EACs presented themselves as legally binding voluntary contracts. This fact justified the substitution of one for the other and was of extreme importance for the effectiveness of the intervention.

Two facts prove the relevance of the legally status "correction" of the agreements.

First, the environmental contracts were published in the official newspaper (*Diário da République*) formalising its statute as "specific measure" to each sector and placing them inside of the public administrative law.

Second, even though, both agreements instituted the immediate loss of privileges in terms of the moratorium that had been awarded as punishment for non-compliance, the fact that the EACs were framed in the scope of the "administrative contracts", provided the public regulator with a set of specific powers

In the Portuguese "administrative contracts", the public administrator is given, amongst others, the following particular powers:

- the "power to orientate ": being the one that safeguards the public interest, it is given to the public administrator, not only, the "power to supervise" the fulfilment of the rules, but also, the power to impose the procedures that should be adopted in order to guarantee effectiveness and higher economy efficiency;

- the "power to punish": with the intention of forcing the private agent to fulfil with the objectives, thus safeguarding the public interest, it is allowed for the public administrator to broaden, in a unilateral way, the strength and variety of sanctions;
- the "power to rescind the contract": this power is given to the public administration not only as a sanction but also as a way to preserve the public interest in case there is no public reason for the maintenance of the contract.

2.3. The implementation o the EACs

The EACs served different purposes in the overall of the Portuguese environmental regulations. Three of this purposes were: to allow for a better implementation of the environmental regulation, to promote a higher participation of the private entities in the regulation processes and, finally, to guarantee a more flexible approach to the implementation of the environmental management practices required by law.

To what regards the implementation of the environmental regulation, the conclusions presented by the DGA (DGA, 2000) gave different results depending on the specific EACs, with an overall positive assessment on the accomplishment of the most significant environmental practices by the majorities of the participating companies. Taking the information on the participating companies, some interesting facts are noticeable in table 2 where it is made a summary of the participation in the EACs.

Table 2. *Companies participation in the EACs*

	Units that signed	Participating Units	% Participation per sector*
Vegetable oils	59	45	62%
Dairy products	44	37	62%
Tomato	12	12	38%**
Cork bark	151	98	66%
Lumber	598	494	41%
Textiles	131	122	15%
Footwear	354	336	76%
Ceramics	140	121	58%
Paper and card board	37	30	59%
Electric and electronic	67	60	30%
Rubber and tires recapping	59	44	90%
Chemicals***	90	90	35%
Nautical industry	18	17	49%

* The data related to the industrial units for each sector was gathered from the Industrial Statistics Reports for 1997 and 1998 (INE, 1997 and 1998), considering the economic units for each sector and taking into account the Portuguese Classification of economic units. The information for two EACs, decorative stones and printing and paper transformation, was not coherent with the number of participating companies and for that reason those contracts were disregard.

** To calculate the percentage participation for this sector it was considered the number of economic units for the sector of food transformation and conservation.

*** Includes three EACs: Chemicals, Printing and paper transformation, Plant protection.

The difference between signing and participating units is explained by the exclusion of some companies. The reasons for the exclusion were: not fulfilling the EACs requirements (e.g., companies declare not to be under the environmental regulation being considered); change, closure or deployment of the economic unit; insufficient information delivered or non-provision of requested information; exclusion requested by the companies.

There were economic units excluded in all EACs. In the particular case of the chemicals EACs, that the ones that adhered during its implementation compensated the four excluded companies.

The two largest exclusions of companies in absolute terms, in the cork bark and the lumber EACs, were justified by the fact that the companies were not under the contract framework.

Taking the percent participation ratio for each sector we find that, in the majority of cases the 50% participation was overcome. The sectors with the highest degree of participation are the rubber and tires recapping, footwear, cork bark EACs.

The textile contract seems to present a disappointing result, only 15% of participation, However, this number is justified by the fact that the contract is constrained to the water management objective and, for that reason, most of the companies that initially wanted to participate were excluded (the number of companies initially wanting to participate was of 574, representing 68,41% of the economic units).

The positive participation by companies, expressed in table 2, is reinforced by the extremely elevated compliance rates with the two initial requirements, related to the impact assessment, the elaboration of the intervention plan and the information sharing.

In all cases, the environmental impact assessment and the intervention plan were completed under the assistance of a technical expertise, and, even though there are mixed information in what regards the actual fulfilment of the environmental requirements in terms of pollution restrictions, the obligations to produce an emissions reports related to the different environmental themes were fulfilled.

Internal practices for the assessment of the magnitude and nature of the pollution emissions and the elaboration of reports with the typification of these emissions were created in the large majority of companies. These established practices have important consequences.

Legal requirements are now being fulfilled:

- in the framework of the waste management law the companies are now providing a periodic report characterizing the types of waste that they generate;
- in compliance with the licensing requirements related to the captation of water and water emissions, companies have now environmental information that allows them to choose the correct procedures and to implement the correct pollution reduction technologies.

At the same time, the information that has been created allows private and public entities to have an overall view of the environmental impact of each sector. This information diminishes the information asymmetries between companies and regulator, contributing to a more efficient public intervention, and supports a better co-operation and co-ordination between the public and private environmental policies.

In the subject of the impact assessment and the creation of the environmental plan it is important to notice the participation of the "external technical expertise". In all EACs the industrial associations accepted the obligation to contract the collaboration of a company with expertise knowledge in pollution management. This came to respond to the technical insufficiencies from which the majority of companies in the different sector were experiencing.

The help provided by the expertise company was important in a first moment for the creation of the environmental impact assessment, but is was also crucial in the definition of the intervention plan and supervision of its implementation.

Finally, in the exploitation of the flexible characteristics of the negotiated contracts, the most significative aspect was the personalized implementation plan established for each sector.

This plan, created as a consequence of the assessment reports, was adapted to the needs of each sector, setting up a personalized schedule for the progressive fulfilment to the environmental norms.

In terms of environmental objectives, only the Chemical sector chose to institute partial objectives that accompanied the environmental plan and obliged the companies to fulfil with intermediate pollution requirements. It is, however, significant that, in many cases, the initial evaluation on the sectorial impact verified that some activities were outside the scope of the specific environmental laws.

A final important aspect on the flexibility was the opportunity opened to the co-operation and co-ordination between social, economic and environmental agents.

The regulator was able to implement the environmental norms required by the EU directives, giving financial support to the companies in the framework of the PEDIP measures and compelling them, ant the same time, to use the technical expertise help.

The companies, accepted a more responsible behaviour in relation to the environment because they were, for once, obliged to decide either to participate or not, and to make an economic assessment of their environmental protection activities (with important results, as in the case of the textile EACs where the equity between companies became an issue).

The industrial associations (a participating agent whose role, views and objectives is still not considered in the literature of voluntary approaches), took this opportunity to:

- increase the value of the "club good" that they provide to the companies, assuming the responsibility for the negotiations with the regulator;
- develop an expertise in the environmental field that allows them to offer an extra service to the companies;
- and increase their public visibility and capacity to intervene.

3. CONCLUSION

In conclusion, the Portuguese experience with voluntary approaches through the implementation of EACs was a creative way to combine the necessities and available means in order to achieve public objectives on environmental quality.

This was accomplished in a way that it was flexible enough to guarantee a gradual compliance to the legal norms without putting to much pressure on the private economic agents.

In terms the EU experience in the implementation of the negotiated agreements, the Portuguese case must be considered as a significant one, not only because of the number of contracts that were negotiated, but especially because of the participating sectors and the ratio of participating companies.

As we saw, the most significant sector in terms of contribution to the Portuguese economy chose this instrument to intervene in the "environmental preservation": the textile EAC, footwear EAC and the cork bark EAC.

At the same time, the voluntary approach was also the choice of sectors that have a significant environmental impact, like the three EACs that we have included inside the chemical sector.

The interest for this approach and its continuous use is assured by the new environmental contracts: the Contracts for Continuous Improvement of Environmental Performance (CCIEPs).

These new contracts are being implemented with the objective of supporting the participating companies' in their effort to adopt environmental management systems in accordance to the EMAS regulation. The over compliance features of the CCIEPs and its "performance target" structure make them an innovative approach to the "environmental preservation" policy. The success in its implementation will be important for the future of the public support for the private initiatives in the implementation of environmental management systems.

4. NOTES

1 The Portuguese environmental agency is now the Environmental Institute (*Instituto do Ambiente*). Information regarding the current use of voluntary approaches is available at the internet site: www.iambiente.pt.

2 It is important to notice that, after integrating the EEC in 1986, different industrial activities suffered a regulation boost in consequence of the obligations to adopt the European directives.
In the field of Environmental regulations the Portuguese Government responded to the European demands with the Law for the Environment (*Lei N.° 11/87 de 7 de Abril*) which should be considered as extremely innovative in the framework and the instruments proposed. The major problems came with the creation of the specific regulation to implement that law and the enforcement of the norms.

3 The olive oil environmental contract end up following an isolated development, and for that reason it was placed outside of the EAC framework.

4 Environmental Management and Audit System, Council Regulation (EEC) N° 1836/93 of 29 June 1993 later substituted by the Council Regulation (EC) N° 761/2001 of 7 November 2001.
The EMAS is formally considered a Public Participation schemes by the European Commission.

5 For a survey in the European implementation of voluntary approaches see EEA (1997) and Börkey and Lévêque (1998).

6 The ECAs for vegetable oils includes industrial units dedicated to the manufacture of food products and industrial units related to the production of cosmetic and hygiene products. For this reason, we have divided these CEAs between the chemical and food manufacture activities taking into account the core business of each unit (60% on the industrial units signing the EACs were integrated in the chemical sector and the other 40% in the food processing sector)

7 Following the characterization presented by Börkey and Lévêque (1998, pag.20) the EAC make companies individually liable in opposition to the EAAs were the formal contract was only signed by the Industrial Association that, assuming collective liability, was afterwards responsible for the participation of the individuals companies.

5. REFERENCES

Arora S. and Cason T. (1995), "Why do Firms Over comply with Environmental Regulation? Understanding Participation in EPA's /50 Program", *Resources for the Future*, September, *Discussion Paper* 95-38.

Börkey P. and Lévêque F. (1998), "*Voluntary Approaches for Environmental Protection in the European Union*", ENV/EPOC/GEEI(98)29/FINAL

Carraro, C. and Lévêque, F., (1999), "*Voluntary Approaches in Environmental Policy*", Feem –Kluer Academic Publishers, Dordrecht Hardbound.

Croci E. and Pesaro G. (1998), "*Voluntary agreements in the environmental sector – the Italian experience*", CAVA working paper n.98/11/5 January 1998

Croci E. (2003), "Voluntary Agreements for CO2 Emissions Reduction: Evaluation and Perspectives", *Energy & Environment*, Volume 14 Number 5 September 2003

EEA, European Environment Agency (1997), "*Environmental Agreements - Environmental Effectiveness*", Environmental Issues Series No.3 - Vol. 1 and 2, Copenhagen.

DGA (2000), "*Contratos de Adaptação Ambiental: relatório final de balanço*", internal document of the DGA, Lisboa

INE (1997), "*Estatísticas da Produção Industrial*", INE, Lisboa

INE (1998), "*Estatísticas da Produção Industrial*", INE, Lisboa

Glachant, M. (1994), "Voluntary Agreements between Industry and the Government: Bargaining and Efficiency", *Business Strategy and the Environment*, Vol. 3, Part 2, Summer, pages 43 - 49.

Segerson, K. and Miceli, T. (1997), "Voluntary Approaches to Environmental Protection: the Role of Legislative Threats", *Fondazione Eni Enrico Mattei, Nota di Lavoro* 22.97.

Storey, M. (1996), "*Policies and Measures for Common Action: Demand Side Efficiency: Voluntary Agreements With Industry*", OECD Environmental Department, Working Paper, Experts Group on the UN FCCC, December.

DESIGNING ENERGY CONSERVATION VOLUNTARY AGREEMENTS FOR THE INDUSTRIAL SECTOR IN CHINA: EXPERIENCE FROM A PILOT PROJECT WITH TWO STEEL MILLS IN SHANDONG PROVINCE

L. PRICE, E. WORRELL, J. SINTON

Energy Analysis Department, Environmental Energy Technologies Division, Lawrence Berkeley National Laboratory, Berkeley, California, USA

Abstract. China faces a significant challenge in the years ahead to continue to provide essential materials and products for a rapidly growing economy while addressing pressing environmental concerns. China's industrial sector consumes about 70% of the nation's total energy each year and is heavily dependent on the country's abundant, yet polluting, coal resources. Industrial production locally pollutes the air with emissions of criteria pollutants, uses scarce water and oil resources, emits greenhouse gases contributing to climate change, and produces wastes. Fostering innovative approaches that are tailored to China's emerging market-based political economy to reduce the use of polluting energy resources and to diminish pollution from industrial production is one of the most important challenges facing the nation today. The use of Voluntary Agreements as a policy for increasing energy-efficiency in industry, which has been a popular approach in many industrialized countries since the early 1990s, is being tested for use in China through a pilot project with two steel mills in Shandong Province. The pilot project was developed through international collaboration with experts in China, the Netherlands, and the U.S. Designing the pilot project involved development of approaches for energy-efficiency potential assessments for the steel mills, target-setting to establish the Voluntary Agreement energy-efficiency goals, preparing energy-efficiency plans for implementation of energy-saving technologies and measures, and monitoring and evaluating the project's energy savings.

1. INTRODUCTION

The use of Voluntary Agreements as a policy for increasing energy-efficiency in industry, which has been a popular approach in many industrialized countries since the early 1990s, is being tested for use in China through a pilot project with the steel industry in Shandong Province. China faces a significant challenge in the years ahead to continue to provide essential materials and products for a rapidly growing economy while addressing pressing environmental concerns. China's industrial sector consumes about 70% of the nation's total energy each year and is heavily dependent on the country's abundant, yet polluting, coal resources. Industrial production locally pollutes the air with emissions of criteria pollutants, uses scarce water and oil resources, emits greenhouse gases contributing to climate change, and produces solid wastes. Fostering innovative approaches that are tailored to China's emerging market-based political economy to reduce the use of polluting energy resources and to

E. Croci (ed.), The Handbook of Environmental Voluntary Agreements, 221 - 235

diminish pollution from industrial production is one of the most important challenges facing the nation today.[1]

2. INDUSTRIAL ENERGY USE IN CHINA

Globally, the industrial sector accounts for about 40% of primary energy use. In 1998, developing countries used an estimated 48 EJ for industrial production, over one-third of world total industrial primary energy use (Price et al., 1998). Industrial output and energy use in developing countries is dominated by China, India, and Brazil. China alone accounts for almost 30 EJ, or about 23% of world industrial energy use (National Bureau of Statistics, 1999).

China's industrial sector is extremely energy-intensive and accounts for about 70% of the country's total primary energy use. Industrial energy use in China has grown about 5% per year since 1980 (Sinton et al., 1996; National Bureau of Statistics, 1999). This growth is five times faster than the average growth that took place in the industrial sector worldwide during the same time period.

The industrial sector can be divided into light and heavy industry, reflecting the relative energy-intensity of the manufacturing processes. In China, about 80% of the energy used in the industrial sector is consumed by heavy industry. Of this, the largest energy-consuming industries are building materials, ferrous metals, and chemicals (Sinton et al., 1996).

3. INDUSTRIAL ENERGY EFFICIENCY POLICY IN CHINA: 1980-2000

Energy-efficiency policy in China has evolved greatly since the People's Republic was established in 1949. At that time and for the next few decades, rapid growth in energy supply was the main energy policy in China. Energy prices were subsidized, a central allocation system provided energy primarily to the industrial sector, and little attention was paid to the environment or energy efficiency. As a result, China's rapidly growing energy system was extremely inefficient (Levine, 2000).

In 1980, following a meeting of more than 100 non-governmental energy experts who declared that China's energy policy was in a crisis situation and required radical restructuring, the Chinese government implemented an extensive series of reforms beginning with the Sixth Five-Year Plan which took effect in 1981 (Levine, 2000). The government announced that it would place equal emphasis on development of energy supply and energy conservation in order to ensure an adequate supply of energy, emphasizing energy conservation in the near term (Lu, 1993). Many energy-efficiency policies and programs were developed and implemented by the central government. Most of these programs were directed toward the industrial sector.

Energy management offices, departments, and agencies were established at all levels of government to implement, manage, monitor, and enforce the numerous rules, standards, and programs related to energy conservation. The Office of Energy Conservation Work in the State Council oversaw all of the efforts. Ministries for specific industrial sectors, such as the Metallurgy Ministry, focused on sector-specific issues. The China Energy Conservation Association, the National Supervising Center of Energy Conservation, and the Energy Conservation Testing Technology Service Centers, along with provincial energy conservation agencies, were also established

Energy efficiency and energy conservation management for the industrial sector during this period involved controlling energy intensity and energy supply through the use of quotas. Energy conservation goals were set in the form of physical energy intensity standards for various manufacturing processes. Other standards addressed industrial equipment such as boilers and motors. Success in attainment of the standards was considered when allocating energy supply quotas for industrial enterprises (Sinton et al., 1998; Liu et al., 1994). Other energy management efforts included dissemination of energy-efficient technologies and products, retiring energy-intensive mechanical and electrical devices, restricting energy-wasting production practices, and monitoring enterprise energy conservation.

Low interest loans for energy conservation projects, tax breaks for energy-efficient products, and monetary energy conservation awards for enterprises were all used to encourage investment in energy efficiency. Funding for energy-efficiency investments was provided by the newly established China Energy Conservation Investment Corporation (CECIC). During this period, energy-efficiency funding for capital construction, retrofits, and transformation projects was equivalent to $16.5 billion (1995 US$) (Sinton et al., 1998).

Information on energy use and intensities was gathered through the national resources conservation and comprehensive utilization network and statistics were compiled by the energy statistical reporting system. National, local, and sectoral energy conservation technology service centers were also established. Education and training programs included the establishment of energy conservation training centers. Over 200 energy conservation centers were established during this period to provide energy monitoring and efficiency services, develop and promote energy-saving technologies, and perform feasibility studies (Liu et al., 1994).

An analysis of the energy savings that resulted from these energy-efficiency efforts found that if energy intensity had remained frozen at 1977 levels, then China would have used 80 EJ in 1995, more than twice as high as the actual consumption of 36 EJ that year (Sinton et al., 1998). Decomposition analyses have shown that most of the energy savings during this period were due to reductions in energy intensity, not structural shifts toward less energy-intensive industry (Huang, 1993; Lin, 1992; Palmer, 1992; Sinton, 1996; Sinton and Levine, 1994; Worrell et al., 1997).

In 1993, the Chinese government enacted a number of significant financial reforms, initiating China's transition to a market-based economy. Energy price reforms included deregulation of coal prices, increases in oil prices, and partial deregulation of electricity prices. A simplified tax code introduced in 1994 eliminated tax rate reductions and tax breaks on energy-efficiency technology development and investment projects. Some banks also began to reduce low-interest lending for efficiency projects.

In 1997, the Chinese government passed the Energy Conservation Law (ECL) which provides broad guidance for the establishment of energy-efficiency policies in China. Article 20 of the ECL requires substantial improvement in industrial energy efficiency in 7200 key energy-consuming industrial facilities in China. This portion of the Law states that "the State will enhance energy conservation management in key energy-consuming entities."

A number of provincial administrations have formulated implementing regulations in accordance with the ECL: Shandong, Shanghai, Beijing, Zhejiang, Jiangsu, Shanxi,

Gansu, Sichuan, Yunnan, and Hubei. Although a review of the Shandong, Zhejiang, and Shanghai implementing regulations characterized them as vague (Wang, 1999), they are still an important step toward providing provincial governments with the tools required to implement energy conservation programs within their jurisdictions.

During this period, energy quotas were eliminated and monitoring of energy intensity levels declined. In 1998, most industrial ministries were demoted to the bureau level and placed under the authority of the State Economic and Trade Commission (SETC). Industrial bureaus were merged into a single Industrial Management Department within SETC in 2000. Quality of statistical collection diminished as state control over enterprises weakened.

In 1999, SETC issued a catalogue of "Outdated Technology Processes and Products" initiating an effort to phase out non-competitive processes or products that consume too much energy or are polluting. The two volumes of this catalogue address 11 industrial sectors (China Environmental Review, 2000). SETC also mandated closure of some inefficient petrochemical plants as well as hundreds of small cement and glass plants, mainly in northern China, small refineries, coal mines, sugar mills, and paper mills for financial, energy efficiency and environmental reasons. This campaign was extended in 2000 to include over 200 small iron and steel plants (China Daily, 2000a and 2000b; Nengyuan, 2000), and similar initiatives continue today.

The 10th Five-Year Plan, which became effective in March 2001, includes a renewed focus on energy end-use efficiency and productivity improvement, development of supporting regulations for the ECL at the local and sectoral levels, formulation of annual energy conservation plans to improve energy utilization efficiency and productivity, formulation of preferential economic policies to support energy conservation demonstration and dissemination projects, enhanced energy management of key energy-using enterprises, and harnessing of grass-roots social forces to save energy.

In 1999, the U.S.-based Energy Foundation, through its China Sustainable Energy Project, funded the China Energy Conservation Association (CECA), under the State Economic and Trade Commission (SETC), and Lawrence Berkeley National Laboratory (LBNL) to began a project with the goals of: 1) developing implementing regulations and relevant standards for the ECL to promote industrial energy conservation and improvement of energy efficiency and 2) promoting the new planning in energy-intensive sectors for energy efficiency to reduce energy consumption of key enterprises (CECA, 2000). CECA, along with a Policy Review Team made up of Chinese energy analysts, experts from Tsinghua University, as well as former staff of China's Ministry of Metallurgy, worked with LBNL and international experts to determine the best approach for reaching these goals. These efforts include analyzing international industrial energy efficiency policies and programs and their adaptability to China, analyzing the status and opportunities for energy conservation in key energy-intensive industrial sectors, reviewing existing energy conservation regulations and policies, and making recommendations for new regulations and policies that work well under a "market-based" economy. The result of these efforts was to recommend further evaluation of the use of Voluntary Agreements for increasing energy efficiency in the industrial sector.

4. INTERNATIONAL EXPERIENCE WITH VOLUNTARY AGREEMENTS

Voluntary Agreements to meet specific energy-use or energy-efficiency targets are used in the industrial sector in many countries around the world (Bertoldi, 1999; Chidiak, 1999; Hansen and Larsen, 1999; Mazurek and Lehman, 1999; Newman, 1998; Paton, 2002). Such agreements can be viewed as a tool for developing a long-term strategic plan for increasing industrial energy efficiency that fully engages not only the engineers and management at industrial facilities, but also includes government, industry associations, financial institutions, and others. An agreement or target can be formulated in various ways. Two common methods are those based on specified energy-efficiency (or energy intensity) improvement targets and those based on absolute energy use or greenhouse gas emissions reduction commitments. Either an individual company or an industrial subsector, as represented by a party such as an industry association, can enter into such agreements.

Voluntary Agreements typically have a long-term outlook, covering a period of five to ten years. The agreements focus the attention of all actors on energy efficiency or greenhouse gas emissions reduction goals. The key elements of Voluntary Agreement programs are the assessment of energy-efficiency potential of the participants as well as target-setting through a negotiated process with all parties. Supporting programs and policies, such as audits, assessments, benchmarking, monitoring, information dissemination, and financial incentives, all play roles in assisting the participants in meeting the target goals.

In its review of 350 voluntary actions and programs, the International Energy Agency found that "past and present experiences with voluntary actions show that, properly designed and implemented, they can achieve stated objectives, sometimes even exceeding those of minimum regulatory standards, and help integrate economic and environmental goals" (IEA, 1997). Another analysis of seven Voluntary Agreement programs found that the programs could be credited with about 50% of the observed energy-efficiency improvement or emissions reductions. In addition to these so-called direct effects of the programs, there are also important medium- and long-term impacts including: changing attitudes and awareness of managerial and technical staff regarding energy efficiency; addressing market, institutional, and regulatory barriers to technology adoption and innovation; fostering market transformation to establish greater potential for sustainable energy-efficiency investments; promoting positive dynamic interactions between different actors involved in technology research and development, deployment, and market development; and facilitating cooperative arrangements that provide learning mechanisms within a sector or industry to combine knowledge and develop new competencies in industry (Dowd et al., 2001; Delmas and Terlaak, 2000).

Based on experience to date, the "Seven Golden Rules" for these type of agreements are: 1) make sure they are negotiated agreements based on assessments of energy efficiency potentials that are more than "business-as-usual", 2) set clear, well-defined targets and specific timetables for achieving those targets, 3) ensure long-lasting government support in the form of policies and programs that assist industries in implementing energy-efficiency improvements, 4) focus on large, energy-intensive industries to start with because this is where the greatest savings are found, 5) establish clear monitoring guidelines, 6) evaluate progress using physical energy intensity measurements, and 7) provide for independent verification of progress (Blok, 2000).

5. ENERGY CONSERVATION VOLUNTARY AGREEMENT PILOT PROJECT IN SHANDONG PROVINCE

The Energy Conservation Voluntary Agreement Pilot Project with two iron and steel enterprises in Shandong Province was modeled after successful international industrial Voluntary Agreement programs, taking China-specific conditions into consideration (Price et al., 2003a). The main participants in the pilot project are two iron and steel enterprises in Shandong Province – Jinan Iron and Steel (Jigang) and Laiwu Iron and Steel (Laigang), the Shandong Economic and Trade Commission (ETC), the State Economic and Trade Commission (SETC, now NDRC), and the China Energy Conservation Association (CECA).[2]

The two steel mills participating in this project have both invested in energy-efficient technologies in the past, but are different in terms of scale and management approach. The Jigang steel mill was built in 1958 and currently produces about 3.0 million tonnes of steel. This plant has four coking furnaces with coke dry quenching, six blast furnaces, two basic oxygen furnaces, one electric arc furnace, and slab and billet continuous casting machines. Jigang is located outside of Jinan, the capital city of Shandong Province, and benefits from a close relationship with the provincial government. The Laigang steel mill was built in 1970 and produces about 2.5 million tonnes of steel per year. The blast furnace, electric arc furnace and basic oxygen converter are all small scale. The product finishing lines are very modern and use state-of-the-art technology. Laigang uses a modern management style and they have been able to attract non-governmental capital from the Asian Development Bank and the Shandong's World Bank-sponsored Energy Management Center to make investments in energy-conservation equipment.

For the Voluntary Agreements, the two iron and steel enterprises are responsible for assessing the current energy-efficiency potential of their enterprises, developing energy-efficiency targets and energy conservation plans, and implementing these plans in order to achieve the agreed-upon targets. The energy-efficiency sector targets of the Energy Conservation Voluntary pilot policy program are based on physical energy intensity metrics.

Using international Voluntary Agreement schemes as a model, the targets were set through a process in which the government and enterprises negotiate the target level based on detailed evaluations of the potential for energy-efficiency improvement in a given industrial sector. Article 4 of China's ECL provides general guidance for establishment of such a program, stating that "the State Council and the governments of provinces, autonomous regions and municipalities directly under the central government should: strengthen their efforts in energy conservation; restructure industry, enterprises, products, and energy consumption patterns; promote technological progress for energy conservation; reduce energy consumption per unit of economic output and energy consumption per physical unit of product;...and encourage the national economy to develop in an energy-efficient manner" (PRC, 1997).

SETC and the Shandong ETC fulfilled the government role in the pilot project and determined which supporting programs to include in the pilot to assist the enterprises in reaching their energy-efficiency targets. An expert Technical Team assisted in evaluating the enterprises' targets and will evaluate the project progress annually.

6. METHODOLOGY FOR ASSESSMENT OF ENERGY-EFFICIENCY IMPROVEMENT POTENTIAL

Assessment of the energy-efficiency improvement potential for each participating enterprise is an essential element in the design of an Energy Conservation Voluntary Agreement Pilot Program because it provides all parties to the Voluntary Agreement with the same information regarding the current energy consumption at the enterprise as well as the options available to reduce energy consumption. This information was needed for negotiating an ambitious, yet realistic energy conservation target. The energy-efficiency assessment was an essential first step for development of energy-efficiency targets and detailed Energy Conservation Plans to document the actions to be taken to reach the Voluntary Agreement targets.

LBNL developed a *Methodology for Assessment of Enterprise Energy-Efficiency Potential* for the steel industry Energy Conservation Voluntary Agreement Pilot Project that incorporates key elements of the various methods used in other countries and in China to determine the energy-efficiency potential of an enterprise. LBNL also developed a user-friendly computer spreadsheet tool to assist the pilot enterprises in implementing this methodology. The energy-efficiency assessment methodology involves determination of the enterprise's physical energy consumption based on total energy consumption for production of iron and steel at that enterprise (subtracting the offsite energy and energy used for non-production purposes).[3] This "total production energy intensity" is calculated for each major process step at the enterprise.

Once the total production energy intensity was calculated, the technical and achievable energy-efficiency potentials for each enterprise were determined. The technical energy-efficiency potential was calculated by comparing the total production energy intensity for each pilot enterprise with benchmark energy intensities that represent internationally available state-of-the-art iron and steel production processes. The achievable energy-efficiency potential was determined by identification of inefficient processes within each enterprise and identification of technologies and measures that could be implemented to improve the energy efficiency of the enterprise, based on the feasibility and cost-effectiveness of implementing technologies. The potential energy intensity reductions associated with implementation of these technologies and measures were estimated to determine the achievable energy-efficiency potential, which was in turn used to set the Energy Conservation Voluntary Agreement Pilot Project targets.

The energy-consuming processes of iron and steel enterprises can be analyzed based on the energy used by fuel type for each process step in the production of iron and steel. The information required includes all energy inputs, recovered energy and energy used for self-generation, as well as the data needed to calculate the process-step energy intensity of the enterprise.

In order to determine the technical energy-efficiency potential for the pilot iron and steel enterprises, the enterprise process-step total production energy intensity must be compared to the process-step energy intensity of a benchmark "state-of-the-art" iron and steel enterprise. Such benchmarks can be constructed using either a hypothetical energy-efficient steel plant[4] or benchmarking to an actual energy-efficient steel plant.[5] For the Energy Conservation Voluntary Agreement Pilot Project an energy-efficiency assessment spreadsheet tool was developed that provides benchmark energy-efficiency values for each major steelmaking process step.[6]

Once the actual energy intensity and benchmark energy intensity have been calculated for each enterprise, they can be used to construct an Energy Efficiency Index (EEI). The EEI is a measurement of the total production energy intensity of an enterprise compared to a benchmark energy intensity. For the Energy Conservation Voluntary Agreements, the EEI is used to calculate enterprise energy-efficiency potential and it is used for evaluating enterprise progress toward the chosen energy-intensity target.

The EEI can be used to calculate enterprise energy-efficiency potential by comparing actual enterprise energy intensity to the energy intensity that would result if the enterprise used "state-of-the-art" technology for each process step. The difference between the actual energy intensity, which is the energy use per ton of product produced, and that of the reference or benchmark technology, is calculated for each of the key process steps of the enterprise and then aggregated for the entire enterprise. The aggregated EEI is calculated as follows:

$$EEI = 100 * \frac{\sum_{i=1}^{n} P_i \cdot EI_i}{\sum_{i=1}^{n} P_i \cdot EI_{i,B}} = 100 * \frac{E_{tot}}{\sum_{i=1}^{n} P_i \cdot EI_{i,B}} \tag{1}$$

Where:

EEI	= energy efficiency index
n	= number of process steps to be aggregated
EI_i	= actual energy intensity (EI) of process step i
$EI_{i,B}$	= benchmark energy intensity (EI) of process step i
P_i	= production quantity for process step i
E_{tot}	= total actual energy consumption for all process steps

The EEI provides an indication of how the actual total production energy intensity of the enterprise compares to the benchmark energy intensity. By definition, a plant that uses the benchmark technology will have an EEI of 100. In practice, all plants will have an EEI greater than 100. The gap between actual enterprise energy intensity at each process step and the reference level energy consumption can be viewed as the technical energy-efficiency potential of the plant. The EEI is an initial screening tool that helps to identify which processes are most efficient and which are most inefficient compared to state-of-the-art conditions and which are most likely to have a substantial potential for energy-efficiency improvement.

The information developed using the *Methodology for Assessment of Enterprise Energy-Efficiency Potential*, including the enterprise total production energy intensity by process step, EEI, technical energy-efficiency potential by process step, and achievable energy-efficiency potential by process step are all fundamental for determining the enterprise Voluntary Agreement target.

7. DEVELOPMENT OF SUPPORTING POLICIES

While the enterprises were evaluating their energy-efficiency potential, the government entities developed policies to offer in support of the Voluntary Agreement. Supporting policies are the key motivational element to encourage enterprises to participate fully in the Voluntary Agreement program. Supporting programs and policies, such as government facilitation of the Voluntary Agreement negotiation and implementation process (including provision of technical assistance and information dissemination programs), enterprise audits and assessments, financial assistance and incentives, and government and public recognition all play an important role in assisting the participants in meeting the target goals. Supporting policies also include elimination or reduction of taxes or environmental regulations for participants.

Existing Voluntary Agreement programs use a variety of supporting policies to motivate and assist industry in reaching its energy efficiency or greenhouse gas emission reduction goals. Table 1 provides an overview of the supporting policies and measures in several Voluntary Agreement programs.

Table 1. *Overview of Supporting Policies and Measures in Selected Voluntary Agreement Programs*

		Supporting Policies and Measures				
Country	*VA Scheme*	*Government Facilitation of VA Process*	*Audits and Assessments*	*Financial Assistance and Incentives*	*Government and Public Recognition*	*Exemption from Regulation and Taxes*
Australia	Greenhouse Challenge	X			X	
Canada	Canadian Industry Program for Energy Conservation	X			X	
Denmark	Agreements on Industrial Energy Efficiency	X	X	X		X
Netherlands	Long Term Agreements	X	X	X	X	X
Sweden	EKO-Energi	X	X		X	
UK	Make a Corporate Commitment, Climate Change Agreements	X			X	X

8. TARGET-SETTING

After the assessment of enterprise energy-efficiency potential was completed and supporting policies were established, all parties to the Voluntary Agreement turned to target-setting. Target-setting is an essential element of Voluntary Agreements. Targets provide all parties to the agreement with a quantitative goal to be reached within the period of the Voluntary Agreement. An important precondition for realistic yet ambitious target-setting is that all parties have the same information regarding the enterprise energy-efficiency potential as well as the governmental supporting policies to assist the enterprise in implementing energy-efficiency technologies and measures.

Using information developed through the assessment of enterprise energy-efficiency improvement potential, as well as information on historical and planned energy intensity reductions at each plant, CECA and the Technical Team worked with representatives at the enterprises and the local government to set achievable yet challenging targets for energy-efficiency improvement within the Pilot Project. During the assessment of enterprise energy-efficiency improvement potential, the calculations of current total production energy consumption and energy intensity by process were made for each of the pilot enterprises. The potential energy intensity in 2005 and was also calculated for both a "business-as-usual" and a "with Voluntary Agreement" scenario. These values, combined with information on historical and planned energy-intensity reductions at each enterprise, were used by all parties to the Voluntary Agreement to determine the targets for the Energy Conservation Voluntary Agreement Pilot Project.

9. SHANDONG PROVINCE VOLUNTARY AGREEMENT WITH LAIGANG AND JIGANG STEEL MILLS

The Voluntary Agreement between Shandong ETC and Laigang and Jigang steel enterprises was signed in April, 2003. At the signing ceremony, both national and provincial decision makers highly praised the sector target Voluntary Agreement approach and stated that they would like to widen implementation within Shandong Province and throughout China in the future.

The signed Voluntary Agreement outlines the targets for both steel enterprises. The parties to the Voluntary Agreement decided to use an energy-intensity target based on two standard Chinese metrics instead of the EEI for this pilot.[7] In addition, the Energy Conservation Rate (ECR), based on comparable energy intensity, will be used to measure the change in intensity over time. In keeping with the national Five-Year Plan schedule, the base year for the targets will be 2000. The parties to the Voluntary Agreement chose a target for 2005 and expect to choose a further target for 2010 in 2005.

Supporting policies for the Energy Conservation Voluntary Agreement Pilot Project include incentive policies provided by the Shandong provincial government as well as the NDRC. The policies to be provided by Shandong ETC are:

- Give priority consideration to the two pilot enterprises under existing preferential policies.
- Coordinate the provision of guarantees for loans and other financial activities required for energy-efficiency projects at the pilot enterprises.

- Use various media to publicize the energy-conservation achievements and contributions of the pilot enterprises.
- Organize intermediary organizations to provide the pilot enterprises with policy, technical, management, and other advice and services.
- Upon evaluation, exempt the pilot enterprises from monitoring of the status of energy utilization.

In addition, NDRC will be requested to provide the following supporting policies:

- For energy-conservation benefits realized through energy-conservation projects, and in accordance with resources comprehensive-utilization policy, investigate and propose recommendations for preferential policies to encourage energy conservation.
- Give priority support to projects undertaken by the pilot enterprises that fulfill the criteria set by national preferential policies.
- Grant a portion of research and development costs for projects undertaken by the pilot enterprises that have significant results in energy and resource conservation and comprehensive utilization, short payback times, and outstanding economic and social benefits, to support enterprises to carry out energy-conservation research and development.
- Give priority to the pilot enterprises when bringing in foreign investment capital.
- Award pilot enterprises the honorable title of "China Energy-Efficiency Voluntary Agreement Pilot Enterprise".

10. LESSONS LEARNED

Preliminary lessons learned from the development of the Shandong Province Enterprise Energy-Efficiency Voluntary Agreement Pilot are that while the general concepts of negotiated agreements and of the value of energy-efficiency improvement in industry were easily comprehended and accepted by the Chinese involved in this project, more specific components of the successful Voluntary Agreements from around the world were not immediately understood or ultimately adopted.

Historically, Chinese industry has operated under annual quotas for energy consumption that were accompanied by fines and penalties if exceeded. Voluntary Agreements, on the other hand, move away from this concept to a focus on long-range planning where annual energy-efficiency progress may fluctuate but ultimately the targets are met in the long run. In the Shandong Province pilot, however, this concept is missing; the pilot plan outlines annual targets for 2003 and 2005, accompanied with the threat of fines for exceeding the quotas. The pilot plan does not include targets for 2010 but rather allows for the development of further long-term pilots of 5 and 10 years duration after completion of this pilot. International advisors also advocated the use of a single, comprehensive energy-efficiency measurement system (the EEI). The Shandong Province pilot, however, uses two standard Chinese energy intensity metrics, neither of which adequately measure actual energy intensity trends at the enterprises.[8] In addition, the pilot plan requests each enterprise to provide twelve additional measurements of energy-efficiency annually. Finally, while some of the supporting policies were established prior to the development of the targets, other supporting policies were not set but rather were advocated as possible policies that the central government could provide. This gives the enterprises no certainty regarding these polices and leads to a situation where relatively weak targets were set for 2005.

On the positive side, in addition to understanding the general concept of Voluntary Agreements and including all of the essential elements in the pilot project design, the Shandong pilot project has extended the energy conservation Voluntary Agreement into the area of environmental pollutants, requesting the enterprises to provide data on their sulfur dioxide and carbon dioxide emissions annually. This is an important link to be made in China, where the drivers for adoption of energy-efficient technologies include the reduction of local, regional, and global pollutants as well as reduction of energy consumption. Also, introducing a new policy mechanism can be a long process when dealing with multiple levels of government as well as members of industry that are unclear on the benefits they will accrue from the policy. The educational and motivational aspects of such an effort cannot be understated. The Shandong Voluntary Agreement pilot has provided the basis for the development of 15 Voluntary Agreements between a diverse array of enterprises and the municipality of Tsingdao (Qingdao) (Qingdao News Net, 2003) and the use of Voluntary Agreements with 8 pilot enterprises in the *Energy Conservation and GHG Emissions Reduction in Chinese Township and Village Enterprise* project of the United Nations Development Programme, the Global Environment Facility, United Nations Industrial Development Organization, and the Chinese Ministry of Agriculture (UNIDO, 2004). Finally, the mechanism of Voluntary Agreements is included as an industrial policy initiative in the upcoming, multi-year China End-Use Energy Efficiency Program of the NDRC, United Nations Development Programme, and the Global Environment Facility (SETC/UNDP/GEF, 2002).

11. ACKNOWLEDGMENTS

This project has been conducted by the State Economic and Trade Commission (SETC), now National Development and Reform Commission (NDRC). This project was sponsored by the China Sustainable Energy Program of the Energy Foundation. Significant technical support for this project was provided by Lawrence Berkeley National Laboratory. Support for Lawrence Berkeley National Laboratory was provided by the China Sustainable Energy Program of the Energy Foundation under contract number DE-AC03-76SF00098. The authors also acknowledge the invaluable assistance of Zhang Ruiying of the Energy Foundation, Jiang Yun of the China Energy Conservation Association, Wil Nuijen of Novem in the Netherlands, as well as Kornelis Blok and Dian Phylipsen of Ecofys in the Netherlands throughout this project. Finally, the authors would like to thank the members of the Chinese Policy Review Team for its guidance during this project.

12. NOTES

[1] This paper contains excerpts from Price et al., 2003a; Price et al., 2003b; Price et al., 2001.

[2] The SETC was recently disbanded and many of its functions, including oversight of the Shandong Province Voluntary Agreement pilot, were transferred to the new National Development and Reform Commission (NDRC).

[3] Offsite energy use includes energy used to heat homes and buildings that are not directly part of the iron and steel plant, as well as energy used for transport outside the plant. Internal transport energy use is included (Price et al., 2002).

[4] Data for a construction of a hypothetical energy-efficient steel plant are available from the International Iron and Steel Institute in IISI, 1998. This document provides data for both a hypothetical "All-Tech" plant that includes technologies that may not be currently economical but lead to significant energy savings and a hypothetical "Eco-Tech" plant that is based on the use of technologies and measures that are considered economical. These values can be used to construct a benchmark "All-Tech" or "Eco-Tech" comparable energy intensity. The difference between this benchmark value and the total production energy intensity values for each pilot enterprise could be considered to represent the technical energy-efficiency potential.
[5] Another source of data for a "state-of-the-art" benchmark are values for an actual energy-efficient steel enterprise, such as the Shanghai Baosteel plant. Data from this plant or other world-class energy-efficient steel enterprises could be used to calculate a "state-of-the-art" benchmark comparable energy intensity.
[6] The simple computer spreadsheet tool that has been developed for use in the Energy Conservation Voluntary Agreement Pilot Project is based on the IISI "Eco-Tech" plant (IISI, 1998).
[7] The two Chinese intensity metrics are comparable (*kebi*) and comprehensive (*zonghe*) energy intensity. The comparable energy intensity normalizes production relative to the ratio of iron to steel produced in order to provide a metric to compare steel plants within China and to plants in other countries. The comprehensive energy intensity metric includes all plant energy use, including uses not directly linked to the production of steel such as for employee homes and schools, as well as other on-site facilities.
[8] The enterprises and Shandong ETC are currently re-visiting the possibility of using the EEI for monitoring progress.

13. REFERENCES

Bertoldi P. (1999), "The Use of Long-Term Agreements to Improve Energy Efficiency in the Industrial Sector: Overview of the European Experiences and Proposal for a Common Framework," *Proceedings of the 1999 American Council for an Energy-Efficient Economy Summer Study on Energy Efficiency in Industry*. Washington, DC: ACEEE.

Blok K. (2000), "Experiences with Long Term Agreements on Energy-efficiency Improvements in the European Union," Presentation at the *Workshop on Learning from International Best Practice Energy Policies in the Industrial Sector*, May 22-23, 2000, Beijing.

Chidiak M. (1999), "Voluntary Agreements for Energy Efficiency in Five EU Countries," in *Energy Efficiency and CO_2* Reduction*: The Dimensions of Social Change: 1999 European Council for an Energy-Efficient Economy Summer Study*, May 31-June 4, Mandelieu, France.

China Daily (2000a), "China to Close Down 50 Small Steelworks." 29 June. http://www.chinadaily.com.cn.

China Daily (2000b), "Polluting Firms to be Closed Down." 9 May. http://www.chinadaily.com.cn.

China Energy Conservation Association (2000), "China Sustainable Energy Program: Progresses on the 'Developing Chinese Regulatory Infrastructure' Project," document prepared for the Meeting of the Management Team and Policy Research Team, November 13-15, 2000.

China Environmental Review (2000), *Legislation on Redundant Technology in China*, Summer 2000.

Delmas M. and Terlaak A. (2000), "Voluntary Agreements for the Environment: Innovation and Transaction Costs," CAVA Working Paper 00/02/13, February.

Dowd J., Friedman K, and Boyd G.. (2001) "How Well Do Voluntary Agreements and Programs Perform At Improving Industrial Energy Efficiency," *Proceedings of the 2001 ACEEE Summer Study on Energy Efficiency in Industry*. Washington, DC: American Council for an Energy-Efficient Economy.

Hansen K. and Larsen A. (1999), "Voluntary Agreements in Industry: A Comparative Description of the Process and a Normative Analysis," *Proceedings of the 1999 American Council for an Energy-Efficient Economy Summer Study on Energy Efficiency in Industry*. Washington, DC: ACEEE.

Huang J-P. (1993), "Industry Energy Use and Structural Change: A Case Study of the People's Republic of China," *Energy Economics* 15(2): 131-136.

International Energy Agency (1997), *Voluntary Actions for Energy-Related CO_2 Abatement*. Paris: OECD/IEA.

International Iron and Steel Institute (1998), *Energy Use in the Steel Industry,* Brussels: IISI.

Levine M.D. (2000), "Energy Efficiency Laws and Policies in China," Presentation to the *Earth Technologies Forum*, Washington DC, October 31, 2000.

Lin X. (1992), "Declining Energy Intensity in China's Industrial Sector," *The Journal of Energy and Development*, 16(2): 195-218.

Liu Z.P., Sinton J.E., Yang F.Q., Levine M.D. and Ting M.K. (1994), *Industrial Sector Energy Conservation Programs in the People's Republic of China during the Seventh Five-Year Plan (1986-1990)*. Berkeley, CA: Lawrence Berkeley National Laboratory (LBL-36395).

Lu Y. (1993), *Fueling One Billion: An Insider's Story of Chinese Energy Policy Development*. Washington, DC: Washington Institute Press.

Mazurek J. and Lehman B. (1999), "Monitoring and Verification of Long-Term Voluntary Approaches in the Industrial Sector: An Initial Survey," *Proceedings of the 1999 American Council for an Energy-Efficient Economy Summer Study on Energy Efficiency in Industry*. Washington, DC: ACEEE.

Michels K. (2000), *ICARUS-4: Sector Study for the Iron and Steel Industry. Revision 2*. Utrecht, The Netherlands: Utrecht University, Department of Science, Technology and Society, NWS-E-2000-10

National Bureau of Statistics (1999), *China Statistical Yearbook 1999*. Beijing: China Statistics Press.

Nengyuan (2000), Goals for the Reduction in Production and Mine Closures for the Coal Industry in 2000. Nengyuan 2000(3):14. March.

Newman J. (1998), "Evaluation of Energy-Related Voluntary Agreements," in Martin et al., (eds.) *Industrial Energy Efficiency Policies: Understanding Success and Failure: Proceedings of a Workshop Organized by the International Network for Energy Demand Analysis in the Industrial Sector*. Utrecht, The Netherlands, June 11-12, 1998 (LBNL-42368).

Palmer W. (1992), E*lectric Power Shortages and the Chinese Economy*. PhD dissertation. Madison: University of Wisconsin, Department of Economics.

Paton B. (2002), "Voluntary Environmental Initiatives and Sustainable Industry," in ten Brink, P., ed., 2002. *Voluntary Environmental Agreements: Process, Practice and Future Use*. Sheffield, UK: Greenleaf Publishing Ltd.

People's Republic of China (1997), *The Law on Energy Conservation of the People's Republic of China*. Approved at the 28th Session of the Standing Committee of the Eighth National People's Congress.

Price L., Michaelis L., Worrell E. and Khrushch M. (1998), "Sectoral Trends and Driving Forces of Global Energy Use and Greenhouse Gas Emissions," *Mitigation and Adaptation Strategies for Global Change* **3**: 263-319.

Price L., Worrell E., Sinton J. and Jiang Y. (2001), "Industrial Energy Efficiency Policy in China," *Proceedings of the 2001 American Council for an Energy-Efficient Economy Summer Study on Energy Efficiency in Industry*, Washington, DC: ACEEE (LBNL-50452).

Price L,. Sinton J., Worrell E., Phylipsen D., Hu X. and Li J. (2002), "Energy Use and Carbon Dioxide Emissions from Steel Production in China," *Energy* 27: 429-446.

Price L., Jiang Y., Worrell E., Du W. and Sinton J.E (2003a). *Development of an Energy Conservation Voluntary Agreement Pilot Project in the Steel Sector in Shandong Province: Report to the State Economic and Trade Commission, People's Republic of China*. Berkeley, CA: Lawrence Berkeley National Laboratory (LBNL-51608).

Price L., Worrell E., Sinton J. and Jiang Y. (2003b), "Voluntary Agreements for Increasing Energy Efficiency in Industry: Case Study of a Pilot Project with the Steel Industry in Shandong Province," *Proceedings of the 2003 American Council for an Energy-Efficient Economy Summer Study on Energy Efficiency in Industry*, Washington, DC: ACEEE (LBNL-52715).

Qingdao News Net (2003), "15 Enterprises Sign Voluntary Energy Agreements," *Qingdao Evening News*, November 6th, 2003.

Sinton J.E. and Levine M.D. (1994), "Changing Energy Intensity in Chinese Industry: The Relative Importance of Structural Shift and Intensity Change," *Energy Policy* 22(3): 239-258.

Sinton J.E. (1996), *Energy Efficiency in Chinese Industry: Positive and Negative Influences of Economic System Reforms*. PhD dissertation. Berkeley, CA: University of California, Energy and Resources Group.

Sinton J.E., Fridley D.G., Levine M.D., Yang F., Zhenping J., Xing Z., Kejun J., Xiaofeng L. (1996), *China Energy Databook*. Berkeley, CA: Lawrence Berkeley National Laboratory (LBL-32822 Rev. 4).

Sinton J.E., Levine M.D. and Wang Q. (1998), "Energy Efficiency in China: Accomplishments and Challenges," *Energy Policy* **26**(11):813-829.

Sinton J., Levine M.D., Fridley D., Yang F. and Lin J. (1999), *Status Report on Energy Efficiency Policy and Programs in China*. Berkeley, CA: Lawrence Berkeley National Laboratory (LBNL-44605).

State Economic and Trade Commission/United Nations Development Programme/Global Environmental Facility (2002), *China's End-use Energy Efficiency Program (EUEEP): Project Brief.* Beijing.

United Nations Industrial Development Organization (2004),. *Energy Conservation and GHG Emissions Reduction in Chinese TVEs - Phase II.* http://www.unido.org/en/doc/4536

Wang A.L., (1999), *A Comparative Analysis of the 1997 Energy Conservation Law of China and the Implementing Regulations of Shandong, Zhejiang, and Shanghai*. NY: Natural Resources Defense Council.

Worrell E., Price L., Martin N., Farla J. and Schaeffer R. (1997), "Energy Intensity in the Iron and Steel Industry: A Comparison of Physical and Economic Indicators," *Energy Policy* 25(7-9): 727-744.

Worrell E., Martin N. and Price L. (1999), *Energy Efficiency and Carbon Dioxide Emissions Reduction Opportunities in the U.S. Iron and Steel Sector*. Berkeley, CA: Lawrence Berkeley National Laboratory (LBNL-41724).

CHAPTER 5

EVALUATION OF ENVIRONMENTAL VOLUNTARY AGREEMENTS

ON THE ASSESMENT OF ENVIRONMENTAL VOLUNTARY AGREEMENTS IN EUROPE

Lessons to be learned from a comparative case study analysis

M. DE CLERCQ, R. BRACKE

Centre for Environmental Economics and Management, Ghent University, Belgium

Abstract. The aim of this study is to gain insight on the factors leading to success or failure of environmental voluntary agreements. To do this we relied on a comparative case study covering twelve voluntary agreements from six different European countries. First, a general evaluation framework for assessing the performance of environmental voluntary agreements is presented. This framework takes into account three different evaluation dimensions: application, impact and resource development. Second, we focus on the factors explaining the level of performance. Four external preconditions for success were identified: the general policy style, the readiness to use severe alternative instruments in case of non-compliance with the agreement, the potential of the sector to negotiate and act as one collective actor and the potential for market success triggered of by the implementation of the agreement. Next to these external factors related to the institutional-economic context wherein a negotiated agreement is used, the specification of an agreement is considered to be an internal factor influencing the performance. The comparative case study shows that taken individually each of the factors is not as such a necessary condition for the success of an environmental voluntary agreement. Rather it is the combination of these success factors that is ultimately decisive for the performance of an agreement.

1. INTRODUCTION

Interest in the use of voluntary approaches as an alternative to regulatory and economic instruments has grown rapidly in the European Union since the publication of the Fifth Environmental Action Programme in 1992, which advocated broadening the range of environmental policy instruments. Within the broad range of voluntary approaches, particular attention has focused on the use of environmental voluntary agreements. In 1996 the Commission produced a Communication on the use of such agreements, which included a number of general guidelines that were intended to ensure their effectiveness, credibility and transparency. Despite the enthusiasm expressed for this new policy instrument, little attention has been paid to the evaluation of environmental voluntary agreements, either in terms of developing a coherent evaluation framework, or in terms of performing ex post analysis of actual agreements.

In 1999, the OECD made an overview of available information on the assessment of voluntary approaches. It states that the evaluation of voluntary approaches, and particularly of environmental voluntary agreements, is hindered by

E. Croci (ed.), The Handbook of Environmental Voluntary Agreements, 239 - 260.

the newness of the approaches and by the fact that practitioners created them. The latter affects the availability of theoretical analysis on their performances, the former constraints empirical investigation.

Nevertheless, the existing literature on the evaluation of environmental voluntary agreements can be divided into two groups. First of all, some authors focus on the development of a theoretical evaluation framework that allows identifying factors that influence the performance of voluntary agreements (see for example EEA 1997; Segerson and Miceli 1998; OECD 1999; Burritt 2002a and Cabugueira 2002). The common feature and also the merits of these works are that they indicate a number of aspects that must be taken into account when assessing the performance of environmental voluntary agreements. On the other hand there exists a rather limited number of ex post case studies on the assessment of one or a few agreements (see for example Klok 1989; Storey et al. 1997; Lehmenn 2000; Immerzeel-Brand 2002 and Burritt 2002b). Only Klok (1989), who discusses eight voluntary agreements in the Netherlands, and Storey et al. (1997), who analyse five agreements to reduce greenhouse gas emissions, have taken up the challenge to assess in a systematic way the performance of a selected sample of environmental voluntary agreements. Although the aim was a systematic evaluation, a great part of their analysis sticks to the descriptive phase. Elements defined important for the performance of environmental voluntary agreements are discussed for each agreement individually, but an in dept comparison of the different cases is lacking. Storey et al. (1997, p.19) conclude that 'there is a lack of clear and established methodologies for evaluating the performance of environmental voluntary agreements'.

This paper is the result of the European Commission sponsored NEAPOL project, which stands for 'Negotiated Environmental Agreements: Policy Lessons to be Learned from a Comparative Case Study'. Negotiated agreements were defined in this research project as: 'agreements between public (national, federal or regional) authorities and industry, wherein both parties commit themselves to realise the environmental goals stated in the negotiated agreement'. The aim of the project was to introduce a general evaluation framework that tries to answer the following research question: Which specific characteristics of environmental voluntary agreements and which factors within the institutional-economic context within which an agreement is used, influence the performance of environmental voluntary agreements? This knowledge can be very useful to policy makers in two different stages of the policy cycle. First, in the phase of the instrument choice, an assessment of the appropriateness of the institutional-economic context could be very helpful when deciding whether to use an agreement or an alternative policy instrument, like a regulation or a tax. Secondly, when an environmental voluntary agreement is preferred, insight in the specific characteristics of voluntary agreements can enhance the effectiveness when implementing the agreement.

In order to reach relevant policy recommendations, we relied on a comparative case study analysis covering twelve environmental voluntary agreements from six different European countries. The analysis consists of three well-defined stages. First a general framework for assessing the performance is developed which can be applied to a broad range of environmental voluntary agreements. The aggregated performance measure is defined as a mixture of the degree of application, the environmental and economic impact and the development of the policy resource base.

Next five factors influencing the performance of an environmental voluntary agreement were identified: the specification of the agreement, the general policy style, the regulatory threat, the sector structure and the competitive structure. The first one is related to the agreement itself, whereas the other four are related to the institutional-economic context wherein the agreement is negotiated and applied.

Finally a comparative analysis of twelve agreements in six European countries was carried out. The individual case studies are compared with respect to their performance and with respect to the specification and the institutional-economic context. This stage allowed us to test the importance of the factors assumed to influence the performance. Once this was done, an evaluation was made and policy conclusions were drawn. The analysis distinguishes itself from previous work by the systematic evaluation method allowing a comparative analysis of the cases. Moreover, the broadness of the sample, although still limited looked upon from a positive research methodological point of view, is remarkable.

The paper is structured as follows. Next, section two presents the measure of performance used to evaluate the selected environmental voluntary agreements. The factors we believe to influence this performance are discussed in the following section. Finally, section four presents the comparative case study analysis from which conclusions and policy recommendations were drawn.

2. THE PERFORMANCE OF ENVIRONMENTAL VOLUNTARY AGREEMENTS

In this section a general evaluation framework for assessing the performance of voluntary agreements is presented. This framework has two important characteristics. First, it can be applied to a broad range of environmental voluntary agreements irrespective of the parties involved or the environmental problem targeted. Second, the evaluation framework assigns a specific performance score to each agreement. Although it is not the aim to make a decisive report of the cases studied by means of one score, this allows ranking of the agreements according to their performance which will be very helpful for the comparative analysis in section four.

The framework takes into account three different evaluation dimensions: application, impact and resource development. These dimensions are explained below. This section ends with a discussion of the presented performance measure.

2.1. Application

The application of the agreement refers to the compliance of the parties with respect to the targets and obligations specified in the agreement. An agreement is considered successful when on the one hand the environmental targets defined in the agreement are reached and on the other hand when all individual obligations of the parties are fulfilled. A distinction between the ‘targets’ and the ‘obligations’ should be made: a good performance on the environmental targets defined in the agreement can influence the environment, while the performance on the other obligations (such as reporting, monitoring,) can for example, influence the cost-effectiveness and the policy resource base, and not the environment.

2.2. Impact

Because the level of ambition of the environmental targets in an agreement could be rather low or because the targets could be rather vague and qualitative, the degree of application is not the only thing that matters when evaluating the performance of an agreement. This leads us to the 'impact' dimension: did the existence of the agreement lead to a substantial environmental impact? Taking into account the environmental impact when assessing the performance of an agreement is crucial because it is the environmental impact of the agreement that matters in the end. If environmental voluntary agreements are just an institutionalisation of business-as-usual scenarios (see, for example, Börkey and Glachant 1998), this will result in a low score for this dimension.

Contrary to the application dimension, the assessment of the impact of an agreement is a complicated problem. Figure 1 draws a clear picture of the methodological problems that arises (Rietbergen and Blok 1999).

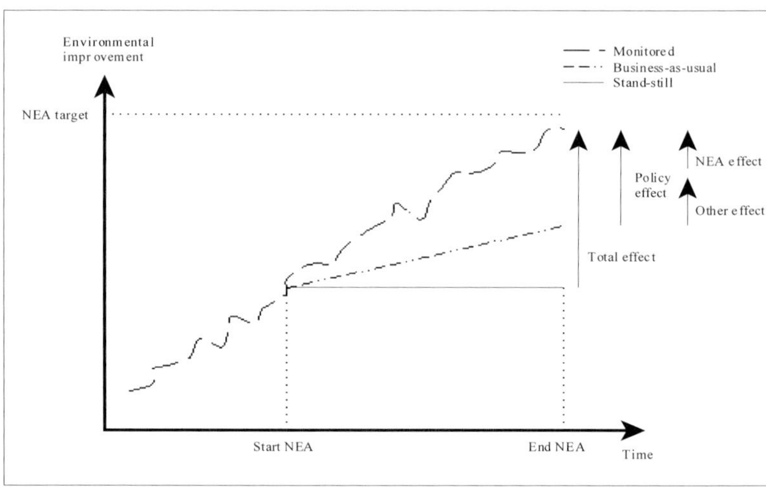

Figure 1. *Methodological problems for assessing the impact dimension.*

First, the environmental improvement due to technological or operational changes in the absence of the agreement and other environmental policy instruments is difficult to determine. In other words a good and reliable business-as-usual scenario is often lacking. Second, other policy instruments can influence the difference between the monitored and the estimated business-as-usual environmental improvement. The way we tried to assess the impact dimension is as follows. First, the situation and evolution of the environmental target before the agreement existed were studied. Next, a critical comparison of this trend and the monitored environmental situation at the end of the agreement was made, taking into account the possible effect of other policy instruments and other structural changes, like output changes due to changing consumer demands. The fact that the sample of agreements studied has different environmental goals, from arranging the collection and recycling of end-of-life vehicles in Germany to regulating the reduction of sulphur dioxide and nitrogen oxide emissions by the power generation industry in the Netherlands, entails an additional difficulty. This is the problem of inter agreement comparison, which will be discussed at the end of this section.

2.3. Resource Development

The resource development refers to the improvements in the policy resource base resulting from negotiating and implementing the agreement. The policy resource base compromises the prevailing institutional network (both formal and informal), the political, economic, legal and cognitive resources of the various actors involved in the policy process, the state of relations between these actors; and the actors' perceptions of the scale of the problem, the need for action and the validity of different policy instruments. In particular, the resource base includes the state of knowledge of the actors (both collectively and individually) encapsulating both the total 'amount' of knowledge, and also the 'distribution' of knowledge among the various actors. As such, it can accommodate the important informational concepts of shared uncertainties and information asymmetries (Aggeri 1999).

Due to the nebulous character of the policy resource base, it is difficult to identify general evaluation criteria for resource development. Nonetheless, there are three aspects of resource development that may be expected to have a general relevance: learning, relations between actors and general awareness and attitudes. Learning can lead to a reduction in the overall cost of achieving the target set in the agreement. An improvement in relations between actors is claimed to be one of the major advantages for voluntary agreements over other instruments (European Commission, 1996). Moreover, the fact that industry is confronted with its environmental impact when negotiating an agreement can bring a positive influence on its general awareness and attitudes along. This dimension clearly reveals the importance of the so-called 'soft-effects' of learning and awareness building (OECD 1999).

Positive resource development thus includes for example: the improvement of relations between actors resulting from increased mutual respect and trust, the generation of new and innovative information about the problem and potential solutions and the dissemination of knowledge amongst the actors. As such, resource development can reflect either an increase in the total quantity of resources (for example a reduction in shared uncertainties), or a decrease in the inequality of existing resource distribution between actors (for example a reduction in information asymmetries).

2.4. The Aggregate Performance Measure

Each of these three evaluation dimensions (the application, the impact and the resource development) is considered relevant for assessing the performance of environmental voluntary agreements. Each dimension plays a certain, but different role in the performance of an agreement. Often, there will be an interaction between them, although this is not necessary the case. For example, the fact that an agreement has a good application with respect to its environmental target does not necessarily mean that there will be an actual impact on the environment. On the other hand, an agreement with a good environmental and economic impact can be considered inferior to an agreement with the same impacts but with a better development of the resource base.

Of the three dimensions, clearly the impact and the resource development in the end determine the performance of an agreement. The application dimension only is too narrow as a judgement base. However, it gives a clear picture of the motivation of the actors and it provides a good estimate of the impact and the resource development that are more difficult to measure.

Therefore, the application, the impact and the resource development of each agreement were measured and aggregated to obtain a total performance score. The assessment of the different evaluation dimensions is done by means of a grading scale technique. Therefore a group of statements was set up for each evaluation dimension (see table 1).

These statements had to be assessed for each agreement by giving them a grade from 1 to 5, showing to what extent the statement is valid. The statements were accompanied by a scoring guide and some explanatory notes to enhance the objectiveness of the assessment. To obtain the average performance of each agreement, we have taken the arithmetic mean of the scores on the three evaluation dimensions: application, impact and resource development.

In order to enhance the objectiveness of assessing the statements, the following procedure, based on the Delphi method, was followed. In each country, a team of two specialists in environmental voluntary agreements was selected with the task of studying the selected agreements in their country intensively, writing a case study report and assessing the statements. First, each specialist assessed the statements individually. Then the scores were compared and discussed in order to give one score on each statement. Next, each case study was passed on to a research partner from another country in order to check the scores on the statements. The German specialists e.g. controlled the French case studies, the German case studies were checked by the Italian specialists and so on. Finally, a discussion session on the assessment of the statements between all project partners was organised.

Table 1. *Statements for assessing the aggregated performance*

Evaluation dimension	*Statements*
Application	▪ Compliance with the environmental performance targets is good. ▪ The target did not break down or eroded substantially during its intended life span. ▪ Compliance with the individual obligations is good.
Impact	▪ There is a significant improvement on the target environmental variable, compared to the business as usual scenario. ▪ The application of the agreement is cost-efficient with respect to compliance. ▪ The administration cost of the agreement is fairly low.
Resource development	▪ The agreement led to an important improvement in the attitudes of the parties concerning environmental issues. ▪ The agreement led to an important improvement in learning. ▪ The agreement has led to substantial innovation in policy making in this area. ▪ The agreement has led to greater trust and more productive relationships between parties. ▪ The agreement has generated product- or process-related innovations and/or market opportunities.

2.5. Discussion

It should be emphasised that the main goal of this study is to determine factors that influence the performance of environmental voluntary agreements in order to draw relevant policy recommendations. To do this we relied on a comparative case study analysis. In order to reach solid policy recommendations, the European Commission obliged us to take diverging agreements in the sample. The selection of the cases was based on five criteria. First, enough information on the agreement was needed. Second, the agreement had to be significant for the environmental problem it deals with. Third, the period during which the agreement has been used should be long enough in order to enable a performance evaluation. Fourth, the sample must consist of both successful and unsuccessful cases. Finally, the economic context of the different industrial sectors involved must be diversified.

Of course, the diverging scope of the sample did not make the task of comparing the performance of these agreements easier. Questions like how to compare a radical shift in the collection and recycling of end-of-life vehicles in Germany to a similar, or why not a minor, reduction of sulphur dioxide and nitrogen oxide emissions from the power generation industry in the Netherlands can not be answered with the given state of knowledge. Consequently, this kind of intra agreement comparison is not taken in account when measuring the performance of the negotiated agreements. Thus when we assessed for example the impact of an agreement we only took into account the improvement compared to a reliable business-as-usual trend without trying to compare this improvement to the impact of another agreement with another environmental target. This sort of inter sector comparison problems can be an interesting subject for future research as it might help policy makers to decide where action is needed most urgently.

3. CRITICAL FACTORS FOR SUCCESS

After assessing the performance of the agreements, we now focus on the factors explaining the level of performance. Knowing these factors and their influence on the performance of environmental voluntary agreements is especially interesting for policy makers as it can help them with the instrument choice and with the actual implementation of agreements. Five factors are identified. Four of these are related to the institutional-economic context wherein an agreement is negotiated and applied, the other one is related to the agreement itself. First we turn to the institutional-economic context.

3.1. The Institutional-Economic context

Four hypotheses were postulated regarding the relation between the different institutional-economic aspects and their expected influence on the performance of the environmental voluntary agreements. All of our hypotheses are supported by other theoretical research on this subject (see, for example, Alberini and Segerson 2002; Arora and Gangopahdyay 1995; Garvie 1999; Glasbergen 1998; Hansen 1997; Klok and Kuks 1994; Maxwell at al. 1998; Segerson and Miceli 1998; Van de Peppel and Herweijer 1994). Below, the hypotheses are presented accompanied with some explanation.

3.1.1. The Policy Hypothesis

"The fact that public environmental policy evolves in a tradition and climate of consensus seeking, joint problem solving, mutual respect and trust is a crucial positive factor for the performance of environmental voluntary agreements."

Environmental voluntary agreements can be seen as a sort of a transaction. Transactions thrive best in a climate of trust. Such a climate is built on positive experiences in the past. The authorities need to trust that the industry will not erode the environmental objectives or cheat by not complying with the agreement. The industry must trust that the authorities will not resort to additional regulations.

3.1.2. The Instrumental Hypothesis

"The fact that public policy makers show the readiness to use alternative policy instruments, as a stick behind the door to deal with the environmental problems, in case the negotiated agreement fails, is a crucial positive factor for the performance of environmental voluntary agreements."

This hypothesis concentrates on the readiness of the policy makers to use an alternative instrument in case of non-compliance to the agreement by the private parties. The readiness of the policy makers however has to be combined with the severity of this alternative when applied. When the threat of the alternative instrument is credible and this instrument has more stringent or costly consequences for the companies involved, they should have a bigger incentive to make the agreement succeed. In this regard, 'speak softly and carry a big stick' is an old adage.

This hypothesis can be considered as a reformulation of the regulatory gains argument, also called the 'stick' approach, which is used in most of the literature for explaining the existence of voluntary approaches (see, for example, Segerson and Miceli 1998). The advantage from the signing of an environmental voluntary agreement would consist of the avoided costs of a public regulation aimed at addressing the same environmental problem.

3.1.3. The Sector Hypothesis

"The fact that the industry involved is homogeneous, has a small number of players and is dominated by one or two players, or has a powerful industry association that can speak for all its members, is a crucial positive factor for the performance of environmental voluntary agreements."

The structure of the target group can influence the performance of an agreement in the negotiating and execution phase. Negotiations will be more efficient if there is a big company or an industrial organisation that can speak for the whole sector. A small number of players can also increase the degree of application because monitoring will be easier and the possibility of free-riding companies diminishes.

3.1.4. The Competitive Hypothesis

"The fact that industries are close to the final market is a crucial positive factor for the performance of environmental voluntary agreements, due to consumer pressure."

Besides the negative incentives arising from the instrumental hypothesis, polluters can also recognise positive incentives to strive for more environmental protection. Literature on voluntary agreements calls this the reputation enhancing argument, or the 'carrot' approach (see, for example, Börkey and Glachant 1998). Voluntary agreements would be an answer to the demand pressure for firm's environmental performance leading to higher demand and profit. The central idea behind this hypothesis is that an agreement will be more feasible when the companies have a certain competitive incentive vis-à-vis the other companies in the area covered by the agreement, to distinguish themselves, for example through a green image. As such, voluntary agreements might be considered a part of firms' public relation activities (Arora and Cason 1996).

Another way of looking at this hypothesis focuses on the increased risk of bad impacts for industries performing badly when there is a high degree of closeness between the final markets and consumers. Due to the salience of the produced brand, the chance of being criticised increases, with the possible effects on both sales to consumers and on the toughness of the provoked responses by authorities.

3.1.5. Assessing the Institutional-Economic context

To gain information on the institutional-economic context, we have carried out an analysis using the same technique as for the performance evaluation. Different statements on each hypothesis were postulated (see table 2) and then judged for each agreement, by giving them a one to five score.

***Table 2**. Statements for assessing the institutional-economic context*

Hypothesis	*Statements*
Policy hypothesis	▪ Environmental policy evolves in a tradition of consensus seeking and joint problem solving apart from the conclusion of the agreement. ▪ Apart from the process leading to the conclusion of the agreement, policy making in the area covered by the agreement is characterised by a climate of mutual trust. ▪ Apart from the process leading to the conclusion of the agreement, the private sector(s) covered by the agreement show(s) a clear readiness to self-responsibility with respect to the environmental problem.
Instrumental hypothesis	▪ The chances that public authorities will use an alternative instrument in case of non-success or non-conclusion of the agreement are high. ▪ If applied, the alternative instrument has more severe consequences for the target group than those resulting from the application of the agreement.
Sector hypothesis	▪ There is already a dominant interest of a major player/a small number of players or a powerful and representative industry association in the area covered by the agreement. ▪ The private parties to the agreement belong to the same industrial sector. ▪ The potential for significant free riding between the members of the targeted sector covered by the agreement, is low.
Competitive hypothesis	▪ Buyers can distinguish the difference in environmental quality performance of the firms in the participating sector(s). ▪ Buyers value environmental sound products in the area covered by the agreement.

Aggregation resulted in a score for each hypothesis. By these scores, we tried to measure how favourable each of the four institutional-economic aspects was with respect to the agreement's performance. A high score for a certain hypothesis meant that the conditions described in the hypothesis were valid for this agreement and thus we expect a good performance.

3.2. The Specification of an agreement

Even when all institutional-economic factors are favourable to the conclusion and execution of an agreement, success is by no way automatically guaranteed. Success indeed depends also upon the creation of a number of internal preconditions. Those internal factors of success are captured under the heading of specification. Special attention should be given to the evaluation criteria for the specification of negotiated agreements presented in table 3. In practice, it may not be feasible for an agreement to compare well against all of the criteria. However, the criteria provide a useful benchmark against which to assess the specification of a particular agreement.

***Table 3**. Evaluation criteria for the specification of negotiated agreement*

Evaluation criteria	*Description*
Target	The inclusion of a clearly defined and quantified target is crucial for the success of an environmental agreement.
Burden sharing rule	An agreement that relies on individual actions by firms to meet a collective target is more likely to be successful when accountability is developed to individual firms. This is best achieved by the inclusion of an explicit 'burden sharing' rule, or a mechanism for appointing the collective target.
Monitoring mechanisms	The inclusion of adequate monitoring mechanisms is crucial for measuring the performance against the target. Ideally, the performance of an agreement would be monitored using information collected and collated by an independent body.
Additional guarantees or sanctions	The inclusion of additional guarantees or sanctions regarding the achievement of targets will considerably enhance the credibility of an agreement.
Contractual form	By providing a clear legal framework, that is enforceable through court decisions, a binding contract adds considerable force to a negotiated agreement.
Legal compliance	In addition to complying with the provisions of the national law under which the agreement falls, it must also comply with the requirements of the EC Treaty and its derived legislation.
General provisions	In order to avoid potential confusion and disputes during the operation of the agreement, it is important that a number of basic issues are clarified like the parties and their respective obligations, the duration of the agreement, the conditions under which it can be revised or terminated, the competent jurisdiction.

Well-specified agreements are important because they lead to a higher rate of application, impact and resource development. A clear description of the obligations of the parties combined with a sanction in case of non-compliance leads to a higher degree of application. Better application and more demanding objectives improve the impact on the target variables. The policy resource base also will be developed more when a credible monitoring mechanism and other reporting activities are included in the agreement. This leads us to postulate the following specification hypothesis:

"*The fact that the agreement is well-specified, containing all important elements of table 3, is a crucial positive factor for the performance of environmental voluntary agreements.*"

Again statements were postulated (see table 4) and then judged for each agreement. A well-specified agreement leads to a high score on these statements, indicating that a successful implementation and execution is expected.

***Table 4**. Statements for assessing the specification*

Specification dimension	*Statements*
Environmental performance	▪ The agreement contains a well-defined environmental performance objective. ▪ The objective represents a meaningful improvement in environmental performance ▪ The agreement contains a credible mechanism for achieving the environmental performance objective ▪ The agreement contains a credible system for monitoring performance against the specified objective.
Learning	▪ The agreement contains a clear objective with respect to learning. ▪ The agreement contains a credible mechanism to support and encourage learning. ▪ The agreement contains an adequate monitoring system for co-ordinating learning activities.
Economic efficiency	▪ The agreement contains a burden-sharing mechanism that is consistent with a cost-efficient outcome. ▪ The agreement contains a credible mechanism to prevent free riding by participants. ▪ The agreement does not create any barriers to new entrants.

4. COMPARATIVE EVALUATION

The final goal of this study was to gain a deeper understanding of the factors influencing the performance of environmental voluntary agreements based on a comparative evaluation of twelve individual case studies. To provide data for this comparative analysis, two agreements were selected in Belgium, The Netherlands, Italy, France, Germany and the UK, giving us a sample of twelve cases (see table 5). Although still limited looked upon from a positive research methodological point of view, it is the greatest sample of environmental voluntary agreements ever analysed in a systematic way. Moreover, the agreements were concluded in different countries and their respective environmental targets cover a wide area of environmental problems. This high diversity among the selected agreements gives a higher degree of validity to the conclusions drawn from the comparative case study. Because these case studies had to be cross-compared, they were made up using a common case study design that links the data to be collected to the statements that need to be judged.

***Table 5**. The selected negotiated agreements*

Abbreviation	*Country*	*Description of the agreement*
GBAT	Germany	Agreement to reduce the mercury-content in batteries and to collect used batteries separately.
GELV	Germany	Agreement to maximise the recycling rate of end-of-life vehicles.
FCFC	France	Agreement to eliminate the use of CFCs in the industry.
FECO	France	Agreement upon the collection and recycling of packaging waste.
BBAT	Belgium	Agreement upon the private separate collection and recycling of used batteries.
BELE	Belgium	Agreement to reduce the emissions of SO_2 and NO_X in power plants.
DSO2	The Netherlands	Agreement upon the reduction of the SO_2-emissions of power plants.
DWHI	The Netherlands	Agreement upon the take back of worn household appliances by their producers.
IVIC	Italy	Regional agreement upon the improvement of the environmental quality in the province of Vicenza
IAGI	Italy	Agreement upon the improvement of gasoline quality
EFAR	UK	Agreement upon the collection from farms of waste plastic films used in the production.
EEFF	UK	Agreement to improve the energy efficiency in the chemical industry.

Having assessed all statements and doing the necessary aggregations, we obtained a score on the average performance, the specification, the policy style, the threat of an alternative instrument and the sector and competitive structure. These scores allowed testing whether there is in fact a positive relationship between the favourability of the conditions and the performance of the agreements in our sample. This has been done using a graphical representation. We already mentioned that the specification and the aspects of the institutional-economic context we studied could be a precondition for the performance of voluntary agreements. These aspects therefore functioned as independent variables that explained the dependent variable, that is the performance of an agreement. Thus the vertical axis measures the performance, the horizontal axis represents the different scores on the aspects considered to influence this performance. A trend line showing the relation between the independent variable and the performance is included in the figures. The hypotheses postulate that a high score on the independent variables should lead to a successful agreement. So a hypothesis is supported if the trend line has a positive slope.

A second way of assessing the hypotheses was done by using Spearman's rank-order correlation test. Spearman's rank-order correlation coefficient, defined between –1 and 1, gives an indication of the relationship between two variables. If the coefficient is negative, a negative relationship between the two variables exists. With a positive coefficient, a positive relationship exists. A coefficient close to zero indicates that there is no discernible relationship between the two variables. The greater the value of the coefficient, the more pronounced the relationship between the two variables. It is stressed that Spearman's rank-order correlation coefficient gives an indication on the strength of the relationship between two variables, but does not allow making conclusions on the causality.

In the correlation test, two hypothesises are postulated and judged. The null hypothesis says there is no correlation between the two variables, the alternative hypothesis states that a correlation exists. This way, the alternative hypothesis corresponds to the postulated hypothesises concerning the influence of the factors for success. In this research project, the null hypothesis was supported if the correlation coefficient is below the critical two-tailed rs value at the 0.05 level of significance (rs .05=0.587). If on the other hand the correlation coefficient was above 0.587, the null hypothesis has been rejected and the hypothesis tested was supported. However, one should keep in mind that because of the limited sample outliners can have a significant effect on the value of Spearman's rank-order correlation coefficient and thus on the results of this quantitative analysis.

4.1. The Policy Hypothesis

Figure 2 shows that except for the BBAT and the IAGI agreement, we can see a quite positive relation between the degree of consensus seeking, respect and trust in the policy making process and the performance of the agreement. The positive slope of the trend line is strongly affected by these two agreements. This brings Spearman's rank-order correlation coefficient close to zero (rs=0.059), indicating that our hypothesis is rejected within a 95 per cent confidence interval.

However, the fact that there are no scatter points in the lower right corner, rather confirms our hypothesis. This shows that there are no agreements concluded in a favourable policy style, which had a low measure of performance. Agreements situated in the upper left corner might be agreements that, despite the unfavourable policy climate, are successful because of other beneficial institutional-economic aspects. It is clear that the policy style is certainly not the only precondition for a successful implementation of environmental voluntary agreements. For that reason, other important features for successful agreements must exist.

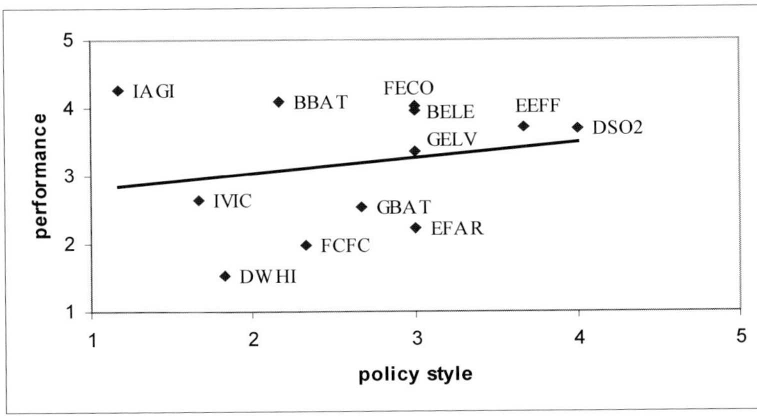

Figure 2. Relation between the policy style and the performance of the studied agreements.

4.2. The Instrumental Hypothesis

The trend line in figure 3 shows a clear positive relationship. Accordingly, Spearman's rank-order correlation coefficient is quite high (rs=0.668), indicating that the instrumental hypothesis is supported.

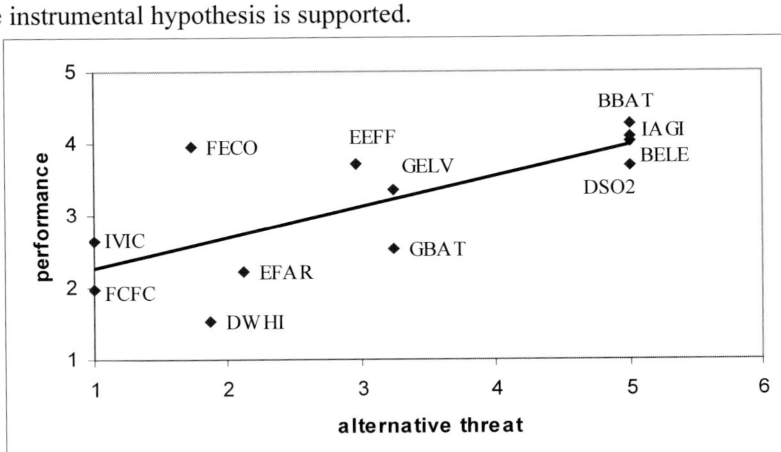

Figure 3. The relation between the existence of an alternative threat and the performance of the studied agreements.

Four agreements were concluded in a context where there was a very strong and severe alternative threat (DSO2, BELE, BBAT and IAGI). All those cases were also evaluated as rather successful ones. Particularly these agreements support the validity of the instrumental hypothesis. Besides these successful agreements there are also two cases, which are assessed with the lowest possible grade (FCFC and IVIC). Accordingly, their performance score is lower than average. Again it is important to notice that the lower right part of the scatter graph remains almost empty. Here, this means that there are no low-performing agreements in the sample when a strong alternative treat was present. In the upper left area, we can detect some agreements that again contribute their high performance to another aspect. We can conclude by saying that, while a strong alternative threat is not necessary, it can clearly contribute to the performance of an agreement.

4.3. The Sector Hypothesis

Again, a positive trend line and thus a positive relationship between the homogeneity of the sector and the performance emerge from figure 4. This hypothesis is also supported by Spaerman's rank-order correlation coefficient test (rs=0.607). Only two agreements break this positive trend, which are the British energy efficiency agreement (EEFF) and the French Eco-Emballages agreement (FECO). All other agreements seem to be in line with expectations.

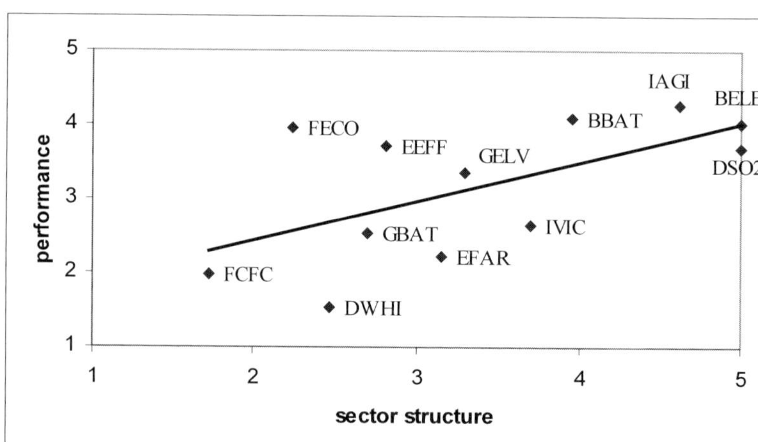

Figure 4. *The relation between the sector structure and the performance of the studied agreements.*

4.4. The Competitive Hypothesis

Whereas the previous three hypotheses seemed to be confirmed, there is less clarity here: the scatter points on figure 5 are dispersed throughout the entire graph. Not surprisingly, Spaerman's rank-order correlation coefficient is very low (rs=0.135) and the competitive hypothesis is rejected. The negative slope of the trend line indicates that the theoretical idea that firms will be prone to a good environmental performance when there is demand pressure from green consumers is not confirmed by our agreements. On the one hand, we have a few agreements concluded with firms in sectors were there is demand pressure, that performed badly (DWHI, EFAR, FCFC), and on the other hand, we have agreements with a rather good performance in markets where demand pressure was not strong (DSO2, BELE, EEFF, GELV).

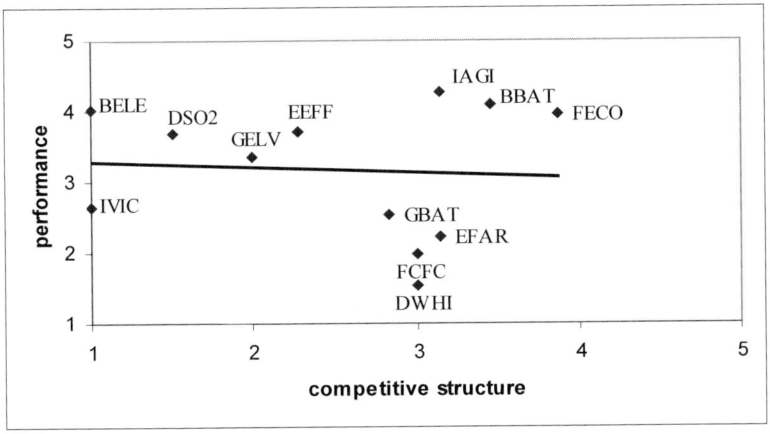

***Figure 5.** The relation between the competitive structure and the studies agreements.*

We can conclude that although the theoretical assumption beyond this hypothesis is quite convincing, this is not supported by our analysis. This could be an indication that when firms voluntary undertake actions to improve their environmental record, they usually do not go much further than business-as-usual. The industry's information advantage over the environmental problem, the alternative abatement strategies and their associated costs, enable companies to fool consumers and government in believing that they are conscious of environmental problems whereas in reality they are only saying, but not doing this ('window dressing') (Alberini and Segerson 2002).

4.5. The Specification Hypothesis

Figure 6 clearly shows a positive relationship between the specification of an agreement and its performance. Spaeman's rank-order correlation coefficient is high (rs=0.839) and supports the specification hypothesis. There are no agreements situated in the upper left corner or in the lower right corner. This shows what we already expected: the degree of specification is an important internal precondition for the performance of an agreement.

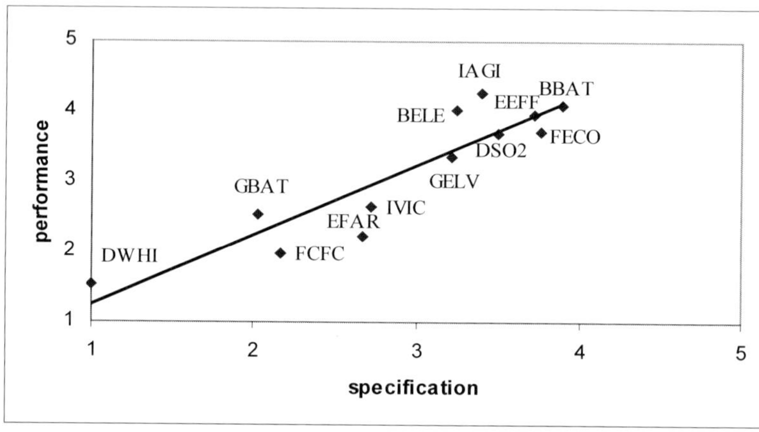

Figure 6. *The relation between the specification and the performance of the studied agreements.*

4.6. The Combined Institutional-Economic Context

When looking at the different hypotheses separately, we already mentioned that the absence of the expected relationship between one institutional-economic aspect and the performance of an agreement can be due to the fact that this performance is positively or negatively influenced by another aspect, diluting the influence of the first. The same holds for the specification of an agreement. Looking at the different hypotheses simultaneously can bring us insight in the possible existence of a 'combined (un)favourable institutional-economic context'. This has been done by defining the combined context as the arithmetic mean of the scores on the five hypotheses. Again a graphical representation with the trend line is presented in figure 7.

A clear positive relationship emerges. Also Spearman's rank-order correlation coefficient (rs=0.783) supports the idea that a favourable institutional-economic context is positively correlated with the performance of negotiated agreements. This leads us to conclude that the favourability of each of the institutional-economic aspects we studied is not a necessary condition for the performance of an environmental voluntary agreement. Rather it is the combined context that determines the performance of the agreements studied. The negative influence of an unfavourable factor can be totally outweighed by the positive influence of another aspect of the socio-economic context.

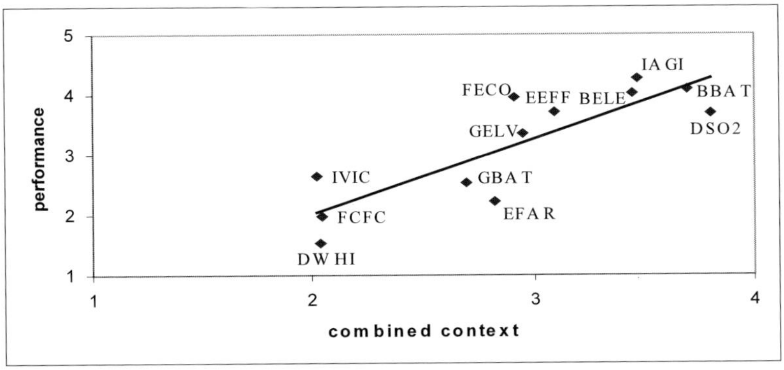

***Figure** 7.The relation between the combined institutional-economic context and the performance of the studied agreements.*

5. CONCLUSION

The aim of this study was to gain insight on the factors leading to success or failure of environmental voluntary agreements. To do this we relied on a comparative case study covering twelve agreements from six different European countries.

First, a measure for the performance of voluntary agreements was developed. In our view, the performance of an agreement is a mixture of the degree of good application of the agreement, the degree of impact the agreement has on the environment and on the economic efficiency and the degree of resource development that occurs while negotiating and implementing the agreement. It is emphasised that taking into account only the application of the agreement would result in a very narrow definition of performance. Moreover, it is the impact on the environment and the development of the policy resource base the agreement brought about that matters in the end.

The theoretical as well as the empirical research point to a number of internal and external factors that influence this performance. Four external preconditions for success were identified: the general policy style, the readiness to use severe alternative instruments in case of non-compliance with the agreement, the potential of the sector to negotiate and to act as one collective actor and the potential for market success triggered of by the implementation of the agreement. Next to these external factors related to the institutional-economic context wherein a voluntary agreement is used, the specification of an agreement is considered to be an internal factor influencing the performance.

In the cases studied, we found evidence that all factors could be important for enabling the success of an agreement. Only the evidence for the competitive hypothesis is less convincing for the twelve cases studied, although the theoretical arguments in favour of this hypothesis are quite robust. Notices that this could be an indication of the low (or wrong) motivation companies have when announcing voluntary actions to strive for a better environmental performance.

It should be emphasised that taken individually each of the factors is not as such a conditio sine qua non for the success of an environmental voluntary agreement. Rather it is the combination of the success factors that is ultimately decisive for the performance of an agreement. This is important because some of the success factors – the sector structure and to a large extend the competitive structure – are independent factors that cannot be manipulated by the government. These factors should play a crucial role in the instrument choice of policy makers. But even if these factors are not favourable to expect a successful agreement, the use of this instrument should not be ruled out in advance. This because the other three factors – the general policy style, the specification and certainly the alternative instrument – are under the control of the policy maker and can thus be manipulated to create a favourable environment for a voluntary agreement. These factors should play a crucial role when negotiating and implementing an agreement.

So, next to the ex post analysis carried out in this paper, the questions used in this study could be used as a quick checklist to ex ante assess whether or not the environment is potentially favourable for the conclusion of a successful environmental voluntary agreement.

6. ACNOWLEDGEMENTS

This paper is part of a broader research exercise under the NEAPOL-project (Negotiated Environment Agreements: Policy Lessons to be Learned from a Comparative Case Study). NEAPOL is a research project financed by the European Commission – DG XII, and is part of the EC Environment and Climate Research Programme (1994-1998) – Research Theme 4: Human Dimensions of Environmental Change (project number ENV4- CT97-0560). For further information see Negotiating Environmental Agreements in Europe: Critical Factors for Success, 2002, Edward Elgar Publishing Ltd. edited by M. De Clercq.

7. REFERENCES

Aggeri F. (1999), Negotiated agreements and innovation: a knowledge-based perspective on environmental policies. In CEEM, Neapol project: theoretical report (pp. 73-96). Ghent: Ghent University.

Alberini A. & Segerson K. (2002), Assessing voluntary programs to improve environmental quality. *Environmental and Resouce Economics*, 22, 157-184.

Arora S. & Gangopadhyay S. (1995), Towards a theoretical model of voluntary overcompliance. *Journal of Economic Behaviour and Organisation*, 28, 289-309.

Arora S. & Cason T.N. (1996), Why do firms volunteer to exceed environmental regulations? Understanding participation in EPA's 33/50 Program. *Land Economics*, 72, 413-432.

Börkey P. & Glachant M. (1998), Les engagements volontaires de l'industrie: un mode original de réglementation environmentale. *Revue d'économie industrielle*, 83(1), 213-224.

Burritt R.L. (2002a), Voluntary agreements: effectiveness analysis, tools, guidelines and checklist. In P. ten Brink, *Voluntary environmental agreements: process, practice and future use* (pp. 267-283). Sheffield: Greenleaf Publishing.

Burritt R.L. (2002b), Application of effectiveness analysis: the case of greenhouse gas emissions reduction. In P. ten Brink, *Voluntary environmental agreements: process, practice and future use* (pp. 284-296). Sheffield: Greenleaf Publishing.

Cabugueira M.M. (2002), Co-regulation performance factors: lessons from theory and from practice in environmental agreements. In P. ten Brink, *Voluntary environmental agreements: process, practice and future use* (pp. 399-417). Sheffield: Greenleaf Publishing.

European Commission (1996), *Communication from the Commission to the Council and the European Parliament on environmental agreements*. COM (96) 561.final.

European Environment Agency (1997), *Environmental agreements: environmental effectiveness*, Vol. I and II. Copenhagen: EEA.

Garvie D. (1997), *Self-regulation of pollution: the role of market structure and consumer information*. Nota di Lavoro 59.97. Milan: Fondazione Eni Enrico Mattei.

Glasbergen P. (1994), Milieuproblemen als beleidsvraagstuk. In P. Glasbergen, *Milieubeleid: een beleidswetenschappelijke inleiding* (pp. 341-358). The Hague: VUGA.

Glasbergen P. (1998), Partnership as learning process: environmental agreements in the Netherlands. In P. Glasbergen, *Co-operative environmental governance, public-private agreements as a policy strategy* (pp. 133-156). Dordrecht: Kluwer Academic Publishers.

Hansen L.G. (1997), *Environmental regulation through voluntary agreements*. Nota di Lavoro 23.97. Milan: Fondationi Eni Enrico Mattei.

Immerzeel-Brand E. (2002), Assessing the performance of negotiated environmental agreements in the Netherlands. In P. ten Brink, *Voluntary environmental agreements: process, practice and future use* (pp. 384-398). Sheffield: Greenleaf Publishing.

Klok P.J. & Kuks S.M.M. (1994), Het doelgroepenbeleid, In P. Glasbergen, *Milieubeleid: een beleidswetenschappelijke inleiding* (pp. 79-96). The Hague: VUGA.

Klok P.J. (1989), *Convenanten als instrument van milieubeleid*. Enschede: University of Twente.

Lehmann M.A. (2000), *Voluntary environmental agreements and competition policy: the case of Germany's private system for packaging waste recycling*. Nota di Lavoro 78.2000. Milan: Fondazione Eni Enrico Mattei.

Lindblom C.E. (1980), *The policy making process*. Englewood Cliffs: Pretentice Hall.

Maxwell J. & Lyon T. & Hackett C. (1998), *Self-regulation and social welfare: the political economy of corporate environmentalists*. Nota di Lavoro 55.98. Milan: Fondazione Eni Enrico Mattei.

Nyborg K. (2000), Voluntary agreements and non-verifiable emissions. *Environmental and Resource Economics*, 17, 125-144.

OECD (1999), *Voluntary approaches for environmental policy: an assessment*. Paris: OECD.

Rietbergen M. & Blok K. (1999), *The environmental performance of voluntary agreements on industrial energy efficiency improvements*. NW&S Report No 99068. Utrecht: Department of Science, Technology and Society.

Segerson K. & Miceli T.J. (1998), Voluntary environmental agreements: good or bad news for environmental policy?. *Journal of Environmental Economics and Management*, 36, 109-130.

Storey M. & Boyd G. & Bowd J. (1997), *Voluntary agreements with industry*. Nota di Lavoro 26.97. Milan: Fondazione Eni Enrico Mattei.

Van de Peppel R. A. & Herweijer M. (1994), Het communicatieve sturingsmodel. In P. Glasbergen, *Milieubeleid: een beleidswetenschappelijke inleiding* (pp. 189-207). The Hague: VUGA.

ENVIRONMENTAL VOLUNTARY AGREEMENTS IN THE DUTCH CONTEXT

H.T.A. BRESSERS, T.J.N.M. DE BRUIJN

University of Twente, The Netherlands

Abstract. This paper describes and analyses the use of environmental voluntary agreements, or covenants, in Dutch environmental policy. Covenants have become a widely used policy instrument in the Netherlands. This trend reinforces the strong neo-corporatist traits of Dutch society with its tendency towards bargaining and cooperation with interest groups. Over the years an authoritarian and distant policy style with a negative attitude towards target groups has changed into a new approach designed to encourage self-regulation. Instead of simply imposing legislation, the Dutch government often concludes agreements with relevant sectors of industry regarding the implementation of environmental objectives. Through negotiations between sectors of industry, the Ministry of the Environment, and regional governments, agreements are sought concerning the contribution of specific industrial sectors to the goals of the National Environmental Policy Plan. These goals aim for 50-90 percent emission reductions for specified pollutants. Since 1989 many such agreements have been reached. In 2002/2003 we carried out a study on the effectiveness of the covenants, commissioned by the Dutch ministry of the Environment (VROM). The focus in the project was the identification of success and fail factors. Our central conclusion on the use and effects of the covenants is quite positive, although we have also identified several constraints. Most importantly, we found the implementation context highly relevant for covenant success. In this chapter we focus on this context in order to understand the workings of environmental voluntary agreements. We describe the background in which the covenants are used as well as the resultant effects. Furthermore we highlight some guidelines for future use.

1. INTRODUCTION

The Dutch Target Group Policy, introduced to implement the first National Environmental Policy Plan (NEPP 1989), is the central element in the current Dutch system of industrial environmental regulation. Since NEPP represented an ambitious new step, it was obvious that these targets could not be reached by conventional policy-instruments only. The publication of the aspiring NEPP coincided with a growing lack of confidence in traditional policy approach with its emphasis on direct regulation. The combination of new ambitions with a lack of confidence in traditional approaches thus called for a new strategy and style. The new strategy would aim more specifically at eliciting private initiative and 'shared responsibility'. Instead of setting technology-forcing standards unilaterally the approach builds on close collaboration with target groups. Environmental voluntary agreements are thought to be the key instruments in this approach. Through negotiating processes and the agreements that result from these, the Dutch government tries to improve the effectiveness of the environmental policy system. In this paper we describe and analyse the use of

E. Croci (ed.), The Handbook of Environmental Voluntary Agreements, 261 - 281

covenants in Dutch environmental policy.

The set-up of the paper is as follows: the next section is an elaboration on the development and use of covenants in Dutch environmental policy. Given the importance of context, we pay ample attention to the developments that lead to the increasing use of covenants as well as the underlying motivation. Section 3 describes the effects of the covenants. Here we build on an extensive evaluation study carried out in 2002-2003, commissioned by the Dutch environmental ministry VROM. Section 4 presents the policy context in which the covenants are implemented and identifies two crucial elements as factors of success. Further positive factors are presented in section 5. In section 6 we draw up our conclusions.

2. THE DEVELOPMENT OF THE USE OF COVENANTS IN DUTCH ENVIRONMENTAL POLICY

Present Dutch environmental policy holds a somewhat unique position in the world. Though the positive outcomes of the policies are highly regarded, it is the approach and subsequent implementation of the policies that demand real scrutiny. The Dutch National Environmental Policy Plans and the way these are implemented through national target group consultations, covenants, the activation of various intermediary organizations and the stimulation of self-regulation have raised a lot of attention abroad. Many of these ideas are manifest, at least in part, in other countries' policies, but nowhere else does this approach exist so fully developed as the Netherlands. Actually, this policy approach was not followed from the very beginning of Dutch environmental policy, but implies a major shift away from the environmental policies of the seventies (Bressers 1991). Here we will focus on environmental policy in the narrower sense, namely the policy followed by the General Directorate of Environmental Policy and its forerunners.

The seventies: construction of the legal framework. In the seventies, Holland liked to see itself as a pioneer in the field of environmental policy. However, this referred more to the speed with which legislation was passed and the strictness of policy-makers' intentions rather than to any special nature of the policy approach. This approach may be characterized as relatively conventional, not so different from other countries in the same period. When various environmental problems had been recognized by politicians and the first real wave of public attention arose for the environment, the construction of a legal framework was begun. During the seventies, this resulted in a series of laws dealing with specific sectors of the natural environment, such as water and air, and making pollution of this environmental sector in question subject to new licensing.

Around 1980 this stage of policy making was more or less complete. At the same time, the first developments could be seen which were to lead to considerable changes in the policy approach.

The eighties: evaluation and adjustment followed by rapid development. Early policy evaluations of the impact of the new environmental acts (e.g. Twijnstra

Gudde 1981; for a survey see Bressers and Coenen 1989) showed the following: the application and enforcement of the licensing systems showed serious deficiencies: licensing procedures took a long time – perhaps too long – to be processed; adjustment problems, having to do both with content and procedure, arose between the various types of sectoral legislation; inspections proved to be costly; and industry complained about the lack of flexibility.

Various measures were taken in response to these new insights in the course of the eighties:

- The licensing obligation was limited by the gradual introduction of systems of general regulations for many branches of industry. The end result of this was that the vast majority of firms no longer were required to have individual environmental licenses. This reduced the procedural burden on the government and on businesses.
- The capacity to implement environmental acts was increased. First this was done mainly through the stimulation of a planned approach to policy implementation by means of subsidies (Noise Pollution Act and environmental implementation programs). Later, more extensive financial stimuli were applied to policy implementation and combined with obligations imposed on municipalities to cooperate in implementing the environmental tasks. This led to a considerable increase in the number of environmental officials.
- An 'umbrella law' was introduced: the General Environmental Hygiene Regulations Act. This coordinated the procedural aspects of the various sectoral laws and provided additional opportunities for in-depth coordination, such as 'Environmental Impact Reporting'.
- Policy memoranda on environmental policy were increasingly formulated in terms of 'environmental issues', i.e. environmental problems exceeding the boundaries of the various sectors. Eventually, the Environmental Control Act was passed which regulated the licensing systems of most sectoral laws.

More than a hundred policy evaluation studies were performed in the field of Dutch environmental policy (Schuddeboom 1994). There is almost no other area in which the authorities so expressly tried to make it possible to learn from experiences in policy-making. However, the new evaluation studies also showed that in the course of the eighties the impact of these policies did not quite meet expectations. Additionally, new environmental problems appeared on the agenda, particularly in the second half of the decade.

The acceptance in the Netherlands of the principle of 'sustainable development' from the Brundtland report (1987) and its rapid application to the Dutch situation by the State Institute of Public Health and Environmental Hygiene (RIVM 1988) subsequently made inevitable the redefinition of environmental policy as a matter of (inter)national priority.

In response to these reports, the first National Environmental Policy Plan (NMP 1989) was drawn up. This plan did not just redefine the policy view of the environmental problem, but also set new and more ambitious targets: emission reductions of most hazardous substances by 70 to 90 percent by 2010. NEPP distinguished themes for policy action (such as acidification, climate change, waste) rather than the different environmental media and distinguished the various target groups (such as industry, traffic, agriculture) that contribute to the thematic problems. NEPP, which was unanimously accepted in Parliament, was enthusiastically received abroad. Nevertheless, it was criticized at the national level. In addition to remarks made by the environmental movement, which called for even more radical targets, this criticism focused on the fact that in view of the still disappointing results of the licensing system, inadequate attention had been paid by the National Environmental Policy Plan to the way in which these ambitious targets might be achieved. Not only did the evaluation studies show that the effectiveness of environmental policy had its limitations, but also a more general picture gradually emerged of a government that could influence developments in society only to a limited extent, let alone steer these developments. In retrospect we can see this moment as a breaking point in the development of the strategy of Dutch environmental policy. With its greatly intensified objective and awareness of the limited possibilities offered by the current policy approach, environmental policy, at the height of its public support, was urgently in need of a revision.

The nineties: a 'new deal'. It was for this reason that the 'NMP+' (National Environmental Policy Plan Plus 1990) emphasized the changes that were needed in policy strategy. Where the first NMP established the goals, NMP+ was more concerned with instrumentation and implementation. Collaboration offers the opportunity to bring together experts from a variety of different disciplines and arenas to fashion solutions that can go beyond the limited perspectives of individual stakeholders (Gray 1989; Hartman et al. 2002). The NMP+ pays a lot of attention to the partners that are necessary to realize these goals. A special Appendix on policy instruments, among other things, announced new directions in policy strategy, which were intended to supplement the existing emphasis on licensing and other forms of direct regulation. One of these strategies was to try to induce the target groups to take more responsibility for a clean environment.

This was elaborated in the Dutch target group policy. The objectives of the NMP+ were taken as a starting point for consultations with representatives of, nearly all the main branches of industry. When agreement is reached on the contribution that the branch in question has to make to achieve its objectives, this agreement is usually recorded in a covenant. These covenants are not just intended to directly influence the behaviour of the firms, but to serve as a guideline for licensing at a later stage. The subsidizing of new environmental technologies and other policy instruments takes place more and more in the context of target group policies. It is no exaggeration to say that target group policies have come to dominate the environmental policy agenda when focused on business and industry.

At this juncture the results of these policies seem encouraging insofar as they deal with technical adaptations of production processes (RIVM 1995). In many fields the goals are within reach, many emissions have been reduced and a decoupling of economic growth and environmental pressure has been realized. It is true that there are several fields where target achievement is not 'on course' toward the NMP objective. But this does partly involve the environmental behaviour of 'hard-to-reach target groups', which makes organized consultation more difficult to accomplish and yield results (NMP 2 1993, p. 11), and environmental problems for which technical adaptations are difficult to apply (such as CO2 emissions). Besides, strong economic growth leads to a rapid increase in the environmental burden (RIVM 1996).

Sensible choices or fortunate coincidence? In the preceding discussion we have consistently presented these developments as rational responses to new insights and circumstances. The perception of the environmental problem gave rise to a policy that corresponded to that of our neighbouring countries. Evaluation studies led to attempts to remove the problem areas. New insight into the environmental problem resulted in redefinition. Recognition of the limitations of the existing instrumentation yielded a new approach. This picture is not only outlined above, but is how successive policy-makers have defined their policy actions.

But is the argument that is presented here really correct? Are all these developments really only the result of considered choices? Some doubts arise if we look not only at what has been done, but also at what might have been accomplished.

Together with the development of environmental policy as a complex of, chiefly, licensing systems in the seventies, very different approaches were chosen by other Ministries in other policy fields. Thus, energy conservation was encouraged by the Ministry of Economic Affairs through enlightenment and subsidies (Van der Doelen 1989); water pollution was combated by the Water Boards and the Ministry of Traffic and Waterways by means of government facilities and charges (Bressers, Huitema & Kuks 1995). After it had long refused to take the problem seriously, the Ministry of Agriculture tackled agricultural pollution. When doing this it used the great variety of instruments that is also typical of general agriculture policy (Termeer 1993). The choice of licensing systems as a means to combat pollution by firms, in particular, was common practice if we look abroad, but as shown by the examples it was not always self-evident.

In the early eighties the first attempts to get in touch with the target groups were seen. The environment minister Winsemius (see Winsemius 1986: 61-67) initiated vigorous efforts to persuade environmental policy-makers and target groups to abandon their entrenched positions. Direct individual contacts mitigated any hostile thinking in the individual Ministries, making the Ministries' own policies less dependent on the involvement of other Ministries. This appeared to have been inspired by general ideas on good management rather than being a response to the inadequacy of other policy instruments. In those days there was not yet any question of a 'crisis mood' concerning the environment. Even before the National Environmental Policy Plan (NMP) was passed, then Minister Nijpels concluded several covenants (Klok 1989). Such developments did not take place, or to a far lesser extent, in Germany (viz. Weidner 1996). Already preceding the 'bankruptcy' of the old approach, a great deal of preparation went into the approach which was to bear fruit in the nineties.

When Alders, the Minister under whose leadership the new policy approach really took shape (1989-1993), was appointed, he still stated that no new covenants would be concluded during his term of government. One year later the NMP announced not one, but two new directions that were intended to supplement policy strategy. In addition to the emphasis on cooperation with partners, at least as much attention was given to the need to develop a system of financial stimuli. However, little came of this in practice. Alders even declared that in retrospect he felt that his efforts on behalf of a CO2 charge, which did not make it, was 'his greatest mistake as a Minister', because he gave those who opposed environmental policy the chance to regroup and join forces, and because the proposal cost a disproportionate amount of energy on the part of himself and his Ministry (oral information, 1996). In brief: the choice of the 'consultation strategy' was certainly not an undisputed one, or free of competition by other ideas. Why this addition bore fruit while the others did not, cannot be understood from the perspective of a politician's intentions.

On reflection we find that changes in the direction of this policy sometimes anticipate and sometimes lag behind the developments to which they appear to be a logical response. From a distance it appears that forces other than purely rational responses to the above-mentioned insights and changing circumstances to which they appeared to correspond so beautifully were at work.

In previous articles, one of the authors of this paper examined these developments from the perspective of the relationship between government and target groups in the policy network (Bressers 1998, Bressers & O'Toole 1998). The chance for a certain instrument type to be chosen is explained by the degree to which it corresponds with the ways in which authorities and target groups interact in the policy network. Instrument choice is seen to re-create and even reinforce the ongoing relationships under normal conditions.

Present position of the approach. A fundamental principle underlying the 'covenant' policy approach is that the responsibility for reaching the environmental targets lies primarily with the target group (Suurland, 1994). Apart from negotiated agreements (or covenants), environmental business management systems and their certification (ISO, EMAS), environmental reporting and accounting and liability rules can also be seen in this perspective. The primary reason for branches of trade to join negotiated agreements is the recognition that the future of industrial production is at stake in defining the boundaries of sustainable development. Second, industry influence would only increase compared to the situation at present. Finally, the market increasingly makes demands on environmental conditions (Van den Broek and Korten, 1997). Apart from these, Suurland (1994) recognizes the significant advantage of streamlining licensing and enforcement procedures. He emphasizes the advantages of integration of sectoral industrial and environmental policies and the integration of environmental and strategic company planning.

In 1995 the Dutch Prime Minister presented a document which contained indications for negotiated agreements (Staatscourant, 1995:249). The document posits that whenever a choice is to be made between regulation and a negotiated agreement, regulation should be preferred. However, if greater efficacy and effectiveness is expected from the conclusion of a negotiated agreement, the option could be considered in four cases. First, in anticipation of regulation, a negotiated agreement can reach results in the interim. Second, if regulation is expected to become superfluous in the near future, this can be sped up by concluding a negotiated agreement. Third, a negotiated agreement can serve the goal of exploring possible forms of regulation. Additionally, a negotiated agreement might be able to support regulation. To what extent negotiated agreements are embedded in Dutch environmental policy today is shown by the fact that the Dutch ministry for the environment is again and officially labelled as a 'negatively prioritized policy instrument'. Since subsidies and taxes are usually not an option, this means that negotiated target group policies still have the lead in Dutch environmental policy.

The challenge of environmental policy has shifted from winning corporate co-operation to harnessing corporate creativity. Dutch environmental policy now emphasizes consultation between government and target groups and encourages self-regulation among businesses. Such a policy requires delicate handling. Consultation can succeed only when realization of the environmental objectives is ultimately perceived by all participants to be 'inevitable'; and this perception can only be achieved by means of sufficient social and political pressure. In such a twin-track policy it is vital to achieve an optimal fine-tuning of legislation and enforcement on the one hand, and consultation and self-regulation on the other.

3. THE RESULTS OF THE COVENANTS

In 2002/2003, commissioned by the Dutch ministry of the Environment (VROM), we carried out a study on the effectiveness of the covenants. Through this study we aimed to identify success and fail factors. We focused on environmental covenants that have been implemented between VROM and private parties. Since 1989 many such agreements have been reached, including the broad-based, NMP-wide agreements with industry sectors. We did not include the many agreements that have been concluded between national, regional and local authorities.

It was an extensive study in which we used various methods for data collection. The initial step was telephone interviews with "the most neutral insider" (almost always the professional mediator that is hired to guide and organize the negotiation processes)[1]. During several phases of the study we talked to almost 120 people. We judged the content of the agreements themselves on aspects as the level of ambition, clarity, etc. We also drew in existing evaluation studies, implementation studies and monitoring reports. These documents often hold concrete, quantitative data on target achievement. So, while our own data relied heavily on the perceptions by the actors involved, we also included hard data from other sources. The data were further complemented by workshops that were organized for 8 covenants. In total some 60 people participated in the workshops. Among them were representatives of all relevant parties: various governmental agencies, industry, trade associations, independent mediators, environmental NGOs, etc. The debate within the workshops focused on the surplus value of covenants over other policy instruments. Of course, the information that resulted from the workshops cannot be used in a statistical analysis. It can, however, be used to explain certain tendencies we observed and to provide additional explanations. In this section we report on the most interesting findings.

In our study we identified 20 covenants with private parties that had already expired and 39 that were still valid, see table 1.[2] These agreements cover a broad area. Some concern a specific issue, for instance the kind of paint to use for indoor staircases. Others concerned the full range of goals from the NMPs. Covenants had been used with different motivations. The most important motivation is to anticipate regulation (almost 40% of the cases). In some of these cases governments do not have sufficient knowledge to issue regulation. Through the use of a covenant they hope to gain this knowledge. In other cases it was thought that covenants could be implemented faster than regulation. A second motivation for the use of covenants is that regulation was not considered to be an option (13%), for instance given the complexity of the target group. In these cases regulation presented no real alternative. Furthermore, our respondents identified covenants as symbolic policy measures in 17% of the cases. In these cases the covenant's main function was to highlight an environmental problem.

Table 1. *Overview of covenants*

Intention Statement for the metal-electro industry
Intention Statement Metal Industry
Agreement on the Environmental Policy for the printing industry
Intention Statement Chemical Industry
Policy Statement on Environmental Goals for the building industry
Intention Statement Implementation Environmental Policy for the dairy industry
Intention Statement Implementation Environmental Policy Oil and Gas industry
Intention Statement Implementation Environmental Policy Textile and Carpet industry
Intention Statement Implementation Environmental Policy Paper and Cardboard Industry
Intention Statement for the Meat industry
Intention Statement for the Rubber and plastics industry
Covenant Glasshouses Horticulture and Environment
Advice on the Environmental Policy for the concrete industry
Covenant benchmarking energy efficiency
Packaging Covenant I
Intention Agreement on a Collecting and Disposal Scheme for Used Packaging Glass
Intention Agreement on a Collecting and Disposal Scheme for Used Paper and Cardboard
Packaging Covenant II
Covenant on Mercury Batteries
Covenant on Liquor packaging
Covenant on PETP-bottles

Covenant on Trucks
Covenant on Alkaline Batteries.
Covenant on Sprayers
Covenant on Energy Savings Existing Houses
Covenant Wooden Stairs
Covenant on Youths
Covenant Synthetic Material Waste Agriculture
Covenant on Electricity Production
Covenant Implementation of Long-term Plant Protection
Agreement on Implementation Environmental Policy Bulbs
Reduction plan Hydrocarbon
Sustainable planning project RIJNMOND
Covenant Sustainable Building
Agreement on Diesel
Covenant on Detergents
Covenant on Crates
Covenant on crop protection
Covenant on Asbestos
Covenant on Waste from inland shipping
Covenant VOTOB
Cargo Residues from push tugs
Protocol on treating soil contamination Gas Stations
Covenant Green Label
Covenant on Industrial Plastic Waste
Covenant on paper isolated cable residues
Covenant on treating soil contamination Dutch Railways
Covenant waste incinerators
Sustainability Charter Higher Education
Environmental Covenant for Swimming Pools, Sports and Skating Facilities
Covenant on the treatment of soil pollution North Nieuwelandsepolder-South at Maassluis
Covenant on the reduction of carbon dioxides of carbon-based power facilities
Covenant on cost sharing for cleaning up soils of industrial estates
Management agreement on Gelder Valley
Implementation Plan for the Channel Zone Zeeuwsch-Vlaanderen
Intention Statement for the areas of Gas Factories in the province of Gelderland
Agreement on the Amsterdamseweg at Arnhem
Agreement on the Administration of the handbook on cleaning up technologies
Covenant Waste of Fisheries

Often the covenant was implemented following societal pressure as well as governmental pressure to deal with the problematic situation (see table 2). This means that covenants are often used within a context where both target groups as well as governmental actors find further action inevitable. A majority of the cases the built on existing collaboration with the target group, although this doesn't necessarily imply that the level of mutual trust is also high from the start.

Table 2. *Initial circumstances*

	(Completely) Agree	***(Completely) Disagree***
There was a great societal pressure	74%	26
There was great pressure from the government	77	23
There was already close collaboration previous to the covenant.	61	39
There was great trust preceding the covenant[3]	48	46

The results of the covenants concern many different aspects. Of course, getting direct environmental results is important. As we indicated earlier, covenants are popular in periods when it is obvious that the permit system will not deliver sufficient results and when there is a need to raise the level of ambition. Next to achieving direct environmental results, the development of new technologies, learning effects, changing attitudes and the improvement of collaboration between governments and target groups are also goals. For describing the results of the covenants we have chosen the following four areas:

- *Environmental results:*

Regarding the environmental results, we identified the level of ambition that was laid down in the covenants in relation to existing regulation. We also looked at compliance and the extent to which covenant targets were reached or are within reach. Furthermore, we asked whether the use of the covenant resulted in a fundamental improvement of the target group's environmental performance. In figure 1 we summarize the results for three indicators:

- Target achievement
- Fundamental improvement of performance
- Development of new technologies

- *Side effects:*

In addition to direct environmental results covenants can have several side-effects; such as building capacity for future environmental improvements. An important side-effect concerns changing the attitudes of the target group towards environmental issues. As one of our respondents said: "Covenants are attempts to share the responsibility for environmental affairs". Through the negotiations and informal contacts governments hope that target groups internalize the goals of environmental policy. The many contacts can also lead to increased mutual understanding, which can then result in improved collaboration between governmental actors and the target group. Furthermore, through collaborating and discussing long-term agendas and implementing joint research programs, another possible side-effect is the development of new knowledge with regard to achieving ambitious environmental targets. Close collaboration harnesses industry creativeness for innovative solutions, thus enabling further policy development. Finally, negotiations may lead to an improved policy integration.[4] Several scholars see this as essential for sustainable development (cf. Lafferty 2002).

In figure 1 we summarize the development of side-effects for three indicators:

- Changing attitudes of the target groups
- New knowledge on achieving ambitious targets
- Improved collaboration between governments and target groups

• *Effectiveness:*

In addition to analyzing results an important question is to what extent these results can be attributed to the covenants. This is a difficult task since the covenants are not implemented in isolation. In both telephone interviews and workshops we spoke with the respondents about the role of the covenant in the policy mix, especially in comparison with other instruments such as regulation. We also discussed possible autonomous developments and the way in which the covenants influenced these.

In figure 1 we summarize the effectiveness for two indicators:

- Role of the covenant in getting results
- Regulation would yield less results

• *Efficiency:*

Environmental voluntary agreements are often thought to improve efficiency through distributing costs among individual members of the target group in a way that considers the individual cost characteristics. Another element that might improve efficiency is the phasing of measures by the target group. Often, target groups get some flexibility in integrating the measures into their own investment schemes. In figure 1 we summarize the efficiency for three indicators:

- Minimizing total costs
- Improved efficiency through distributing costs over the target group
- Improved efficiency through phasing of costs

In figure 1 below we summarize the results. The data have been gathered using a five-point scale. The figure shows the average score for all covenants; a score between 1 and 5. When necessary the data has been recoded so that the size of the shaded area indicates positive results (the bigger the size the more positive the results are).

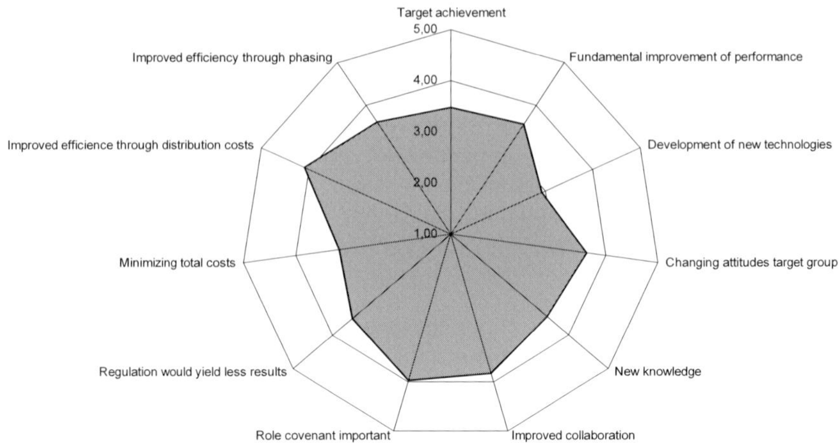

Figure 1. *An overview of the results of the covenants*

Notably, almost all scores are above the average score 3, indicating that the majority of our respondents are relatively positive on the totality of the results. We judged the targets laid down in the covenant as ambitious, although not necessarily much more ambitious as regulation. This is due partly to the fact that covenants in a lot of cases are explicitly linked to regulation. For instance, sometimes the environmental license is used as a fall-back option if individual members of the target group don't live up to the agreement (see also the next section). Maybe this link explains our respondents' positive view of compliance, despite the inherent problem of free-riders.

Most importantly, the judgment on target achievement is positive. Our respondents saw covenants as a positive contribution to the results of the last decade. We should note that this is not just perception. Many evaluation studies on individual covenants during the last years as well as progress reports confirm these results. In cases where the covenant did not reach its goals this was mostly due to an unrealistically high level of ambition and lack of motivation by the target group, next to phasing in the investment plans by the target group, free riders, and a lack of support by governmental agencies.

The results also point out that covenants have significant other benefits beside direct environmental results:

- The covenants have led to improved collaboration between governments and the target group (mentioned by 80% of the respondents);
- Governments and target groups have gained mutual understanding of their points of view (78%);

- The attitude of the target groups towards environmental issues has improved (74%);
- The covenants have resulted in more knowledge on reaching ambitious environmental goals (71%);
- The covenants have facilitated further developments in environmental policy (63%);
- The covenants have led to innovative behavior of the target group (54%).

Based on these results we may conclude that our respondents have a fairly positive judgment on almost all side-effects. They also see positive contribution to policy integration. The fact that different levels of government and different governmental agencies are present during the negotiations leads to a more coherent policy approach towards a target group. External integration is much harder to accomplish, although our respondents still see a positive contribution by the covenants.

Most of our respondents are fairly positive about the role of the covenant in reaching the mostly positive results. A quarter of them have doubts, however, whether 'old-fashioned' regulation would have resulted in less results.

The covenants also contribute to an improved efficiency through the possibility of distributing costs. Of course, this benefit is only realized in the case of sector-wide agreements. Remarkably, although minimizing costs is often mentioned as one of the benefits of covenants, it is clear that our respondents see missed opportunities here.

The indicator that is judged most negatively is the development of new technologies, that is technologies that are not already indicated in the agreement. Real technological breakthroughs are not to be expected. This is confirmed by other studies on Dutch covenants as well (see for instance Van der Woerd et al 2002). This is a highly relevant finding since originally covenants were implemented with the ambition to lead to fundamental innovation. The latest NMP mentions the need for so-called transitions. For this next step in environmental policy the hope is once again put on covenants. Our study provides little evidence to support this hope.

Covenants, thus, have strong benefits next to some deficits. The results as presented above illustrate this. To get an overall judgment we also asked for a response on the two following statements:

- "Covenants are valuable policy instruments";
- "Covenants can facilitate the achievement of ambitious environmental targets".

The results are shown in figure 2 and 3.

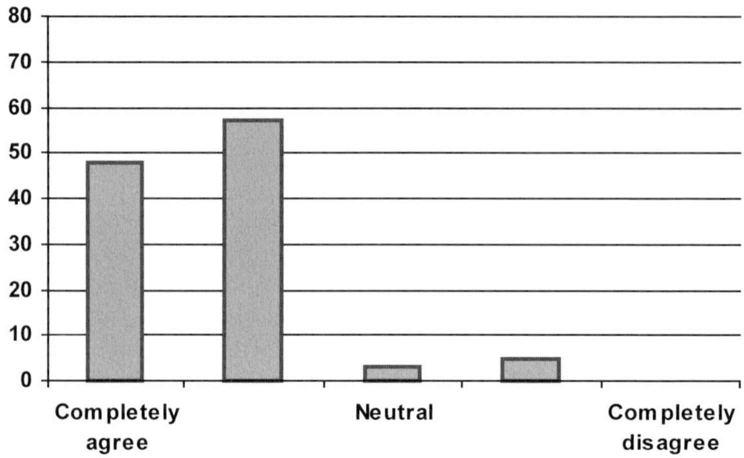

Figure 2. "Covenants are valuable policy instruments" (number of respondents)

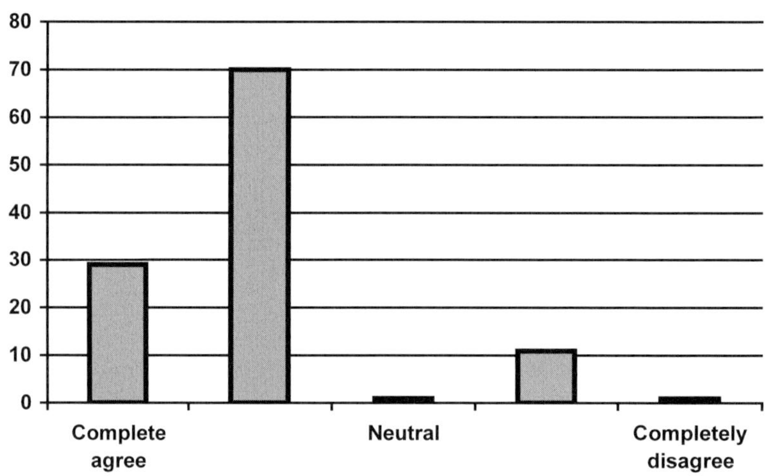

Figure 3. "Covenants can facilitate the achievement of ambitious environmental targets" (number of respondents)

Based on the figures 2 and 3 we conclude that the overarching picture is positive. Our respondents view covenants as important policy instruments. Although they see both pros and cons, they value the contribution covenants make. Most strikingly is that respondents from governments, industry and NGOs alike are fairly positive about covenants.

4. EXPLAINING THE SUCCESS OF COVENANTS: THE BROADER POLICY CONTEXT

It is not only the design and implementation of the instrument itself, but also the broader context in which a covenant is used that determines success or failure. Target groups do not respond to a specific instrument in isolation; rather their response depends largely on the environmental policy system as a whole. Thus, instruments that are not designed to fit with and complement the other elements of a nation's environmental policy system are less likely to be successful. One of the main characteristics of the Dutch approach is the integration and coherence of the diverse strategies and instruments.

Covenants are not the sole policy instrument that influences the behavior of the target group. Other instruments such as regulations and financial instruments are implemented simultaneously. Some instruments cannot be seen without the use of covenants, since their use originates from the agreements and they are an integral part of the negotiating processes. Some examples are subsidies and technology programs that are implemented to support the covenants. There is a close link with other instruments.

In contrast to other countries' practice the use of covenants is deeply embedded in the environmental policy system of the Netherlands. In this system public regulation, co-regulation, and voluntary action are all important, mutually sustaining, ingredients (cf. Lévêque 1996). The government decides on overall targets (as laid down in the NEPP); target groups get a say in what measures can best achieve the targets. Although there is always discussion about the specific goals for a target group, the range of the goals is a given. The gains for target groups are increased flexibility and efficiency. In the Dutch system of co-regulation, free riders will be forced by the regulators to meet the same standards in the end as participating companies. In practice, of course, regulators run into the well-known limits of the permit system and the link is looser than ideal. Implementation of the covenants proves to have the most benefits for the knowledge of local regulators. Direct effects on the permit system are far less visible, although governmental agencies admit that there is a convergence in permit requirements for comparable companies (cf. Hoek en Van de Peppel 2002). Not only the threat of other instruments is important though. Covenants need to be actively linked to other instruments. These instruments can support the covenant, and vice versa. Direct regulation can deal with free-riders. Subsidies can help lift technological barriers. The sole use of covenants will be less effective than the design of a complete package containing many instruments.

The current Dutch policy strategy leans heavily on integrating more collaborative approaches into the rest of the policy system, thus bringing environmental policy making more into line with the mediating national style

(Liefferink 1997). Over the years the traditional hierarchical stance with its distant, negative attitude towards target groups has evolved into a new approach designed to encourage collaboration, voluntary action, and self-regulation (Bressers and Plettenburg 1996). The use of covenants is the prime example of the new approach and marks a major shift in the philosophy of environmental governance and regulation. Consultation and collaboration, central characteristics of the use of covenants, are more likely to flourish in a corporatist context characterised by pragmatic bargaining and consensus building between administrative and societal actors. The use and effectiveness of negotiated agreements is, therefore, more easily realised in some countries than others.

5. A CHECKLIST FOR THE USE OF COVENANTS

Positive results notwithstanding it is obvious that some covenants are far more successful than others. This is not a surprise given the diversity in covenants in Dutch environmental policy. Covenants differ in terms of scope, target group, level of ambition, etc. Our analyses point out that not all situations are equally suited for the use of covenants. The next step in our analyses, therefore, focuses on the identification of success and fail factors. In the previous section we identified two success factors for the use of covenants: integration in the policy system and fit with the dominant policy style. Through interviews as well as in-depth discussion in the workshops we were able to distinguish more success and fail factors that concern the design of the instrument itself.

5.1. Success and fail factors

What did we learn? Throughout the study it was obvious that the covenant itself is very important for attaining positive results. There is a clear need for concrete and quantified targets in covenants with agreements on deadlines, responsibilities, etc. A covenant needs to send out a signal for action. Therefore, it needs to be clear who is addressed. Obviously, in some cases it is much easier to arrive at such agreements than in others.

The most important function of environmental voluntary agreements is the sharing of responsibilities with target groups. A fundamental principle underlying NEPP is that the responsibility for reaching the ambitious environmental targets lies primarily with the target group itself (Suurland 1994). Through the negotiating processes and the covenant target groups can show that they are ready and willing to take up this responsibility. A prime success factor, therefore, is the readiness by the target group to truly take responsibility. If this readiness is lacking the use of covenants is quite useless.

There are also some structural characteristics of the target group that are important. The most obvious concerns the representative of the target group. Governments need a strong partner to negotiate with. This partner should have a close link with the target group and be able to actually negotiate on behalf of the target group. In practice, this is not as obvious as it sounds.

Next, covenants need a climate of trust. Although trust can be built through the processes of negotiation, partners need to start with a basic level of trust in one another. On the one hand covenants do need to have clear and quantified targets in order to be effective. Having flexibility during implementation is, on the other hand, one of the main arguments for working with covenants. Flexibility can promote first movers by freeing the private sector to think creatively about how to achieve improved environmental performance, rather than responding in a more remote fashion to meeting regulations with proven and familiar technology, e.g. best available technology (BAT), maximum achievable technology (MACT), etc (Ashford 1993; NAPA 2001; Porter and Van der Linde 1995).Agreements shouldn't nail down every detail. Some aspects should, therefore, be left at the partner's discretion. This can only happen if partners start of with a basic level of trust in each other.

Even as our statistical analysis revealed the importance of having a stick behind the door, the workshop participants added the notion of having a stick *before* the door. Target groups need to have a clear argument for joining the 'voluntary' negotiations. This stick can have many forms. In some cases it was the government threatening to introduce tough regulations that drove industry to the negotiating table. In other cases it was public opinion that made industry realize change was inevitable. We also had the example of an environmental group that was able to develop so much pressure and mobilize the community, that the industry felt forced to act. Of course, the same holds true for the government actors, sometimes driven by the same (societal) forces as industry. Participants summarized it by saying that there needs to be a 'sense of urgency' by all partners.

Finally, our analyses reveal the importance of the negotiating processes for the effectiveness of the instrument. Contrary to our expectation, processes that could be qualified as tough or even quarrelsome proved to lead to better results than processes that seemed to run smoothly. Without negotiations in which all major disagreements have been discussed, it is hard to arrive at an agreement that is clear and feasible and supported by all partners. Potential conflicts need to be settled during the negotiations, otherwise the covenant will turn out to be a form of 'conflict displacement' (Mayntz 1976). A more or less independent participant that directs the processes can be useful in this respect. In order to discuss all relevant points partners also need to have frequent contact. Where the statistical analysis showed the importance of the socio-economic context for the results that covenants deliver, the workshops pointed at the crucial importance of having the negotiating processes themselves develop well.

As an (important) sideline, during the workshops it became obvious that assessing whether a covenant has been successful is a difficult task. In some cases, the partners can not agree fully on what to measure, the baseline for comparison, the value of parameters, etc. This shows the importance of continuous monitoring and the need to lay down in the covenant how this monitoring will take place, preferably via third party involvement.

5.2. Getting to the essentials

The statistical analyses based on the telephone interviews, together with the workshops, gave a good insight into the success and fail factors for covenants. We have combined the outcomes of both parts of the research into a checklist. This checklist distinguishes four phases:
1 Initial choice for the instrument
2 The negotiating processes
3 The covenant itself
4 Implementation.
For each phase the success and fail factors are summarized. Taken together they determine the level of success of the instrument to a large extent. Of course, not all factors are evenly important. Within the rather long list of factors we distinguish five central conditions:

- ✓ *The sectoral structure should enable the use of covenants;*
  A strong representative partner is a necessity, one that can really negotiate on behalf of the target group. This is probably easier realized in cases where the target group is not too large or too heterogeneous.
- ✓ *There is a clear stick before and after the door that keeps the target group motivated.*
  The covenant needs to be embedded in the policy system as a whole. In the end, covenants are more dependent upon other elements of this system for their effectiveness;
- ✓ *All major disagreements are solved during the negotiating processes;*
  Parties need to discuss potential major issues since these will eventually come to the forefront. If not dealt with during the negotiations, they are likely to disrupt the implementation of the agreements.
- ✓ *The covenant holds concrete, quantitative goals with responsibilities assigned to partners;*
  From reading a covenant it should be obvious who is supposed to do what and when, and to what end.
- ✓ *There is ample attention for monitoring, progress and evaluation throughout the processes.*
  We have seen some examples of processes in which in the end parties disagree on what has been established or this is simply unknown. As is true for all policy processes, monitoring is a crucial element for evaluating the value of the covenant.

When these conditions are met, chances are that a covenant will be successful. The use of covenants is, however, far from easy as they require governments to play different roles simultaneously. Our study shows the importance of employing different instruments and different strategies. Governments need to collaborate and negotiate, put pressure on target groups, punish and listen to arguments. Combining these strategies is a tall order.

6. CONCLUSIONS

Since the end the 1980s we have seen a worldwide trend of an increasing use of voluntary approaches and covenants (OECD 1999). The Netherlands is one of the countries in the forefront of this movement. The use of covenants is inchoate. The question of their efficacy is to some extent still unanswered, as is the question of identifying success and fail factors. We tried to systematically contribute to answering to those questions. Our main focus was analysing the value of covenants in the total policy mix. The main research question focused, therefore, on identifying the situation in which the use is most appropriate and situations in which use should be dissuaded from.

All in all, our judgment is quite positive on the contributions that covenants can have to successful environmental policy. Next to the positive environmental results that have been realized, the main benefit of covenant building is found in the concomitant processes. Through these processes mutual trust is strengthened, new knowledge is developed, and partners have the option of building their relationship in a constructive manner. There are some constrains: technological breakthroughs do not evolve easily from covenants and covenants are not a panacea for all situations. The level of information proves to be important for the surplus value. Without some basic understanding of the environmental problem at hand, establishing clear targets is very difficult. If all partners know beforehand what solutions there are to an environmental problem, then it is questionable whether a covenant is the most efficient manner in which to proceed. We have also come across vicious processes that can carry on for years. The question of whether these efforts are justified is legitimate. This means that covenants are best used in a certain phase of a policy cycle, namely for dealing with problems that need further exploring before solutions are found. When these solutions become more obvious, the question is justified whether the covenant can be succeeded by regulation.

With our analyses we have shown that covenants are dependent upon other, accompanying policy approaches (sometimes based on direct regulation or subsidies) for their effectiveness. In the end the real question, therefore, is not whether the covenant is effective or not; it is whether the policy system as a whole is effective or not. Given the results of our studies we feel that covenants can play an important role in this system.

7. NOTES

[1] In case there was no such 'neutral insider' we held two interviews, one on the side of the authorities and one with a representative of business. This was done in 15 cases.

[2] Furthermore we identified 73 covenants between public authorities, 41 non-environmental covenants and 35 covenants on energy. These agreements were excluded from our analysis, since our focus was on agreements between public (VROM) and private parties.

[3] Six percent of our respondents gave a neutral answer to this proposition.

[4] We distinguish internal integration (coherence of diverse policy approaches aiming at a specific sector) and external integration (coherence between environmental programs and programs on other policy fields).

8. REFERENCES

Ashford N. (1993), Understanding Technological Responses of Industrial Firms to Environmental Problems: *Implication for Government Policy. Environmental Strategies for Industry*. K. Fischer and J.Schot. Washington D.C., Island Press.

Brand E.L., H.Th.A. Bressers and J.J. Ligteringen (1998), *Policy science approach and Dutch aspects of negotiated agreements*, Enschede: CSTM.

Bressers H.Th.A. (1985), *Milieu op de markt* (Environment on the market), Amsterdam, Kobra.

Bressers H.Th.A. and Klok P.J. (1988), Fundamentals for a theory of policy instruments, in: *International Journal of Social Economics*, Vol. 15, No. 3-4, 1988, pp. 22-41.

Bressers H.Th.A. (1989), Naar een nieuwe cybernetica in de beleidswetenschap (Towards a new cybernetics in the policy science), inaugural speech, Enschede: University of Twente.

Bressers H.Th.A. (1998), The choice of policy instruments in policy networks, in: B. Guy Peters and F.K.M. van Nispen (eds.), *Public policy instruments,* Cheltenham: Edward Elgar, pp.85-105.

Bressers J.Th.A., L. Plettenburg (1997), The Netherlands, in: M. Jänicke and H. Weidner (eds.), *National environmental policies: a comparative study of capacity-building*, Berlin: Springer, pp. 109-131.

Bressers H. (2003), Understanding the implementation of instruments: How to know what works, where, when and how, in W.M. Lafferty (Ed.), *Governance for Sustainable Development: The Challenge of Adapting Form to Function.* Cheltenham: Edgar.

Bruijn T.J.N.M. de and. Lulofs K.R.D (1996), *Bevordering van milieumanagement in organisaties* (Stimulating environmental management in organisations), Enschede: University of Twente, dissertation.

Bruijn T. de and Norberg-Bohm V. (2001). *Voluntary, Collaborative, and Information-Based Policies: Lessons and Next Steps for Environmental and Energy Policy in the United States and Europe.* Cambridge, MA, Harvard University: 63.

Bruijn T., Bressers h., Lulofs K.& van der Veer A. (2003), *Evaluatie Milieuconvenanten: Eindrapportage (Evaluation Environmental Covenants: Final Report)*, Enschede: CSTM.

Clercq M. de, Seyad A., Suck A. and Ameels B. (2000), *A comparative study of environmental negotiated agreements*, Ghent: CEEM.

Clercq M. de (Ed.) (2002), *Negotiated environmental agreements in Europe: Critical factors for success*, Cheltenham: Edward Elgar.

Environmental Environment Agency (1997), *Environmental Agreements: Environmental effectiveness*, Copenhagen.

Glasbergen,P. (1998), Partnership as a learning process, environmental covenants in the Netherlands, in: Glasbergen P. (ed.), *Co-operative environmental governance, public-private agreements as a policy strategy*, Dordrecht: Kluwer Academic Publishers, pp. 133-156.

Gray B. (1989), *Collaborating: Finding Common Ground for Multiparty Solutions*. San Francisco,

Jossey-Bass.

Hartman C. L., Hofman P.S., et al. (1999). "Partnerships: a path to sustainability." *Business Strategy and the Environment* 8(5): 255-266.

Klok P-J.(1989), *Convenanten als instrument van milieubeleid* (Covenants as instruments of environmental policy), Enschede: University of Twente.

Klok P-J. (1991), *Een instrumententheorie voor milieubeleid*, Enschede: University of Twente, dissertation, (Partly summarised in English: Bressers H.Th.A and Klok P-J, *Explaining implementation and effectiveness of policy instruments*, paper OECD conference on the effectiveness of economic instruments in environmental policy, Paris, June 1995).

Mayntz R. (1976), *External Pressures and Conflicts in the Formation and Implementation of Environmental Policy.* ECPR, London.

NAPA (2001), *Leading Change: Advancing Effective Governance in the 21st Century.* Washington, DC, National Academy of Public Administration.

OECD (1999), *Voluntary Approaches for Environmental Policy; an Assessment.*

Peters B. G. (1993), *American Public Policy: Promise and Performance.* Chatham, New Jersey, Chatham House Publishers, Inc.

Porter M. and Van der Linde C.(1995), "Toward a new conception of the environment-competitiveness relationship." *Journal of Economic Perspectives* 9(4): 97-118.

Soet M.C. de (1990), Omgaan met milieuconflicten in de besluitvorming: de consensusbenadering voor win/win-uitkomsten (Dealing with environmental conflicts in policy formation: the consensus approach for win/win outcomes), in *Milieu*, vol. 5, no. 1, pp. 8-13.

Staatscourant (1995), *Aanwijzingen voor convenanten* (Indications for negotiated agreements), p.249.

Suurland J. (1994), Voluntary agreements with industry: the case of Dutch covenants, in: *European Environment*, vol. 4 (4), pp. 3-7.

Tweede Kamer der Staten-Generaal (1988-1989), *Nationaal Milieubeleidsplan: kiezen of verliezen* (National Environmental Policy Plan: to choose or to lose), 21 137, no. 1-2, The Hague.

Tweede Kamer der Staten-Generaal (1989-1990a), *Nationaal Milieubeleidsplan plus (National Environmental Policy Plan-Plus)*, 21 137, no. 20-21, The Hague.

Woerd K.F. van der, van der Grijp N.M., De Boer J. (200"9, *Effectiveness of sectoral voluntary agreements*, IVM VU, reportnr. E-02/01, maart

ANALYSING THE EFFECTIVENESS OF AN ENVIRONMENTAL VOLUNTARY AGREEMENT: THE CASE OF THE AUSTRALIAN NATIONAL PACKAGING COVENANT

R.L. BURRITT[#], H. LEWIS[##], K. JAMES[##]

[#]*School of Business and Information Management, Australian National University, Canberra, Australia*
[##]*Centre for Design, RMIT University, Melbourne Australia*

Abstract. In Australia, corporate signatories to the National Packaging Covenant (NPC) agree to engage in a collaborative approach with government to improve the management and environmental performance of packaging in an effective manner. The NPC, signed on 27 August 1999, is subject to review after five years and a set of criteria are needed to establish its effectiveness. Based on the notion of 'effectiveness analysis' this paper examines whether the NPC can be seen to be effective in relation to a set of criteria that address the following aspects: the confining of corporate action; institutional structuring of agreements; and checking of performance outcomes. Comment is made on the results of applying effectiveness analysis to the NPC and suggestions are made about potential changes to any future Covenant that is negotiated.

1. INTRODUCTION

The packaging industry is under increasing pressure to take greater responsibility for reducing environmental impacts of its products throughout their life cycle. This principle is often called 'product stewardship'.

In Europe, Japan and parts of the US and Canada, there are directives and legislation in place that require the packaging industry to implement product stewardship in some form, while Australian governments tend to have a preference for 'voluntary' approaches. This is reflected in the National Packaging Covenant (NPC), a product stewardship program that was negotiated between state, territory, federal and local governments and companies in the packaging supply chain in Australia. While participation is supposedly voluntary, it is supported by mandated alternative regulation (i.e., the National Environment Protection Measure for Used Packaging) to encourage voluntary compliance.

A range of policy instruments exists for encouraging environmental protection, including command and control regulation, self-regulation, voluntarism, educational and informational strategies, economic incentives and laissez-faire environmentalism (Gunningham and Grabosky 1998: p.38). Whereas command and control regulation directly prohibits or restricts activities that harm the environment, self-regulation may be defined as a process whereby an organized group, such as an

E. Croci (ed.), The Handbook of Environmental Voluntary Agreements, 283 - 306

industry association, regulates the behaviours of its members. In contrast, voluntarism is based on the corporation agreeing unilaterally to undertake actions that will protect the environment. No coercion from government is involved, however government may initiate, facilitate or co-ordinate the agreement. In general, voluntary agreements are non-mandatory partnerships between government and individual businesses where the incentives for action to improve environmental performance are related to mutual benefits rather than sanctions.

Voluntary environmental agreements are used by government to change the behaviour of corporations without direct intervention. The agreements are couched in terms of environmental stewardship of products, processes and activities. When the natural self-interest of corporations is to protect the environment and where participation of those responsible actions taken is encouraged, voluntary agreements can be effective as long as environmental protection becomes part of the collaborative norm (Gunningham and Grabosky 1998; Schaltegger *et al.* 2003). For their effect, voluntary environmental agreements rely more on education and persuasion than on incentives and power imposition, however, all of the characteristics of an enforcement pyramid play a part in producing effective voluntary environmental agreements, including the potential threat of negative incentives and mandated requirements if the voluntary mechanism does not produce the desired result (Burritt 2002). Figure 1 presents an enforcement pyramid associated with the regulatory mix.

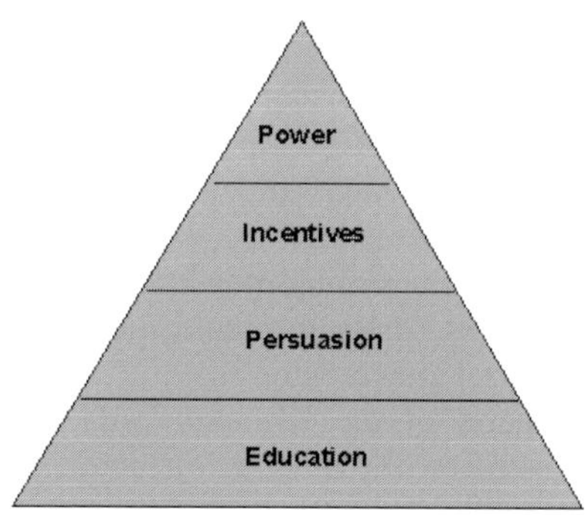

Source: Based on (Ayers and Braithwaite 1992).

Figure 1. *Enforcement pyramid associated with the regulatory mix*

Voluntary environmental initiatives are '*private or public sector efforts to improve environmental performance beyond existing legal requirements*' (Paton 2002: p.37). According to Paton, they can be divided into four categories: unilateral initiatives, private codes, voluntary challenges and negotiated agreements. Table 1 provides examples of all four categories in Australia.

Table 1. *Examples of voluntary environmental initiatives*

Type	**Definition**	**Australian examples**
Unilateral initiatives	Company-specific actions to improve environmental performance within the firm, and in some cases, its business partners.	Ford Australia's requirement for suppliers to have an EMS
Self-regulation (private codes)	Initiatives by industry associations, NGOs and standards organisations.	Chemical industry's *Responsible Care* and plastics industry *Plascare* programs; ISO 14000 series; Environmental Code of Practice for Plastic Shopping Bags; PVC industry's Product Stewardship program.
Voluntarism - Voluntary challenges	Government sponsored programs that create opportunities for firms to receive public recognition (no contracts signed).	EcoRecycle Victoria's *Waste Wise* program.
Voluntarism - Negotiated agreements	Contracts reached between government and industry to improve environmental performance.	Australian Greenhouse Office's *Greenhouse Challenge* program; National Packaging Covenant.

Source: Based on (Gunningham and Grabosky 1998; Paton 2002: p.39).

While the NPC includes specific requirements for governments (e.g. implementation of best practice kerbside recycling programs), this paper focuses on the effectiveness of the NPC in changing behaviour of participants in the packaging supply chain.

The remainder of the paper is structured as follows. Section 2 provides an overview of packaging and the environment as seen in its international context, and development of the NPC. Section 3 develops the research questions, while the following section addresses the research method. Section 5 presents the results of the analysis, while section 6 summarises the paper and draws several conclusions for policy makers, corporations and others interested in assessing the effectiveness of the NPC or other voluntary programs. This includes reference to a recent evaluation of the Covenant funded by the Covenant Council, and how it differed to the review undertaken for this paper.

2. PACKAGING AND ITS INTERNATIONAL CONTEXT

The packaging industry is under pressure internationally to reduce the environmental impacts of its products. Public awareness about packaging and its environmental impacts have their roots in the 1960's and 1970's, when vocal parts of the environment movement began to express concern about the impact of modern consumption patterns - the 'throwaway society' - on the environment, e.g. (Nader 1965; Toffler 1970; Meadows et al. 1972). A particular concern was the replacement of refillable beverage bottles with single-use bottles, making them a highly visible component of the waste and litter streams (Ackerman 1997: p.125). As a result of this pressure from the environment movement, packaging began to be defined as an 'environmental problem' that needed to be addressed.

In the United States, the growing litter problem was a particular focus of attention, resulting in the introduction of 'bottle bills' or Container Deposit Legislation (CDL) in many states, which places a deposit on bottles which could later be redeemed when the bottle was returned to the point of purchase. According to Rathje and Murphy (1992), these bills were opposed by the major beer and soft drink companies, who were trying to establish national markets based on single-use bottles. One of their strategies to combat further bottle bills was to establish recycling centres across the country as an alternative means of combating the litter problem - *'the upshot of all the contentiousness of bottle bills is that through one means or the other (or both) a variety of recycling programs got off the ground almost everywhere'* (Rathje and Murphy 1992: pp.198-199).

The plastics and packaging industries came under intense pressure in the 1970's and 1980's because '*they had become a politically incorrect symbol of the threat to the environment*' (Byars 1995, p.12). The symbolic role that plastics played in the emerging environmental movement was very real:

> "By definition the plastics industry was everything ecological activists wanted to expunge from the American experience. Since the early twentieth century, promoters of industrial chemistry and synthetic materials had boasted of transcending age-old limits of transitional materials by extending scientific control over nature. During the 1920's predictions of an expanding stream of inexpensive artificial goods had suggested material abundance as the basis for a utopian democracy. By the final third of the century that transcendence threatened to drain natural resources and pollute the society that supported it by generating a flow of irrecoverable, inassimilable materials - garbage, society's excrement" (Meikle 1995: p.264).

Kerbside recycling programs for packaging and paper have been introduced in most developed economies, and Australia is one of the most successful in achieving high recycling rates despite the absence of any regulation requiring participation. In a survey of residents in Victoria, Australia, 97% of respondents agreed that kerbside recycling is an essential service, and 92% agreed that municipal rates should continue to be used to finance recycling (EcoRecycle Victoria 1998). An estimated 3 million tonnes of packaging is used each year in Australia (Lewis *et al.* 2002), and approximately 800,000 tonnes of domestic packaging waste is recycled from kerbside (Nolan ITU *et al.* 2001). In addition some unknown amount of commercial and industrial packaging waste is recycled. While packaging has impacts upon the environment, it also performs many functions within the supply chain which need to be considered. Functions of packaging include the containment of the contents; protection and preservation of the product from external factors such as micro-organisms, water and oxygen; communication of content (ingredients and nutritional); and convenience in use when holding, opening and dispensing the product (James 2003). Packaging also plays a role in waste reduction (Kelsey 1989), particularly in the food distribution chain. For example, efficient packaging ensures fresh fruits and vegetables are delivered to the point of sale with minimal loss of product therefore reducing the environmental burdens associated with the product loss (Erlov *et al.* 2000).

3. INTERNATIONAL POLICY APPROACHES TO PACKAGING

International policy approaches to packaging range from command and control regulations through to voluntary agreements and self-regulation.

In many European countries a highly regulated approach to product stewardship has been adopted. Germany introduced the *Ordinance on the Avoidance of Packaging Waste* in 1991 that set mandatory recycling rates for packaging materials, and requires companies to take back and recover their own packaging from consumers. Recovery and recycling programs are funded through a collective industry initiative, the 'Green Dot' system.

The European Directive on Packaging and Packaging Waste initially adopted in 1994 sets recovery and recycling rates that member countries are obliged to meet. Company responsibilities are not clearly specified; rather they are required to meet the specific legal requirements in each jurisdiction. On 11 February 2004 the Directive was amended for a further 5 years phase.

Despite this overarching framework established by the Directive, approaches taken within individual jurisdictions varies considerably. In Belgium, Austria and Germany, trade and industry have total responsibility for packaging waste management. Elsewhere in Europe, a 'shared responsibility' has been adopted whereby responsibility is shared between industry and local government. While systems and responsibilities vary, product-recycling fees have been imposed in 13 countries.

The Netherlands has adopted a different approach, and one that became one of the models for the NPC. The Dutch Government introduced a Covenant in 1991 that required the packaging industry to introduce recycling programs, under the threat that if they were not successful then back-up legislation would be enforced. Responsibility is shared: local government is responsible for collection of waste and incineration of contaminants, while industry is responsible for delivering collected materials to the re-processor. The program is also supported by a ban on packaging materials going to landfill (Immerzeel-Brand 2002).

The Dutch Packaging Covenant is one of many covenants negotiated between the Dutch Government and industry since the late 1980's. A report published by the Dutch Government found that while there had been many successes arising from this approach to environmental policy, only the 'low-hanging fruit' had been harvested to date while difficult issues such as producer responsibility and eco-design had not been dealt with (Glazbergan, 1998, cited in Immerzeel-Brand, (2002: p.384).

The UK has been much further behind some of its European neighbours in both legislation and industry programs, and they are now struggling to meet their EU responsibilities. The *Producer Responsibility Obligations (Packaging Waste) Regulations* were introduced in 1997 and recently revised in 2003. This is a shared responsibility approach - local government retains responsibility for waste collection and disposal, but producers are required to recover and recycle certain percentages of packaging waste. Regulations have been introduced to divide responsibility for achievement of targets between different players in the supply chain in a clear way, either through individual action or collective activity (Castle 1997; NEPC 1999).

The United States has tended to rely on voluntary rather than regulated product stewardship programs, although some states have CDL or recycling targets for beverage packaging. The US EPA runs a voluntary program called *Waste Wise*, through which companies make a commitment to reduce solid waste. A similar approach is evident in Canada, with voluntary recycling targets at a national level, and CDL in some States.

In Japan the *Packaging Source Separation and Recycling Law* (1997) makes producers responsible for recycling of plastic containers, glass, paper cartons and paperboard boxes. They pay a fee to the Japan Container and Recycling Association to cover costs and the government sets targets based on recycling capacity.

Australia has approached the matter in a different way, through the introduction of the NPC.

4. THE NATIONAL PACKAGING COVENANT

In November 1996 the Australian and New Zealand Environment and Conservation Council (ANZECC) decided to commence negotiations with local government and the packaging supply chain on 'a national packaging agreement based on the principle of shared responsibility' (ANZECC 2000: p.1). This was followed by three years of negotiation involving ANZECC officials, the Australian Local Government Association (ALGA) and industry associations representing the food, beverage, supermarket, packaging and plastics industries.

At the formal launch of the National Packaging Covenant on the 27 August 1999, 8 Commonwealth, State and Territory Ministers, 2 local governments, 9 industry associations and 13 industry organisations/companies signed. By the end of October 2000, there were a total of 131 signatories and by February 2004 the total had increased to 629 (Figure 2).

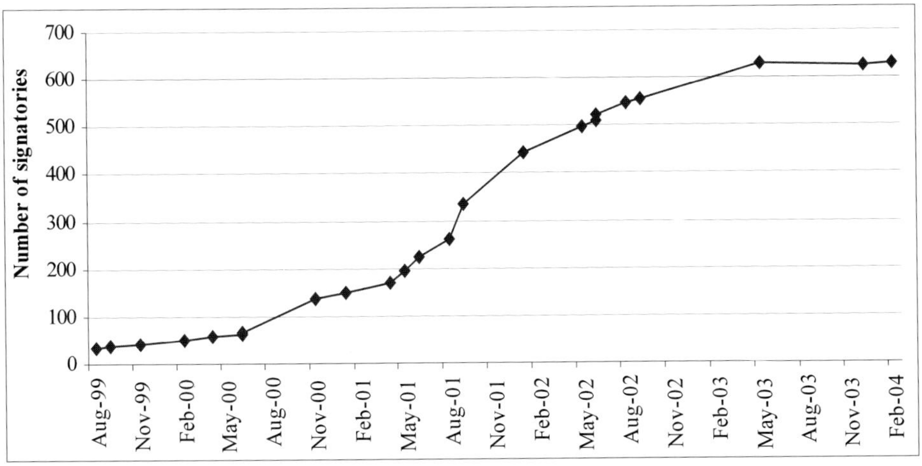

Figure 2. *National Packaging Covenant Signatories*

The stated objectives of the NPC are to (ANZECC 2000: p.3) (*emphasis added*):

- 'Establish a framework based on the principle of shared responsibility for the *effective lifecycle management* of packaging and paper products including their recovery and utilisation.
- Establish a collaborative approach to ensure that the management of packaging and paper throughout its lifecycle and the implementation of collection systems including kerbside recycling schemes, produces real and sustainable environmental benefits in a *cost effective* manner.
- Establish a forum for regular consultation and discussion of issues and problems affecting the recovery, utilisation and disposal of used packaging and paper, including costs'.

It also states that the NPC 'is based on the principle of product stewardship' and that, 'consequent on this principle, all participants in the packaging chain – raw material suppliers, designers, packaging manufacturers, packaging users, retailers, consumers, all spheres of government, collection agencies – accept responsibility for the environmental impacts associated with their sphere of activity' (ANZECC 2000: pp.4-5); see also (Schaltegger *et al.* 2003) regarding spheres of activity.

Signatories to the NPC are required to prepare an Action Plan 'for evaluating and improving environmental outcomes, as appropriate, in their production, usage, sale and/or reprocessing and recovery of packaging materials' (ANZECC 2000: p.6). By June 2003, 507 Action Plans had been processed by the National Packaging Covenant Council (NPCC 2003).

The NPC is a 'voluntary' commitment by signatories to reduce the environmental impacts of packaging. It is however, supported by regulation. The National Environment Protection Measure (NEPM) for Used Packaging Materials (NEPC 1999) was signed in August 1999 and is gradually being enacted through State & Territory Government regulation. Management and administration of the various elements of the NPC is shared between a number of organisations (Figure 3).

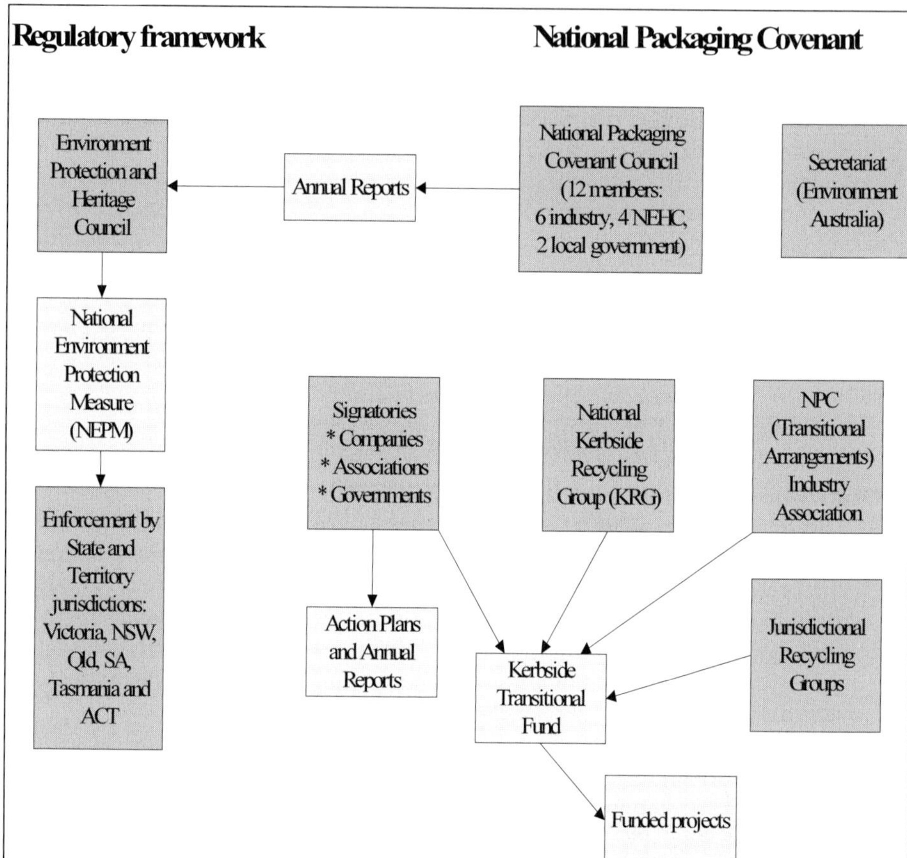

Note: New South Wales (NSW), Queensland (Qld), South Australia (SA), Australian Capital Territory (ACT) are states within Australia along with Victoria and Tasmania.

Figure 3. *Management and administration of the NPC*

5. RESEARCH QUESTIONS

In 1999, the NPC was established as a voluntary environmental agreement for a five-year period. Policy options beyond this date are under consideration by the parties involved. Past effectiveness of the Covenant will have an influence upon its future direction.

Two formal evaluations of the effectiveness of the NPC have been undertaken: (i) A review of Action Plans and Annual Reports (GHD 2002) and (ii) An evaluation of the National Packaging Covenant (Nolan ITU *et al.* 2004). The first review of action plans and annual reports aimed to identify whether key elements of the Covenant had been incorporated into Action Plans, and to provide an indication of the written level of understanding and commitment of signatories. While the review was generally positive, it identified a number of important gaps, for example:

- Most action plans failed to include details on data collection, measurable targets or the allocation of resources for implementation;
- Some companies had copied industry association templates without looking at specific implications for their own operations; and
- Many action plans involved information gathering and '*motherhood statements*' (GHD 2002: p.32) rather than detailed actions to implement Covenant commitments.

The second formal evaluation of the NPC, is currently (February 2004) being considered by the National Packaging Covenant Council, although the Packaging Council of Australia has reported that the three major recommendations from the report are (PCA 2004):

- The Covenant with its regulatory safety net should be retained for a minium of another three years;
- The operational elements of the existing Covenant/NEPM should be substantially improved; and
- The Covenant should focus on achieving measurable quantitative outcomes for consumer packaging on a nationally consistent basis.

The *first research question* is **has the NPC been effective in achieving its stated objectives?**

Effectiveness of voluntary environmental agreements can be analysed in various ways and a range of criteria have been suggested (see for example, (Allars 1990; EEA 1997; Industry Canada and Treasury Board 1998; OECD 1999; Industry Canada 2000)). Burritt (2002) combines the elements discussed in previous approaches into a comprehensive and useful checklist designed to assess gaps between expected and actual performance of voluntary agreements and to identify potential areas for improvement. His checklist draws upon a number of spheres of influence on voluntary agreements – political, socio-cultural, technological, economic and legal (Burritt 2002; Schaltegger *et al.* 2003) – which are integrated into thirty-three different characteristics of effectiveness. These characteristics are explored in Tables 2 to 7 below.

The *second research question* is **what are the gaps that are identified by the application of effectiveness analysis to the NPC?**

6. RESEARCH METHOD

Effectiveness is viewed in terms of the nearness of actual results to the achievement of expectations; for example, the term eco-effective has been used to capture the difference between the desired ecological footprint of a business and its actual footprint (Gray 2001). Effectiveness analysis is a tool that can act as a guideline and checklist for the assessment of the effectiveness of voluntary agreements (Burritt 2002: 367). The tool recognises that effectiveness is a multi-dimensional concept because the expectations of different stakeholders vary, can be in conflict and all expectations need to be addressed if overall effectiveness is to be assessed. The emphasis is on providing a pragmatic, or purpose orientated, test of performance (Chambers 1993; Schaltegger and Burritt 2000).

Key assumptions behind the usefulness of this approach to assessing the effectiveness of voluntary environmental agreements include: the perception of the organisation as a sometimes unstable, coalition of stakeholders (Simon 1949); recognition of the organisation as a separate legal entity with the power to sue and be sued in its own right; acknowledgement that managers act on behalf of the organisation, rather than on behalf of any one group of stakeholders (Chambers 1993: p.15); and the need for relevant performance information to be gathered and communicated to the signatories to any voluntary agreement in order to confirm or deny the success of the venture. Effectiveness analysis provides this touchstone.

Effectiveness analysis is a performance tool that has specifically been developed to identify gaps between the objectives, targets and goals established for voluntary agreements and actual results. Effectiveness can be analysed in terms of the number and variety of stakeholders (e.g. government and the business) that lead to a set of objectives whose effectiveness is being assessed. Effectiveness can also be assessed by considering the spheres of influence (e.g. environmental, political, economic, socio-cultural, legal, etc.), which lead to particular objectives being established for the voluntary agreement (Schaltegger *et al.* 2003). The aim of voluntary environmental agreements between government and the organisation is to affect behaviour of the organisation in a favourable way towards its environmental impacts, the organisation being seen here as a legal entity, separate from any of the stakeholders supporting it (e.g. managers, shareholders, local communities, environmental groups, government). In this situation, three characteristics of the organisation are thought to be of particular interest – confining, structuring and checking behaviours (Allars 1990: p.11).

Confining. Managers who sign an agreement, which specifies certain actions to reduce environmental impacts, confine organizations that would otherwise be free to follow their own course of action. Success in achieving specific objectives will depend on explicit commitment of the leaders who can act as champions for the initiative, as well as on the buy-in or 'ownership' of other parties. In addition, governments may adopt mechanisms designed to persuade managers of organizations to sign agreements and comply with them.

The organization may be offered a range of positive and negative incentives to sign, one of which is that legislation will not be introduced, or will be held in abeyance, if the voluntary agreements successfully help the government towards its own goals. Positive drivers include: the provision of technical services or training; publicity; subsidies; a seat at the table in policy decisions; provision of credit for early action; cost reduction; increased community support; awards and use of labels; and other competitive advantages. Negative drivers include: penalties for non-achievement; administrative costs of involvement; threats that other levels of the enforcement pyramid will be used if the voluntary agreement fails; increased transparency of performance; negative publicity such as 'naming and shaming' as used in Japan and the UK. Through these mechanisms the agreement constrains the power of an organization to act, while allowing government to reduce its visibility as the wielder of the big stick of environmental legislation, unless the voluntary initiative fails.

Structuring. Organizations are also constrained by *structure* imposed on them in the agreements. When voluntary agreements are signed managers acting on behalf of organizations accept an obligation to follow certain administrative procedures and processes. For example, these obligations may include organizational structures that develop and implement specific action plans, or the institutionalization of educational processes and continuous improvement in order to help the organisation towards its agreed targets. Size and scope of signatory organisations included in the agreement are also measures of effectiveness of the reach of an agreement.

Checking. Finally, organizations may be monitored or checked as part of the accountability process when they claim to have achieved certain targets, or undertaken certain agreed actions. Information gathering plays an important part when establishing the existing situation at the time an agreement is signed, identifying expected changes in the situation through actions taken, and when identifying any gaps in performance and developing awareness of responsible persons in the organisation in order to ensure accountability. Checking may be by self-assessment and reporting or through internal verification, or by independent second or third party verification, and, through reporting of performance, provides important 'transparency' and 'participatory' components to the generation of an effective dialogic accountability process.

The information used to evaluate the NPC for this paper included:

- A review of key policy documents such as the Covenant and the NEPM, as well as supporting documentation provided by the NPC Council (such as Action Plan Guidelines); and
- The two formal evaluations of the NPC undertaken to date, including a review of Action Plans and Annual reports by two of the authors (Lewis and James 2003).

Lewis and James (2003) undertook a review of action plans and annual reports from signatories to the National Packaging Covenant (NPC) as part of the performance review and benchmarking aspects of the evaluation of the NPC (Nolan ITU *et al.* 2004). The review was undertaken during October and November 2003. A total of 95 action plans and 39 annual reports were reviewed from a total of 54 signatories. Signatories were drawn from all sectors of the packaging and retail industry, including raw materials suppliers, packaging manufacturers, fillers/brand owners, retailers and government. The project team did not review action plans prepared by local government signatories, as these were dealt with under a separate evaluation by the Australian Local Government Association. The selection criteria for the action plan assessment were:

- Companies selected were drawn from each sector in the packaging chain in proportion to their signatory numbers.
- Companies selected were a mix of small (one sixth), medium (one third) and large corporations (one half).

- Companies selected were drawn from all states in proportion to signatory numbers.
- Companies selected represent the breadth of each sector e.g., retailers were drawn from food, hardware and general merchandise sectors.
- Two industry association action plans were reviewed to assess how these meshed with company specific plans.
- A significant proportion of companies selected have produced plans and at least one annual report.
- Across all sectors, states, and company sizes, no area was neglected overall.
- Where possible companies that have a national packaging involvement were selected (rather than a single state focus).

The methodology used to evaluate the action plans and annual reports included the use of the semi-quantitative scoring system developed by GHD (2002) for the first evaluation of action plans in 2002. This system was used to obtain a general overview of the degree to which signatories had addressed the different elements of the Covenant. General observations were made about the content, transparency and detail within action plans and annual reports, with examples provided of positive and negative findings.

7. USING EFFECTIVENESS ANALYSIS TO EVALUATE THE NATIONAL PACKAGING COVENANT

7.1. Confining: support and objectives

Table 2 presents the evaluation of the NPC against the effectiveness criteria of confining – support and objectives. There are no explicit requirements for Chief Executive Officer (CEO) support of the Covenant, although the review of action plans found that 'CEOs/senior management are involved' (GHD 2002: p.iii). The list of Covenant signatories includes many CEOs, but also marketing, sales and finance managers.

The objectives of the Covenant are many and varied. Most are very broad and ambiguous, such as '*effective life cycle management of packaging*'; collection systems for packaging that produce '*real and sustainable environmental benefits*'; and '*improving environmental outcomes*' at each stage of the packaging life cycle.

Other supporting documents refer more specifically to reducing waste, for example '*Your Action Plan sets out what actions you will take to contribute to packaging waste reduction and management*' (Environment Australia undated-b: p.2) or '*The National Packaging Covenant is an agreement to reduce waste from packaging: wasted material, wasted energy and money and waste going to landfill*' (Environment Australia undated-a: 1), but there is no benchmark data, no targets and no reference to how this will be measured. One of the four goals mentioned in one document is '*to ensure that the voluntary process continues*' (Environment Australia undated-a: p.3).

The ambiguity of the language used throughout the Covenant probably reflects the long and difficult negotiation process between the parties over a three-year period, and the many compromises made to reach agreement about wording of the final document. Nevertheless it is problematic because the terms are not defined and are therefore open to widely different interpretations.

***Table 2.** Confining: support and objectives of the National Packaging Covenant*

Checklist number	Generic description of characteristic	National Packaging Covenant characteristic
1	Confining: support and objectives	
1.1	*CEO support (champion) required*	There is no restriction on who can sign the Covenant on behalf of an organisation.
1.2	*Rank and file and other stakeholder buy- in*	Not required
1.3	Objectives	The Covenant's stated objectives are: • Effective life cycle management of packaging and paper; • Collection and management of packaging that produces real and sustainable environmental benefits in a cost effective way; and • Improving environmental outcomes in production, use, sale and/or reprocessing and recovery of packaging. The Action Plan Guidelines refer to: • Packaging waste reduction; and • Improvement in efficiency and sustainability of kerbside recycling

7.2. Confining: drivers

The main driver for signing the Covenant appears to be a negative one, i.e. the wish to avoid alternative legislation. Over 70% of companies surveyed in the food packaging supply chain claimed to have signed the Covenant for this reason (James 2003: p.163). The Covenant Action Plan Kit states that 'in the longer term, if industry fails to check the use of virgin materials and the amount of waste packaging going to landfill, future governments will become tougher. Local Government may also seek to recover more of the costs of kerbside collection' (Environment Australia undated-b: p.2).

The NPC Council stated in their 2003 Annual Report that 30 companies were reported to jurisdictions as not being in compliance with the Covenant commitments. Only 9 of these companies remain outside Covenant compliance and are being followed up by jurisdictions (Lewis and James 2003)

There are no obvious positive drivers for participation such as a plaque, logo, financial support or training. Signatories are listed on the Environment Australia web site (http://www.ea.gov.au/industry/waste/covenant/signatories.html), and action plans are published on the Packaging Council of Australia site (http://www.packcoun.com.au/). Apart from this there is very little publicity or exposure for companies that are involved in the Covenant and who are achieving good outcomes. This is in contrast to most other voluntary environmental programs in Australia such as the Commonwealth Government's *Greenhouse Challenge* program (http://www.greenhouse.gov.au/challenge/) and EcoRecycle Victoria's *Waste Wise* program (http://www.ecorecycle.vic.gov.au/). The *Greenhouse Challenge* Program provides participants with workbooks and guides, positive publicity through a public event with the Minister and a logo for use on corporate promotions. Waste Wise participants receive some advice and support from industry advisors or regional education officers, as well as a wall plaque and the right to use the program logo.

Some companies appear to regard the Covenant as an opportunity for competitive advantage. In the survey mentioned above, 18% of respondents mentioned good business, commercial or marketing reasons for participation (James 2003: p.163). VisyPak is probably unique in marketing 'covenant friendly packaging' (http://www.visy.com.au/overview/covenant.asp). Table 3 lists the NPC characteristics under the confining - positive drivers categories and Table 4 presents the confining – negative driver categories.

Table 3. *Confining: positive drivers*

Checklist number	Generic description of characteristic	National Packaging Covenant characteristic
2.1	***Positive drivers***	
2.1.1	Government technical or consultation services, or training, offered	Limited support, e.g. Action Plan Guidelines, Action Plan Kit.
2.1.2	External 'good' publicity	Signatories are listed on the Environment Australia web site. Action plans are published on the Packaging Council of Australia web site.
2.1.3	Direct subsidies for action	None
2.1.4	Influence activities (e.g. roll-on policy development)	Indirect influence through industry associations represented on the NPC Council and other committees.
2.1.5	Credit for future activities	None
2.1.6	Cost savings	There is potential for costs savings, e.g. through implementation of waste reduction

		plans, although not guaranteed.
2.1.7	Greater community support	Unlikely as there is very little promotion or public education about the Covenant.
2.1.8	Competitive advantage	The National Environment Protection Measure aims to protect signatories from competitive disadvantage through 'recover and utilise' requirements on free loaders.
2.1.9	Awards or plaques, logos	None

Table 4. *Confining: negative drivers*

Checklist number	**Generic description of characteristic**	**National Packaging Covenant characteristic**
2.2	***Negative drivers***	
2.2.1	Penalties for non-achievement (e.g. financial or suspension from agreement)	Brand owners that do not sign the Covenant, submit an Action Plan or who do not meet Action Plan commitments can be required to collect and recycle materials through State regulations linked to the NEPM.
2.2.2	Dollar cost of participation	The signatories meet administrative costs. They are also required to contribute to the Transitional Fund for kerbside recycling activities over 3-years.
2.2.3	Threat of alternative regulation	There is a threat of alternative regulation if the Covenant does not succeed.
2.2.4	Loss of confidentiality a concern	Commercially sensitive information can be excluded from public reports, e.g. by inclusion in non-public appendices.
2.2.5	Threat of negative publicity	No provision for this.

7.3. Structuring: implementation

The main tools used in monitoring the implementation of the NPC are Action Plans and Annual Reports submitted by signatories, but evaluation is difficult due to the ambiguous and varied objectives of the Covenant, the absence of any requirement for companies to provide benchmark data on environmental impacts of their activities or to provide quantified data on reductions achieved. This problem is starting to be addressed, for example, in January 2003 Action Plan guidelines were amended to require 'performance indicators, targets and other measurables' (Environment Australia 2003). Table 5 lists the NPC characteristics against the structuring – implementation categories of the effectiveness analysis.

Table 5. *Structuring: implementation*

Checklist number	Generic description of characteristic	National Packaging Covenant characteristic
3.1	*Letter of intent or signed agreement required*	Organisations are required to sign the form that forms part of section 10 of the Covenant and send it to the NPC Council. An Action Plan must be submitted within 6 months of signing.
3.2	*Baseline inventory required*	Baseline data is not required. The Covenant does require: • That Action Plans have measurable performance objectives and mechanisms to monitor their achievement; and • That all signatories maintain records to enable them to demonstrate, on request, that they are meeting their Covenant undertakings.
3.3	*Action plans required for improvement*	Action Plans can be for between 1 and 5 years. Subsequent Action Plans must demonstrate continuous improvement.
3.4	*Basis for forecasts of reductions*	Forecast reductions in packaging waste or other indicators are not required.

7.4. Structuring: scale and scope

There were no specific targets set for levels of participation, although small companies are exempt (those with <1% market share). By June 2003 there were 608 signatories with 507 action plans processed and 324 of these being recommended to Council for registration (NPCC 2003: p.9). Prior to the introduction of the NPC there were three specific agreements in place with respect to paper, steel and PET in which less than 30 companies were covered. With the introduction of the NPC the entire packaging supply chain is covered and for the first time the major supermarket chains are included (Williams 2003).

Table ***6*** presents the NPC characteristics against the structuring – scale and scope categories.

Table 6. *Structuring: scale and scope*

Checklist number	Generic description of characteristic	National Packaging Covenant characteristic
4.1	*Size of organizations targeted*	Companies with <1% market share are exempt.
4.2	*Number of sectors or organizations included*	There are currently over 600 signatories and over 400 Action Plans published. Approximately 64% of company signatories are packaging users/brand owners, 17% are wholesalers/retailers, 15% packaging manufacturers and 3.5% raw material suppliers.

7.5. Checking

Covenant signatories are required to report annually on progress. These reports (and Action Plans) are published on the Internet. The Covenant Council is also required to report annually on issues affecting the Covenant, including performance of Covenant signatories. In its first annual report the Council states that:

> "In addition to the formal reporting procedures required by Ministers, a less formal but equally important requirement for an annual report will be to promote the Covenant and its achievements to the general public and potential Covenant signatories. The report will also assist in creating and maintaining credibility in the self-regulatory instrument which is very much based on cooperation, goodwill and trust between all participating parties" (NPCC 2000: p.4).

The 2003 Annual Report from the National Packaging Covenant Council contained information on the number of signatories, discussion on the validation and assessment of action plans, discussion on the random reviews of action plans and summaries of activities undertaken by the Kerbside Recycling Groups and the key industry associations.

There is no requirement for independent verification. The Covenant does provide for auditing of Plans, either on a random basis, in response to information supplied by a third party or on the Council's initiatives in order to '*determine if they adequately address the undertakings of the Covenant*' (NPCC 2000: p.10). Concerns about transparency and accountability are expressed in the Council's first annual report, which highlighted the need to develop (NPCC 2000: p.14):

- A credible audit process that provides independent verification of signatories achievements';
- Development of a mechanism of accountability, including consideration of a penalty for non-compliance; and
- Development of defendable and relevant measurable performance indicators for each industry sector to be used in the annual assessment of the Covenant.

A Covenant Council Review and Evaluation Working Group, which meets quarterly, has been established to examine the adequacy of organisation's action plans and commitments. The general finding of this working group is that overall, the action plans are improving in quality though there are a number of areas where plans need to be improved – i.e., not consistently meeting all of the assessment criteria and in particular the inclusion of measurable performance indicators and application of the Environmental Code of Practice for Packaging. The working group has randomly reviewed 57 action plans in 2002/2003 (NPCC 2003: p.5).

A more extensive review of action plans and annual reports was undertaken by Lewis and James (2003) for the most recent evaluation of the Covenant (Nolan ITU *et al.* 2004). One of the positive conclusions of the review was that action plans demonstrate that a significant amount of activity is occurring within the packaging supply chain to reduce environmental impacts, from purchasing through to operations and product design. The *Environmental Code of Practice for Packaging* (an important element of the Covenant) is now being incorporated into product development processes by many signatories, particularly by packaging companies and brand owner / fillers; and many brand owners / fillers are developing packaging material databases.

The review identified a number of aspects that need to be improved, for example:

- Many plans include commitments that are expressed in general terms using vague or non-committal language, without providing detailed lists of actions;
- There is an almost universal lack of measurable (quantified) targets in action plans;
- Most action plans do not indicate how data will be collected to measure progress;
- Most action plans do not include resources allocated to implementation of specific activities;
- Many of the Annual Reports do not specifically report against the original actions and measures; and
- Some reports include a list of achievements, but many do not, and some report on activities rather than outcomes (e.g., reports, feasibility studies, meetings etc).

The review made a number of recommendations in relation to transparency and accountability, for example that all signatories should produce publicly available action plans and annual reports. Any confidential information should be included in an appendix for viewing by the Covenant Council only (at present some plans and reports are not published at all because they include confidential information). It also recommended that, following approval by the Covenant Council, all action plans and annual reports should be published on the PCA web site, as soon as possible. The review found that there had been significant delays in publishing plans and reports.

A recommendation with respect to the submission of action plans and annual reports was that additional resources should be allocated to allow for effective monitoring and follow-up by the NPC Council, to ensure that action plans are submitted annually (or for longer periods of time but with annual updates in reports) and that annual reports are provided within 3 months of the end of each financial year. Signatories that do not meet these requirements despite follow-up by the Covenant Council should be considered for follow-up under the NEPM.

With respect to the content of action plans, the review recommended that a standard checklist be produced that provides guidelines on the required content of an action plan and annual report, taking into account different expectations depending on the size of the company and their place in the supply chain (raw material supplier, packaging manufacturer, brand owner / filler, retailer, industry association). The review recommended that all action plans include detailed actions, targets or outcomes, timelines, responsibilities, funding allocations and an indication of how data will be collected to measure performance.

Table 7 lists the NPC characteristics against the effectiveness analysis criteria of checking – reporting.

Table 7. *Checking: reporting*

Checklist number	Generic description of characteristic	National Packaging Covenant characteristic
5.1	*Reporting*	
5.1.1	None, full information or exception reporting	Signatories must report to the Covenant Council on progress against their Action Plan.
5.1.2	By company/by sector	Action Plans can be submitted by groups of companies (e.g. through industry associations) or local councils (e.g. through State municipal associations). Reporting can be done by the same organisations.
5.1.3	Periodicity	Reports must be provided annually.
5.1.4	Changes – in action plans	These must be addressed in annual reports.
5.1.5	Public disclosure (transparency)	Action Plans and Annual Reports for each signatory are published. The Covenant Council is also required to report annually on progress.
5.2	*Independent verification*	
5.2.1	Required/not required	Not required
5.2.2	Method of selecting verifier	NA

8. CONCLUSIONS AND RECOMMENDATIONS

Using Burritt's (2002) notion of 'effectiveness analysis', this paper set out to examine whether the NPC can be seen to be effective in relation to a set of criteria that address the confining of corporate action, institutional structuring of agreements and checking of performance outcomes. The analysis is used to address two research questions: 'Has the NPC been effective in achieving its stated objectives?' and 'What are the gaps that are identified by the effectiveness analysis of the NPC?' Based on this analysis, some general conclusions can be drawn about the effectiveness of the NPC and a list of recommendations are presented to help guide the development of potential improvements that could be made if the NPC is to continue beyond 2004.

8.1. The effectiveness of the NPC in achieving its stated objectives

The general nature of the NPC objectives makes it difficult to evaluate whether the stated objectives have been met.

The first objective, to 'establish a framework based on the principle of shared responsibility for the effective lifecycle management of packaging and paper products including their recovery and utilisation' appears to have been met by the establishment of the NPC policy and implementation framework.

The second objective to 'establish a collaborative approach to ensure that the management of packaging and paper throughout its lifecycle and the implementation of collection systems including kerbside recycling schemes, produces real and sustainable environmental benefits in a cost effective manner' has only been met to the extent that many of the key organisations involved in management of packaging and paper at end-of-life are now cooperating to a much greater extent through the NPC framework. No comprehensive assessment has been made to determine whether collection systems are producing real and sustainable benefits in a cost effective manner.

The NPC Council and the Jurisdictional Recycling Groups are helping to achieve the third objective, which is to 'establish a forum for regular consultation and discussion of issues and problems affecting the recovery, utilisation and disposal of used packaging and paper, including costs'.

It is clear that the most fundamental NPC goals of *effective life cycle management*, *sustainable environmental benefits* and *cost effective collection systems* are virtually impossible to measure without clear performance indicators and monitoring systems.

8.2. Gaps identified through the effectiveness analysis

An effective Covenant should ideally maximise both the environmental outcomes being sought by policy makers and the broader community, and commercial benefits to the packaging supply chain. The most critical areas to be addressed in the NPC are:

- The need for clearer objectives and targets, supported by a comprehensive data collection system to measure and report progress; and
- An increased focus on positive business drivers such as technical support and public recognition.

The potential effectiveness of the NPC in confining corporate behaviour is limited by the lack of clear objectives and targets. The positive impact of this is that it provides organisations with a high degree of flexibility in how they choose to respond, and can therefore choose actions that are the most appropriate and cost effective for their individual circumstances. The downside is that many organisations have failed to address the NPC requirements in a serious way, and some have tended to use 'motherhood statements' in their Action Plans, or simply copy industry association templates (GHD 2002). Such actions are unlikely to change industry behaviour in any substantial way.

The objectives of the NPC need to be re-written to give clear direction to industry on stakeholder expectations. This could start with providing definitions for vague terms such as 'effective life-cycle management of packaging and paper' and 'improving environmental outcomes'. For example, if reduction in the amount of waste packaging going to landfill is one of the agreed intentions of the NPC, then this objective needs to be made more explicit.

While the NPC includes a strong 'negative driver' through the NEPM for Used Packaging, the threat of regulatory action against non-performers needs to be seen to be real. To date there has been no legal action taken for non-participation or non-performance.

State and territory jurisdictions need to enforce the NEPM. Signatories need to understand that if they do not perform to acceptable levels that the 'threat' of the NEPM is real.

'Positive drivers' are lacking from the NPC. Some of the characteristics included in the effectiveness analysis, but absent from the NPC, are technical support and training, good publicity, subsidies for action and roll-over into other policy making. Technical support and training are likely to be particularly important in encouraging the uptake of new technologies and practices. Education and training have been found to be particularly important for small to medium sized enterprises (SMEs) in improving environmental performance (Gunningham and Sinclair 2002). SMEs tend to lack the resources and skills to integrate the environment into their business practices, and regulators may lack sufficient technical expertise to provide specific advice to even larger businesses. (Gunningham and Sinclair 2002)

It is necessary to provide sufficient education and training to signatories in the form of workbooks; information sessions to inform signatories of current status and future plans of the NPC; and an annual national conference where signatories can present their NPC activities to delegates.

While there is a requirement for signatories to prepare Action Plans and report annually, the effectiveness of this is limited by the lack of benchmark data on industry performance, and quantitative reporting against actions and targets. An appropriate framework for data collection is already included in the NEPM, which requires brand owners to annually provide information on units of packaging; weight of material used, recovered, recycled, used for energy recovery or sent to landfill. This data could then be used to calculate recovery rates.

One of the benchmarks for evaluation of the Covenant could be the amount of packaging collected and reprocessed in relation to the amount of packaging consumed in Australia. This could become a mandatory requirement for Annual Reports. Alternatively, the data could be provided to the Environment Protection and Heritage Council on a confidential basis for collation into national material or product recovery rates.

Linked to the need for better data collection, is the need for clear accountability. Reports must be published, so there is a degree of transparency. The large number of signatories and the diversity of organisations, approaches and actions, means that it is extremely difficult for even the most informed observer to evaluate the performance of individual organisations, and the industry as a whole, in achieving the program's objectives.

Develop a uniform system for data collection and reporting by signatories that would allow for a degree of aggregation and analysis by the NPC Council, and dissemination of results to broader stakeholders. This would also allow for easier verification of reported outcomes by the Council.

There are positive signs that the introduction of the NPC has initiated change within the packaging supply chain regarding environmental issues. Future negotiations on the design and direction of the NPC need to ensure that the momentum that has been achieved to date continues.

9. REFERENCES

Ackerman F. (1997), *Why Do We Recycle?* Island Press, Washington.

Allars M. (1990), *Introduction to Australian Administrative Law*. Butterworths, Sydney.

ANZECC (2000), *National Packaging Covenant*, Australian New Zealand Environment Council, *www.packcoun.com.au/NPC.htm*, Last accessed 15 January 2002.

Ayers I. and J. Braithwaite (1992), *Responsive Regulation: Transcending the Deregulation Debate*. Oxford University Press, New York.

Burritt R. L. (2002), Voluntary Agreements: Effectiveness Analysis - Tools, Guidelines and Checklist in *Voluntary Environmental Agreements: Process, Practice and Future Use*, ed. P. ten Brink, Greenleaf Publishing. Sheffield, UK.

Byars M., Ed. (1995), Mutant Materials in Contemporary Design. The museum of Modern Art,.New York.

Castle D. (1997), *Wasting Away*, Printing World, 20 October, www.dotprint.com/pack/waste.htm, Last accessed 27 October 2002.

Chambers R. J. (1993), "Positive Accounting Theory and the PA Cult." *Abacus* **29**(1)1-29.

EcoRecycle Victoria (1998), Public Views 98: Community Attitudes to Waste and Recycling, EcoRecycle Victoria, Melbourne,

EEA (1997), Environmental Agreements. Environmental Effectiveness, Vols 1 & 2, Environmental Issues Series No. 3, European Environment Agency, Copenhagen,

Environment Australia (2003), *What's new January 2003*, *www.ea.gov.au/industry/waste/covenant/whatsnew.html*, Last accessed 24 April 2003.

Environment Australia (undated-a), *The Covenant 'Action Plan' Kit*, *www.ea.gov.au/industry/waste/covenant/kit.html#download*, Last accessed 24 April 2003.

Environment Australia (undated-b), *National Packaging Covenant Action Plan Guidelines*, *www.ea.gov.au/industry/waste/covenant/pubs/action_plans.doc*, Last accessed 24 April 2003.

Erlov L. Lofgren C. and Soras A. (2000), PACKAGING - a tool for the prevention of environmental impact, Report Number 194, Packforsk, Kista, Sweden, June,

GHD (2002), Independent Review of Action Plans and Annual Reports. Final Report to the National Packaging Covenant Council., Gutteridge Haskins and Davey Pty Ltd, Sydney, November,

Gray R. (2001), "Forbidden Fruit." *Tomorrow: Global sustainable business* **XI3**(June)50-53.

Gunningham N. and Grabosky P.N. (1998), *Smart Regulation Designing Environmental Policy*. Clarendon Press, Oxford.

Gunningham N. and Sinclair D. (2002), *Leaders & Laggards: Next Generation Environmental regulation*. Greenleaf Publishing, Sheffield.

Immerzeel-Brand E. (2002), Assessing the Environmental Performance of Negotiated Environmental Agreements in the Netherlands in *Voluntary Industry Agreements: Process, Practice and Future Use*, ed. P. ten Brink, Greenleaf Publishing. Sheffield, Uk.

Industry Canada (2000), An Evaluative Framework for Voluntary Codes, Office of Consumer Affairs, Ottawa,

Industry Canada and Treasury Board (1998), Voluntary Codes: A Guide for Their Development and Use, Office of Consumer Affairs at Industry Canada and Regulatory Affairs Division at the Treasury Board, Ottawa,

James K. L. (2003), *Environmental life cycle costs in the Australian food packaging supply chain*, PhD Thesis, School of Accounting and Finance, Victoria University, Melbourne.

Kelsey R. J. (1989), *Packaging in Today's Society*. Technomic Publishing Company Inc, Lancaster, Pennsylvania, USA.

Lewis H. and James K.L. (2003), Appendix F: Action Plan and Annual Report Review. Report on the National Packaging Covenant Action Plans and Annual Reports from NPC Signatories (November 2003), Centre for Design, RMIT University, Melbourne,

Lewis H., Sonneveld K. , Fitzpatrick L. and Nicol R. (2002), Towards Sustainable Packaging. Discussion Paper, unpublished.

Meadows D. H., Meadows D.L. , Rander J. and Behrens III W. (1972), *The Limits to Growth*. Pan Books, Sydney.

Meikle J. (1995), *American Plastic: A Cultural History*. Putgers University Press, New Jersey.

Nader R. (1965), *Unsafe at Any Speed*. Grossman, New York.

NEPC (1999), Used Packaging Materials: National Environment Protection Measure, National Environment Protection Council, Adelaide, 2 July,

Nolan ITU, Lewis H., James K.L., Robinson B.and Watts B. (2004), Evaluation of the Covenant. Report to the National Packaging Covenant Council, Nolan ITU Pty ltd, East Kew, Victoria Australia, January 2004,

Nolan ITU, SKM Economics and Envirosris (2001), Independent Assessment of Kerbside Recycling in Australia, Report to the National Packaging Covenant Council, Sydney, January,

NPCC (2000), *First Annual Report*, National Packaging Covenant Council, www.ea.gov.au/industry.waste/covenant/publications.html, Last accessed 24 April 2003.

NPCC (2003), National Packaging Covenant Council Annual Report June 2003Canberra, National Packaging Covenant Council http://www.deh.gov.au/industry/waste/covenant/annual-report/index.html.

OECD (1999), Voluntary Approaches for Environmental Policy in OECD Countries: An Assessment, Working party on Economic and Environmental Policy Integration, Environmental Policy Committee, OECD, Paris,

Paton B. (2002, Voluntary Environmental Initiatives and Sustainable Industry in *Voluntary Industry Agreements: Process, Practice and Future Use*, ed. P. ten Brink, Greenlead Publishing. Sheffield, UK.

PCA (2004), *Packaging Expose Volume 01/04 - The Covenant*, Packaging Council of Australia www.packcoun.com.au/expose/0104.htm, Last accessed 5th February 2004.

Rathje W. and Murphy C. (1992), *Rubbish! The Archeology of Garbage*. Harper Collins, New York.

Schaltegger S. and Burritt R. (2000), *Contemporary Environmental Accounting. Issues, Concepts and Practice*. Greenleaf Publishing Limited, Sheffield, UK.

Schaltegger S., Burritt R.L.and Petersen H. (2003), *An Introduction to Corporate Environmental Management: Striving for Sustainability*. Greenleaf Publishing, Sheffield, UK.

Simon H. (1949), *Administrative Behaviour*. MacMillan.

Toffler A. (1970), *Future Shock*. The Bodley Head, London.

Williams G. (2003), *The National Packaging Covenant: The Benefits - The Lessons, Presentation to the South East Queensland Regional Organisation of Council Forum*, http://www.packcoun.com.au/speechpage.html, Last accessed 2 July 2003.

TOWARDS AN INTEGRATED PERFORMANCE INDICATOR FOR (ENERGY) BENCHMARKING COVENANTS WITH INDUSTRY

J. COUDER, A. VERBRUGGEN

University of Antwerp, Belgium

Abstract. Voluntary approaches play an important role in reducing industrial energy use and CO_2-emissions. Benchmarking can provide a starting point for negotiating targets, and are an added value to a monitoring program. Indicators are perfect for identifying the performance gaps, and to track performance over time. However, indicators at the firm level are still characterized by a low degree of standardization. Lack of comparability makes benchmarking very difficult. Indicators measure changes in one aspect (e.g. energy use) as if they were completely independent of changes in other aspects (e.g. waste generation). We argue that integrated indicators, based on micro-economic productivity theory, may one day assume the role of certified tools in the field of flexible policy instruments.

1. VOLUNTARY APPROACHES

Voluntary approaches (VAs) – on their own or along with other policy instruments – have already been widely adopted in EU member countries for improving energy efficiency in the industrial sector, primarily to achieve CO_2 reduction targets

Voluntary approaches is a broad term that encompasses many different kinds of arrangements, such as "... self-regulation, voluntary initiatives, voluntary codes, environmental charters, voluntary accords, voluntary agreements, co-regulation, covenants, and negotiated environmental agreements, to name just a few." Higley *et al.* (2001, p. 5) and Croci and Pesaro (1997; 1999) were among the first to provide a practical taxonomy of VAs in the environmental sector. One now commonly classifies voluntary approaches into three main types: *unilateral commitments*, *public voluntary programmes* and *negotiated agreements*. Unilateral commitments are environmental improvement programmes initiated and undertaken by industry (firms and industry associations) and communicated to stakeholders (consumers, employees, shareholders, neighbours, etc). The contents of the commitment – targets, time schedules etc – are decided by the firms and industry associations involved in the programme. A typical example is the Responsible Care programme by the chemical industry. Public voluntary programmes are schemes initiated by government, which firms can voluntarily join. The public authorities 'set standards as regards some combination of processes and procedures to be followed, or targets to be attained, and participating firms agree to meet these targets.' (Higley *et al.*, 2001, p. 6) Typical examples include the Eco Management and Auditing Scheme

E. Croci (ed.), The Handbook of Environmental Voluntary Agreements, 307 - 332

[EMAS] of the European Union, the US-EPA Energy Star Programme or the Dutch Benchmarking Covenant. Negotiated agreements (NAs) are contracts between some public agency and industry, either individual firms (*individual* agreements) or branch organisations (*collective* agreements). Through negotiations one or more overall targets and a time schedule that industry must meet are fixed. Environmental organizations, who may or may not sign the agreement, may also participate during the negotation phase. (Croci and Pesaro, 1999) In return the public agency usually refrains from introducing more coercive or costly regulatory measures. (Hansen *et al.*, 2000) Examples are the Dutch Long Term Agreements on Energy Efficiency (LTAs) and the Danish Agreement on Industrial Energy Efficiency.

Organisations like EEA (1997) and OECD (1999) formulate recommendations for the design and implementation of VAs in general and NAs for energy and environmental policies in particular. We interpret, with the aid of Hansen *et al.* (2002), the EEA and OECD recommendations as follows:

The *target setting* process is open, i.e. made known to the general public and involves third party interest groups. Transparency provides an incentive to comply with the agreements; increases credibility and avoids or decreases the risk of 'regulatory capture', where the government is persuaded by industry to agree on unacceptably weak targets. The target setting process should include the setting of interim (staged) objectives to avoid the opportunistic use of agreements as a means to retard the implementation of more effective regulation. The targets themselves should be clearly defined, unambiguous and quantitative, so that uncertainty cannot be exploited to avoid obligations;

The agreement establishes estimates of a *business-as-usual* (BAU) trend in order to provide a baseline scenario against which improvements will be measured. The business-as-usual trend defines what the outcomes are likely to be, given natural technical progress within the industry. The agreement targets should be significantly beyond the baseline scenario, to assure environmental effectiveness. A baseline scenario limits the problem of asymmetric information in the target-setting phase, where the government lacks data about firms' technological capabilities and firms use private information to set targets below or very close to the baseline scenario;

The agreement includes reliable and clear *monitoring* mechanisms. Monitoring encompasses all kinds of reporting procedures, indicators and steering groups, which may help increase the regulator's knowledge about the feasibility of the targets, the firm's performance (the distance between the targets and the estimated BAU), unexpected problems, and technological and economic limits. (Aggeri, 2000) If the risk of getting caught in not reaching the agreed targets is very small or non-existent, industry will make little efforts to fulfil its obligations. In the case of collective VAs, monitoring should be made both at the firm and the sector level to tackle 'free riders' (firms that benefit from the collective agreement without making any effort). Monitoring by independent organisations may increase the credibility of the monitoring program;

The agreement includes a credible *regulatory threat*. Sanctions for not reaching the (interim) targets must be set. Assuming that a) industry is boundedly rational and does not act out of altruistic motives, and b) that non-compliance is costless whereas compliance is costly, industry will not make an extra effort if there are no clear consequences for non-compliance, such as the threat of an alternative instrument (e.g. regulation or tax). Sanctions may consist of denying or revoking operating permits, shutting down (part of) an operation, reimbursement of tax rebates, monetary penalties, etc.;

Technical solutions are available in order to reach the agreed targets, and VAs should promote the *diffusion* of knowledge and of cleaner, more energy efficient technologies;

There is an efficient burden sharing of targets among industry, i.e. the costs of complying with the VA are minimised and are relatively similar for all members of the target group (economic *efficiency* or 'cost effectiveness').

Krarup and Ramesohl (2000) and Chidiak (1999) compare five European voluntary agreements for CO_2 reduction and the associated goal of improving energy efficiency: 1) the Dutch Long-Term Agreements on Energy Efficiency (LTAs); 2) the Declaration of German Industry on Global Warming Prevention (DGWP); 3) the French Voluntary Agreements on CO_2-reductions; 4) the Danish Agreement on Industrial Energy and 5) the Swedish EKO-energy. Hansen *et al.* (2002) discuss the adequacy of the OECD recommendations for three cases of NAs related to industrial energy consumption (Netherlands, France and Denmark).

2. BENCHMARKING

2.1. Definition, Types and Methodology

The use of benchmarking has a long tradition as a management tool in business firms for economic reasons (comparison of financial performance, sales, operations, etc.)[1]. The application of energy and environmental benchmarking is a fairly recent phenomenon.

Voluntary approaches can be linked to benchmarking for setting targets (Figure 1). Accepted benchmarks could provide a starting point for negotiating VA targets. Altham (2002, p. 3) states that

> Benchmarking should be seen as "adding value" to a monitoring program; the use of data currently collected in your monitoring program being used to judge your performance against your peers with the aim of identifying where opportunity to improve may be found.

Literature presents various definitions of benchmarking[2]. We define benchmarking as the continuous, systematic process of comparing the current level of performance against a predefined point of reference, the *benchmark*, in order to evaluate and improve performance. There are many types of benchmarking, e.g. internal (compare identical or similar activities within the same firm); external or 'best in class' (compare against outside firms that are known to be 'best in class'); performance, competitive or sectoral (compare against direct competitors drawn from the same sector); functional or industrial (compare against firms from different sectors performing similar activities); and generic or process benchmarking (compare unrelated activities that can be practised in similar ways, regardless of the sector).

There is no uniform standard benchmarking methodology. (See e.g. Bogan and English, 1994; Camp, 1995) Instead, once the benchmarking team determines the objectives of the project, the methodology is customized according to the needs of the project. (Diebäcker, 2000, p. 491).

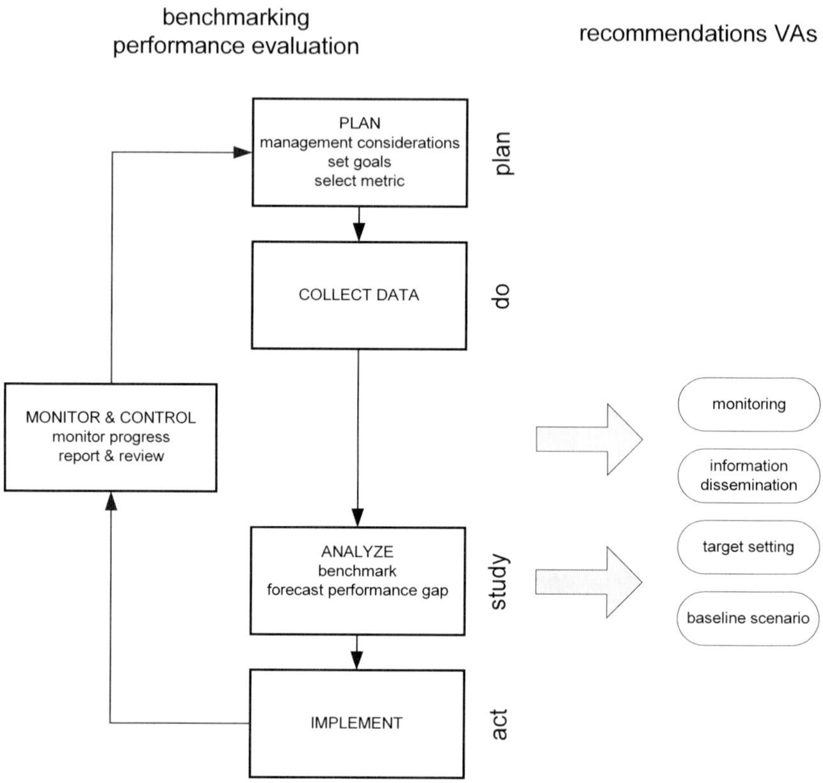

Figure 1. *Voluntary approaches, benchmarking and performance evaluation*

TQLO (Kraft, 1997) surveyed 20 benchmarking models, before developing their own 10-step model based on Demming's (1993) Plan-Do-Study-Act cycle. Using benchmarking of energy performance as an example, the necessary phases and steps may look something like this (for illustrative purposes only):

> Planning. Gain total, visible commitment from the top levels of management; clearly define the goals (e.g. reduce energy consumption) and communicate these goals throughout the firm; assign responsibility and accountability for achieving those goals; understand the critical processes and how they are measured; choose the metrics; establish and document what kind of data is needed, what data are available and how data will be collected;
>
> Internal data collection and analysis. Collect information to understand in detail your own processes. In the case of energy benchmarking: analyse energy use patterns, i.e. find out how energy is used, what types of energy sources are used, how much is used,

when the use takes place, where the energy is used, why it is used, etc. Identify practices that enable and hinder superior performance;

External data collection and analysis (benchmarking). Identify possible benchmarks (e.g. market, economic, technical or theoretical energy saving potential [Phylipsen *et al.*, 1998]) and/or benchmarking partners; collect information to understand in detail the processes of others; gather information on metrics through site visits, questionnaires, literature surveys and expert opinion, or – if available – 'benchmarking services'; compare performance and identify the performance gap with the use of metrics; forecast future performance gaps;

Implementation. Identify exactly how and where performance improvements (e.g. energy saving opportunities) are to be made to close the performance gap; establish acceptable project risks and returns; develop actions plans for each goal, to determine who does what and when, and implement them;

Monitoring and control. Monitor progress; provide feedback into the previous steps: recalibrate, i.e. re-evaluate and update the benchmarks (maintain and keep current the benchmarking database; regularly search for public domain data); adjust the action plans if goals are not being met.

A number of barriers hamper the use of (energy or environmental) benchmarking. Firms (especially SMEs) are unwilling to disclose internal information, for fear their competitors may learn from this information and improve their competitiveness; or they are sceptical because they suspect that government might use the benchmark results for purposes other than improving (energy or environmental) performance. Also, firms do not always understand benchmarking, or are unaware of its potential benefits. Of a more practical nature are the lack of quality data (e.g. because firms lack suitable metering equipment); and the difficulty to identify firms producing similar outputs or output mixes, using similar processes, consuming similar raw materials or resources.

2.2. Energy Benchmarking Covenants in The Netherlands and Flanders

The Dutch Benchmarking Covenant was signed on 6 July 1999 by the Dutch government and the energy-intensive industry (with an energy consumption > 0.5 PJ per year per plant), including the electricity production sector. Participating industries pledge to reach world top in terms of energy efficiency for processing installations by no later than 2012. In return the government will refrain from imposing any additional specific national measures governing energy conservation or CO_2 reduction on the participating companies. For the other industries, a new long-term agreement was developed, referred to as the second generation LTAs. In the Benchmarking Covenant, the companies themselves, by means of an international benchmark, set the target. The first step is to determine the top ten percent worldwide. Once a firm has identified the top international standard, it determines the gap from the top global performers and prepares an Energy Efficiency Plan (EEP). The EEP sets out, per individual plant, the concrete measures to be adopted. The participation rate (84 % of potential benchmark companies, representing 94 % of energy consumption for the industrial sector and 100 % for the power-generating sector) is very high. It is estimated that the covenant's potential

CO_2 reduction for the period from 2000 to 2012 would work out at 3,2 to 4,0 million tons of CO_2 for the industrial sector. (Gerrits, 2003, p. 3-64)

The Flemish Covenant Energy Benchmarking, approved on 29 November 2002, has a working period up to 2012. It is almost an exact copy of the Dutch Benchmarking Covenant. The participating companies undertake to implement a benchmark study, to draw up and execute plans to bridge the gap from the Best International Standard and to stay there, to monitor their progress and report about it. (website: www.benchmarking.be)

3. INDICATORS

3.1. Definitions, Properties, Principles, Measurement and Types of Indicators

Indicators, meaningful metrics reflecting (energy / environmental) performance on a company level (e.g. quantity of energy used per year or per unit of production), are perfectly suited to identify the performance gaps in benchmarking, and to track performance over time (monitoring in VAs and benchmarking).

A metric is a system of measurement that includes the item being measured, the unit of measurement, and the value of the unit. (Geisler, 1999) Gallopin (1996) defines an indicator as a variable, which is an operational representation of an attribute (quality, characteristic, property) of a system. States (or values) of the variable are entities through which the variable manifests itself. We will use the terms metric and indicator interchangeably, although an indicator as a (combination of) metric(s) only gets its meaning within the context of a pre-defined value, often referred to as reference, baseline or benchmark level, threshold or target. (Gallopin, 1997)

Rigby *et al.* (2000) survey the literature on the desirable properties (purposes, functions) of indicators. We prefer the following list: 1) selection of the most significant information; 2) quantification of information, so that its significance is more readily apparent; 3) simplification of complex phenomena; and 4) communication of information, particularly between data collectors and data users. (Piere *et al.* 1995, based on Adriaanse, 1993 and Hammond *et al.*, 1995)

Qualitative indicators are also possible. According to Azzone *et al.* (1996), qualitative metrics describe intangible items that are difficult to quantify, such as the characterisation of the type of environmental strategy carried out by a firm. One can still evaluate qualitative indicators using scoring systems, such as a number scale in connection with an explanation for the evaluation (e.g. 1 = reactive, 2 = proactive).

Indicators can be calculated for different system boundaries, e.g. countries or regions, corporations or organisations, sites or plants, production processes or units. Our discussion focuses on the development of indicators for companies, and will therefore exclude the many initiatives aimed at the global, national, or regional (macro-economic) level of industrial activities. Approaches such as the OECD (2003) Environmental Indicators, Indicators of Sustainable Development by the UN Commission on Sustainable Development (UN CSD, 1996), Environmental Indicators at the European Environment Agency (EEA, 1999), are based on the P-S-

R, D-S-R or D-P-S-I-R framework[3] respectively, and respond mainly to environmental (sustainability) policy follow up.

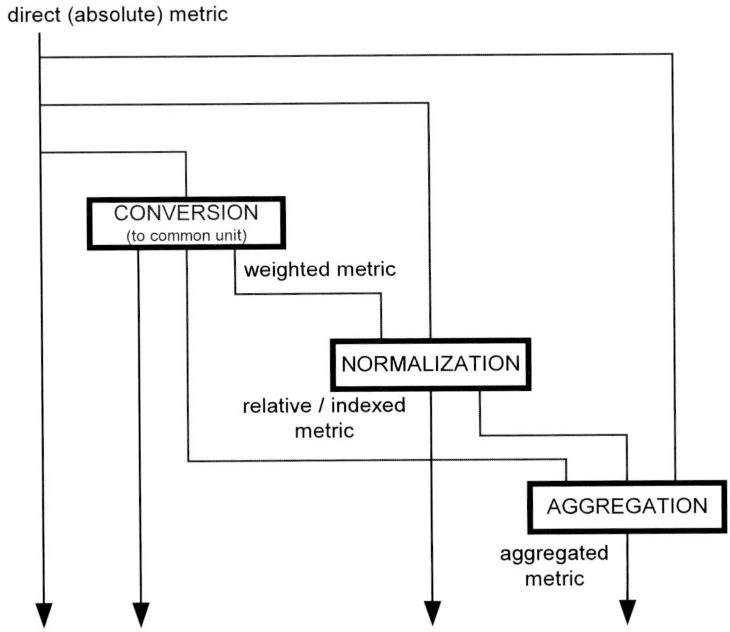

Figure 2. *Conversion, normalization and aggregation of indicators*

Other measurement issues include conversion to standard units (transformation), normalisation and aggregation (Figure 2).

> Conversion to standard units refers to converting the data to a single system of units. For example, all energy inputs are expressed in terms of Joules (J), or greenhouse gas emissions in terms of global warming potential (GWP). Potency factors[4] become relevant when converting environmental performance indicators to core theme indicators such as climate change, acidification, stratospheric ozone depletion, troposphere excess of ozone and other photo-oxidants, toxic dispersion, eutrophication and waste problems. Conversion may thus lead to weighted metrics, describing data 'modified by applying a factor related to its significance'. (ANSI/ISO 14031-1999) For many indicators however, there are no mutually agreed upon methods for transforming data into "common" units. 'Thus, great care needs to be taken when data are transformed, and the method of transformation must be clearly described.' (Keffer *et al.*, 1999);

> Normalisation of data ensures that data is converted to a form that relates a direct (or absolute) metric (e.g. consumption of energy) to a defined baseline in order to achieve comparability despite of fluctuations in activity levels. For organisations possible normalisation metrics may include units of output or service, value added, sales, number of employees, … Normalisation usually leads to relative metrics (e.g. amount of energy used per unit of product) or indexed metrics (e.g. amount of energy used in the current

year as a fraction of the amount of energy used in a baseline year). Different normalisation measures can lead to different performance ranking (Wagner and Wehrmeyer, 2001);

Aggregation collects and expresses data of the same type, but from different sources, as a combined value or aggregated metric. Aggregation can obscure potentially important information about an organisation's performance, as data from different products, processes or geographic areas may be combined, hiding details about the performance of individual units. 'As a result, aggregation of data must be done very carefully, and with complete transparency to the end-user, so that the limitations of the information can be well understood' (Keffer *et al.*, 1999).

The World Business Council for Sustainable Development working group on eco-efficiency metrics & reporting recommends that indicators (at the company level) should conform to the following principles (Keffer *et al.*, 1999):

- Be relevant and meaningful with respect to protecting the environment and human health and/or improving the quality of life;
- Inform decision making to improve the performance of the organization;
- Recognize the inherent diversity of business;
- Support benchmarking and monitoring over time;
- Be clearly defined, measurable, transparent and verifiable;
- Be understandable and meaningful to identified stakeholders;
- Be based on an overall evaluation of a company's operations, products and services, especially focusing on all those areas that are of direct management control;
- Also recognize relevant and meaningful issues related to upstream (e.g. suppliers) and downstream (e.g. use) aspects of a company's activities.

The Global Reporting Initiative (GRI, 2002a) groups reporting principles in four clusters, namely those that:

1. Form the framework for the report (transparency, inclusiveness, auditability);
2. Determine what to report (completeness, relevance, sustainability context);
3. Ensure quality and reliability of the report (accuracy, neutrality, comparability);
4. Govern access to and availability of the report (clarity, timeliness).

Indicators should be dependent on data that are readily available or obtainable at a reasonable cost-benefit ratio, adequately documented, of known quality and updated at regular intervals (UN CSD, 1996).

We already mentioned some types of indicators: quantitative vs. qualitative; weighted, direct (or absolute), relative, indexed and aggregated indicators.

In addition, indicators at the firm level are commonly subdivided into two categories:

- "Core" (GRI), "generally applicable" (WBCSD) or "generic" indicators: relevant or applicable to virtually all organizations and of interest to most stakeholders;
- "Additional" (GRI), "Supplemental" (WBSCD) or "company / sector specific" indicators: based on the needs of a particular organization and its stakeholders.

There are many approaches measuring performance of industry at the firm level, but few address full sustainability. It would be almost impossible to list all types of indicators that have ever been devised. Most studies propose a specific indicator framework (a basic set of principles defining how indicators will be selected and used), and, in the best cases, a broad list of indicators. We will therefore concentrate on three global (GRI, ISO 14031 and WBCSD) and three European initiatives (Anite, Ellipson and MEPI), primarily in function of a firm's (energy / environmental) performance evaluation.

3.2. Global initiatives

GRI and WBSCD, using the ISO terminology, organize information on the same three hierarchic levels: categories, aspects and indicators. Categories are the broad areas of economic, environmental or social issues of concern to stakeholders (e.g. the environmental category). Aspects are general types of information related to a specific category (the what, e.g. energy consumption). Indicators are the specific measurements of an aspect that can be used to track and demonstrate performance (the how, e.g. MJ of energy used).

3.2.1. Global Reporting Initiative (GRI)

The Global Reporting Initiative (GRI), founded in 1997 and until spring 2002 a project of the UN Environment Programme (UNEP) and the US Coalition for Environmentally Responsible Economies (CERES), is now a permanent, independent organisation[5] (http://www.globalreporting.org), whose mission is to develop, promote, and disseminate Sustainability Reporting Guidelines. The GRI's "Guidelines" (GRI, 2002a) represents the first comprehensive framework for reporting an organisation's sustainability performance, based on a global, multi-stakeholder process[6]. GRI defines sustainability reporting as a process for publicly disclosing an organisation's economic, environmental, and social performance, in relation to its operations, products and services. A revised version of the "Guidelines" is expected in 2005. GRI (2002b) is also developing technical protocols on indicator measurement. Each protocol addresses a specific indicator or set of indicators by providing detailed definitions procedures, formulae and references to ensure consistency across reports.

GRI's "Performance Indicators" are the core of a sustainability report, and are divided into economic, environmental, and social performance indicators. These are still largely treated as separate reporting elements. GRI does not identify a standardised set of integrated performance indicators, given '... the unique relationship of each organisation to the economic, environmental and social systems within which it operates' (GRI, 2002a, p. 44) The 2002 "Guidelines" contains five "Energy Consumption Indicators", each belonging to the set of "Environmental Performance Indicators", namely two "core" indicators (Direct Energy Use Segmented by Primary Source, Indirect Energy Use) and three "additional" indicators (Initiatives to use Renewable Energy Sources and Increase Energy

Efficiency, Energy Consumption Footprint of Major Products, and Other Indirect Energy Use not Reflected Above, Such as Travel and Use of Energy-Intensive Materials). The Energy Protocol (GRI, 2002b) provides key definitions and conventions (e.g. primary vs. intermediate energy) and measurement methodologies (e.g. conversion factors, units, the use of an Energy Balance Sheet).

3.2.2. ISO 14031 and Environmental Performance Evaluation

The ISO 14031 standard (ANSI/ISO 14031-1999), part of the ISO 14000 family of voluntary international environmental standards, provides guidance on the design and use of environmental performance evaluation (EPE) within an organisation. It applies to all organizations, regardless of type, size, location and complexity. Whereas the GRI framework focuses on indicators that are most relevant to the stakeholders, ISO 14031 makes no recommendations about reporting or about which indicators an organization should utilize – although it does include, in Annex A, a list of 197 topics from which companies could select indicators for environmental management.

Following ISO 14031's definition EPE is essentially an iterative process to facilitate management decisions regarding an organization's environmental performance by going through three main stages:

1. Planning (management considerations and selecting environmental indicators);
2. Evaluation (collecting data; analysing and converting data; assessing information against environmental performance criteria; reporting and communicating);
3. Periodically reviewing and improving the EPE.

ISO 14031 can thus be seen as being complementary to the GRI framework. We also draw attention to the similarity between the EPE stages and the previously discussed benchmarking phases and steps.

ISO 14031 describes two general categories of indicators for EPE: 1) Environmental Performance Indicators (EPIs) and 2) Environmental Condition Indicators (ECIs). ECI is defined as a "specific expression that provides information about the local, regional, national or global condition of the environment". EPIs are subdivided in Operational Performance Indicators (OPIs) and Management Performance Indicators (MPIs). The latter provide information about management efforts to influence the environmental performance of the organization's operations, whereas the former provide information about the environmental performance itself. OPIs relate to inputs, outputs and the physical facilities and equipment of the organization. Energy (input) indicators clearly belong to the set of OPIs.

3.2.3. WBCSD and Eco-efficiency

Environmental performance indicators (EPIs) do not make any reference to economic indicators, in contrast to "eco-efficiency indicators". The concept of eco-efficiency was first coined in 1992 by the Business Council for Sustainable

Development (BCSD)[7], and has become widely recognized by the business world. Although the eco-efficiency metrics are limited to bringing together the economic and environmental categories, rather than sustainability issues, WBSCD has now begun to address corporate social responsibility (as a separate business concept) as well.

Following WBSCD, eco-efficiency is achieved by 'the delivery of competitively priced goods and services that satisfy human needs and bring quality of life, while progressively reducing ecological impacts and resource intensity throughout the life-cycle to a level at least in line with the earth's estimated carrying capacity.' The basic calculation for eco-efficiency indicators is the ratio of 'product or service value' per 'environmental influence'. Or, in OECD (1998) terminology, the ratio of an "output" (value of products and services produced) divided by an "input" (the sum of environmental pressure generated). In this form[8], an increasing eco-efficiency reflects a positive performance improvement.

$$\text{eco - efficiency} = \frac{\text{product or service value}}{\text{environmental influence}}$$

Examples of product or service value include quantity of units produced, sales or value added[9]. Examples of environmental influence include the use or consumption of resources (land, raw materials, energy and water); pollutants releases to air, water and soil; and generation of solid and liquid waste, including hazardous waste[10]. Both numerator and denominator can be expressed in physical and / or monetary units. To improve eco-efficiency, companies can 1) reduce the material intensity of goods and services; 2) reduce the energy intensity of goods and services; 3) reduce toxic dispersion; 4) enhance material recyclability; 5) maximize sustainable use of renewables; 6) increase material durability; and 7) increase the service intensity of goods and services.

WBCSD proposes a common framework to provide all companies – regardless of their business or geographic base – with a basic foundation or starting point to use the eco-efficiency concept, in a way that meets the needs of its business and their external stakeholders. (Verfaillie and Bidwell, 2000) The WBCSD indicator framework includes the following key steps:

- Get an overview and a better understanding of the eco-efficiency concept, by reviewing a) environmental and value-related indicators classification, b) agreed upon definitions and c) examples;
- Use the limited set of "core indicators", which WBSCD believes a) are more or less global concerns or values, b) have a widely agreed upon measurement methodology, and c) are valid for virtually all businesses;
- Develop and select "supplemental indicators", valid for specific businesses (at the company or sector level);
- Collect data and calculate relevant eco-efficiency ratios, i.e. quantify the relation between economic/value performance and environmental performance using the eco-efficiency indicators concept;

- Integrate eco-efficiency metrics into performance reports to internal and external stakeholders for a better decision-making.

3.3. European Initiatives

3.3.1. Anite and Ellipson

Anite (1999) published the results of a pilot study, commissioned by Eurostat and DG Enterprise, containing a list of eco-efficiency indicators for two industry sub-sectors: chemical industry and metal industry, in two EU member states: the Netherlands and France. Anite defines two core eco-efficiency indicators related to energy: value added/energy consumption (excluding coke) and value added/non-renewable energy use.

Ellipson (2001) published a concept paper to provide guidance on the identification, selection and construction of eco-effiency indicators for enterprises. For the environmental problem of 'depletion of non-renewable energy sources' Ellipson proposes two generic eco-efficiency indicators: non-renewable primary energy input/value added and energy costs/value added.

3.3.2. Measuring Environmental Performance of Industry (MEPI)

We will discuss the EU-funded project – Measuring Environmental Performance of Industry (MEPI) – in more detail. To our knowledge MEPI is the first large scale attempt to empirically construct integrated industrial performance indicators, including indicators based on productivity theory.

One of the key objectives of the European MEPI project was to develop quantitative indicators for comparing the overall environmental performance of industrial firms. To date the MEPI project provides the most comprehensive piloting of developing a standardized approach to environmental performance measurement across several industrial sectors (electricity generation, pulp and paper, fertilisers, book and magazine printing, textile finishing and computer manufacture). (Berkhout *et al*, 2001)

MEPI first and foremost makes a distinction between variables (data on performance) and indicators (normalised measures of performance). Normalising factors included standardised units of production for a given sector; value added, turnover, profit, or number of employees. MEPI further distinguishes physical indicators (concerned with materials and energy inputs and wastes generated from production processes, and reported as ratios of mass or energy per unit of product produced); eco-efficiency indicators or business indicators (concerned with linking physical aspects of environmental performance with key business indicators); and impact indicators (concerned with the actual impact on the general environment of a firms' activities, i.e. contributions to climate change, acidification and so on). MEPI also makes a distinction between generic/core and sector-specific indicators.

The MEPI study includes four main stages:

1. Data collection. Data for 274 firms and 676 production units (firms, business units and sites, mainly in the UK, the Netherlands, Belgium, Germany, Austria and Italy) were collected, screened for quality, transformed into

appropriate units and normalized for comparability. To manipulate the data, a flexible and expandable dedicated database was built;

2. Defining indicators sets (parallel with stage 1). Having consulted with an 'Advisory Panel' (representing policymakers and industry), sets of generic/core and sector-specific indicators for each of the six industrial sectors were defined;
3. Data analysis. The objective was two-fold: a) to analyze statistically the environmental and business data collected in stage 1; and b) to conduct more detailed analysis of a small sample of firms for which data was available, to investigate the causes of variability in recorded performance;
4. Policy and industrial interface and reporting. In the final stage MEPI was concerned with evaluating current and future needs for corporate environmental reporting, the use of benchmarking in environmental management, EPI development in other industrial sectors, and the application of firm-level EPIs for policymaking and evaluation.

The statistical analysis (stage 3.a) consists of three different approaches:

1. The production of standard descriptive statistics;
2. The use of principal component analysis (PCA), to identify variables that best explained the variability of environmental performance. PCA allowed MEPI to isolate a restricted set of variables that were tested in multiple regression analysis. These regressions allowed identifying links between different dimensions of environmental and business performance (e.g., is it true that more eco-efficient firms are also, in general, more profitable than less eco-efficient firms, as the 'Porter-hypothesis' would suggest?);
3. The use of data envelopment analysis (DEA) – a method of establishing rankings of performance of firms by aggregating across a range of variables – in an attempt to introduce a more integrated picture of performance.

The main analytical results of the MEPI project are:

- A small number of indicators give a relatively good representation of the overall environmental performance of a firm;
- The range of environmental performance between firms often spans order of magnitude, mainly but not always explained by the use of different technologies;
- The environmental performance on the site-level is (as expected) strongly related to the process technology being used;
- The size (scale) of the production site appears not to be related to environmental performance;
- There is little evidence of a positive or negative correlation between high profitability and high levels of environmental performance;
- Firms with a certified environmental management system do not appear to perform better than those without (though statistical significance was low)

  > The thinness of the evidence for an 'environmental management effect' is somewhat surprising, given the expectations that voluntary schemes would have a positive impact on performance. (Berkhout *et al.*, 2001, p. iv);

- A strong country effect could not be shown.

4. THE NEED FOR INTEGRATED INDICATORS

The current practice of using indicators is still characterized by a low degree of standardization. There is no consensus on which indicators to use within an enterprise (the 'what'), and the methodology for the construction of indicators (the 'how') varies across enterprises. This has led to a plethora of indicators, and the existent 'jargons' often mix similar terms with different, even contradictory, interpretations. Lack of comparability (within the enterprise, across different enterprises within the same sector or even across different enterprises of different sectors) makes it impossible to compare the performance of one enterprise with another (benchmarking).

Also, most environmental, economic and/or social indicators still measure changes in one category or aspect as if they were completely independent of changes in the other categories or aspects. In particular, energy efficiency improvement opportunities are all too often considered independent of other industrial productivity improvements, such as pollution or waste prevention.

There is an obvious need for a restricted list of standardized performance indicators at the firm level. Ideally, integrated indicators should reflect the interdependent changes, which occur at all levels.

4.1. Benchmarking Indicators Derived from Data Envelopment Analysis (DEA)

Tyteca (1996) adapted an approach to environmental performance analysis derived from the theory of Data Envelopment Analysis (DEA).

DEA is based on the engineering concept of technical efficiency, in the single-input, single-output case defined as the ratio of useful output over input. In a seminal paper Charnes, Cooper and Rhodes (1978) extended this measure of efficiency to a multiple-inputs multiple-outputs framework, defining technical efficiency as a weighted sum of outputs divided by a weighted sum of inputs. The weights normalize for different units of measurement and give an indication of the relative importance of a particular input or output.

$$\text{technical efficiency} = \frac{\text{weighted sum of outputs}}{\text{weighted sum of inputs}}$$

To 'objectively' assign the appropriate weights, the standard DEA method uses an optimization model. This model calculates, for each firm under evaluation, the optimal set of input weights and output weights, such that the efficiency of that particular firm is maximised relative to the efficiencies of all the other firms in the sector. The other firms use the same type of inputs and produce the same type of outputs as the firm under evaluation, but in varying amounts[11]. The efficiencies of the other firms are calculated using the same optimal weights as for the firm under evaluation. Technical efficiency (a dimensionless ratio) is furthermore restricted to lie between zero and one, ensuring that the optimal weights are given so that the highest efficiency any firm can ever achieve is one. Once the efficiency for each firm in a sector has been calculated, using the above method, it is possible to rank the different firms according to their 'efficiency scores'. DEA also allows identifying, for each inefficient firm (efficiency score < 1), its 'best practice' counterparts (efficiency score = 1), thus making DEA very apt as a benchmarking tool. Variants of DEA models can incorporate the effects of returns to scale, additionally allowing the computation of scale efficiency (defined as the ratio of total efficiency calculated on a constant returns to scale technology and one computed on a variable returns to scale technology).

Excellent introductions to the use of DEA can be found in Charnes *et al.* (1995) and Cooper *et al.* (2000).

The standard DEA approach has been extended to incorporate 'undesirable outputs and inputs' or 'bads'. Tyteca (1996) explicitly uses this extended DEA approach to construct a number of environmental performance indicators (EPIs) for industrial firms, namely:

- The Undesirable Output (UO) EPI: minimises the ratio of the weighted sum of undesirable outputs over the weighted sum of inputs and desirable outputs;
- The Input-Undesirable Output (IUO) EPI: minimises the ratio of the weighted sum of undesirable outputs and inputs over the weighted sum of desirable outputs;
- The Normalized Undesirable Output" (NUO) EPI: minimises the ratio of the weighted sum of undesirable outputs over the weighted sum of desirable outputs.

Ball *et al.* (1994) and Yaisawarng and Klein (1994) had previously designed similar environmental performance measurements, without explicitly referring to EPIs as such.

DEA thus allows the construction of EPIs, which are solely based on physical quantities, which are dimensionless (excluding the need to explicitly convert to a common unit), and which take into account the scale at which the firm operates. In general, the DEA framework can be applied to any selection of variables, provided they can be distributed into subsets of those that should be maximized and those that should be minimized.

> For example, undesirable outputs can be substituted by variables describing environmental impacts when appropriated (whose value should be as low as possible to contribute to enhanced environmental performance), or production outputs can be replaced by financial quantities such as sales or turnover (the value of which should be maximized to reflect economic efficiency), and so on. (Tyteca, 2001, p. 12).

This makes DEA methods suitable, not only for the calculation of environmental performance indicators (EPIs), but also for eco-efficiency indicators or even impact indicators. In fact, although one of the strengths of DEA is the "objective" assigning of weights, variants of DEA have been developed to incorporate weight restrictions[12], e.g. to make provision for the actual effects of pollutant discharges on the environment. Other variants of DEA allow the use of categorical variables (variables that take on only a finite number of values), opening a pathway to the construction of qualitative indicators. Yet other DEA variants consider the use of non-discretionary variables (variables not subject to direct management control, e.g.

weather conditions). Finally, DEA can just as easily be used to define "pure" energy efficiency indicators at the firm level (Couder and Verbruggen, 2003).

4.2. Benchmarking Indicators Derived from the Theory of Productive Efficiency

In this section we look at integrated performance indicators derived from the micro-economic theory of productive efficiency. Under certain conditions, these indicators are mathematically equivalent to those derived from the DEA approach. Both approaches are commonly referred to as (best practice) 'frontier methods', the micro-economic approach being easier to interpret graphically and providing a better foundation for the construction of integrated performance indicators. Our discussion draws heavily on Chung *et al.* (1997) and Färe *et al.* (2001).

To construct an integrated performance indicator for industrial firms, the representation of their production technology (or production model) requires the following properties:

- Be able to handle multiple inputs and multiple outputs;
- Be able to handle undesirable inputs and outputs, or 'bads';
- Require only information on physical quantities of inputs and outputs – and 'bads' in particular – since market or shadow prices for bads such as pollutant emissions can rarely be found or estimated;
- Require no pre-specified assumptions of profit or revenue maximization or cost minimising behaviour of a firm;
- (Preferably, but not necessarily) require no pre-specified functional form on the underlying technology. This allows a non-parametric construction of the production frontier, similar to the DEA approach;
- Be able to decompose productivity growth into efficiency change (measuring the relative change of position of a firm with respect to the production frontier), and technical change (measuring the shift of the production frontier).

A production model that fits the above requirements is the 'distance function', pioneered by Shephard (1953). A distance function may have either an input or an output orientation. Without loss of generality, we limit ourselves to output distance functions.

Given N inputs $x = (x_1, \ldots, x_N) \in \Re_+^N$ and M outputs $y = (y_1, \ldots, y_M) \in \Re_+^M$, the output producible sets are given by:

$$P(x) = \{y : x \text{ can produce } y\}, x \in \Re_+^N .$$

Output producible sets represent the sets of all output vectors, which can be produced using the input vectors.

It is assumed that the technology satisfies the standard axioms:

- The output producible sets are convex, bounded and closed for all input vectors;

- Possibility of inaction, i.e. a firm can produce no outputs using no inputs;
- No free lunch, i.e. all production requires the use of some input;
- Strong (free) disposability of (good) outputs, i.e. a firm can reduce one or more of its outputs without changing the consumption of its inputs or the production of other outputs;
- Strong (free) disposability of (good) inputs.

An output distance function looks at how much the (good) output vector may be equiproportionally expanded with the (good) input vector held fixed, and may be defined upon the output producible sets as:

$$\vec{D}_O(x,y) = \inf_\theta \left\{ \theta : \frac{y}{\theta} \in P(x), \theta \in \Re_+ \right\}.$$

By definition, the reciprocal of the value of the output distance function, or the non-negative scalar $1/\theta$, measures the maximum amount or "distance" by which the output vector can be inflated, given the input vector. The output distance function will take a value which is less than or equal to one if the output vector is an element of the output producible set, and a value of unity if the output vector is located on the boundary of the output producible set. An efficiency measure is thus obtained by measuring the distance between the observed input-output vector of a firm and a technically feasible input-output vector, as defined by the frontier of the production possibility set. The efficiency performance indicator is determined, in effect, by determining the projection directly along the output vector, holding the input vector constant.

We now distinguish between good (desirable) outputs $y^g = \left(y_1^g, \ldots, y_G^g\right) \in \Re_+^G$ and bad (undesirable) outputs $y^b = \left(y_1^b, \ldots, y_B^b\right) \in \Re_+^B$, so that $y = \left(y^g, y^b\right) \in \Re_+^{M=G+B}$

We further impose the assumptions of *null-jointness* (introduced by Shephard and Färe, 1974) and *weak disposability* (introduced by Shephard, 1970) between these two types of outputs. The strong (free) disposability assumption is retained for good outputs

- Null-jointness means that no bad outputs are produced only if none of the good outputs are produced;
- Weak disposability of bad outputs means that the reduction of bads for a given level of inputs is allowed only when accompanied by a *simultaneous* reduction of good outputs. This avoids the need of separately modelling abatement activities, consuming resources that could have been used to expand production of good outputs.

Stated mathematically:

$$(y^g, y^b) \in P(x) \wedge y^b = 0 \Rightarrow y^g = 0$$

$$(y^g, y^b) \in P(x) \wedge 0 \leq \theta \leq 1 \Rightarrow (\theta y^g, \theta y^b) \in P(x)$$

Given a direction vector $d = (d^g, -d^b)$ we can now define a directional distance function:

$$\vec{D}_O(x, y^g, y^b; d^g, -d^b) = \sup_\beta \{\beta : (y^g + \beta d^g, y^b - \beta d^b) \in P(x), \beta \in \Re_+\}$$

The value of the directional distance function is the maximum feasible expansion of good outputs and contraction of bad outputs (when the expansion and contraction are identical proportions), for a given level of inputs[13]. This directional distance function is illustrated in figure 3.

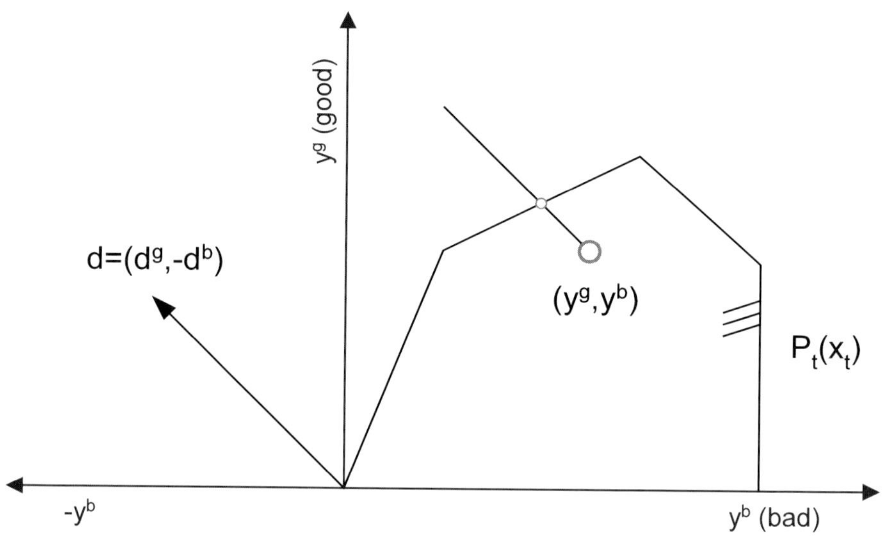

Figure 3. *Directional distance function and projection on the transformation frontier*

In theory, given a sample of observed input-output vectors of firms in a particular sector, one can estimate the transformation frontier[14] using econometric (stochastic, parametric) techniques. These require specification of the transformation function (e.g. the translog functional form) and of the error term[15]. In practice, most analysts prefer mathematical programming (deterministic, non-parametric) methods to calculate the value of the distance function for a firm whose environmental performance is under evaluation. Programming techniques simply "envelop" the observed input-output vectors to obtain a piecewise linear transformation frontier, discarding the need for the specification of any functional form. Conditional on the assumptions, the mathematical programming techniques take the form of linear programming (LP) models, which are the mathematical equivalents (duals) of the LP models used in the DEA approach. LP methods are fairly robust, but can be computationally intensive, considering one has to run the model for each firm in the sample. Unlike the econometric approach, the programming methods suffer from sensitivity to measurement error (not being able to distinguish between statistical noise and technical inefficiency) and the impossibility to test the significance of parameter estimates using standard statistical tests[16].

Färe *et al.* (2000) use two separate distance functions two create what they call an "Environmental Performance Index". The first distance function is an output distance function for the subvector of good outputs, the second an input distance function for the subvector of bad outputs. These distance functions are used to define a quantity index of good outputs (measuring the relative success of a firm in expanding its good outputs while using the same level of inputs and producing the same amount of bad outputs as another firm), and a quantity index of bad outputs (measuring the relative success of a firm in contracting its bad outputs while holding its good outputs and inputs at the same level as another firm)[17]. The environmental performance index is simply the ratio of those two quantity indices (how much good output is produced per unit of bad output), the implicit benchmark being the highest ratio of good to bad outputs.

4.3. Measuring Productivity Change

Monitoring progress is an important characteristic of voluntary approaches, benchmarking and performance evaluation in general. The Flemish Energy Benchmarking Covenant clearly states that "...the industry will annually draw up a monitoring and progress report." The Covenant also states that one has to "...predict the energy efficiency of the Best International Standard until 2012", and this prediction can be based on "... extrapolation of efficiencies from the past". (www.benchmarking.be) Clearly some insight in productivity change is called for.

Fortunately, the productive efficiency approach can be extended to the measurement of productivity change. The values of the distance functions, or "efficiency scores", are used to calculate what are known as Malmquist-Luenberger (ML) indices of productivity change. This is done by benchmarking against a single period's transformation frontier and assuming constant returns to scale.

The choice of which period to use as a reference or benchmark is arbitrary, but can lead to different results. It is therefore customary to specify the geometric mean of two ML indices, one using technology $P_t(x_t)$ of period t as the reference set and one technology $P_{t+1}(x_{t+1})$ of period t+1.

The ML index with technology of period t as reference or benchmark is defined as:

$$ML^t = \frac{\left[1+\vec{D}_O^t(x_t, y_t^g, y_t^g; y_t^g, -y_t^g)\right]}{\left[1+\vec{D}_O^t(x_{t+1}, y_{t+1}^g, y_{t+1}^g; y_{t+1}^g, -y_{t+1}^g)\right]}$$

where the direction vector of the output distance functions is given by $(d^g = y_t^g, -d^b = -y_t^b)$. The ML index ML^{t+1} for period t+1 is specified analogously, simply replacing $\vec{D}_O^t(\bullet)$ with $\vec{D}_O^{t+1}(\bullet)$.

The ML index of productivity change, using output distance functions, is defined as $ML_t^{t+1} = \sqrt{ML^t \times ML^{t+1}}$. One readily sees that an improvement in productivity is signalled by $ML_t^{t+1} > 1$, a decrease by $ML_t^{t+1} < 1$, whereas $ML_t^{t+1} = 1$ means there were no changes in inputs or outputs between the two time periods.

Furthermore, the ML index of productivity change can be decomposed into technical efficiency (ML_TE) and technical change (ML_TC) components, or ML = ML_TE x ML_TC. Although the mathematics remains fairly simple, the equations are rather cumbersome, and we refer to Färe *et al.* (2001) for the details. The main results are given in table 1.

***Table 1.**. Decomposition of the Malmquist-Luenberger index of productivity change*

Index	*Value*	*Meaning*
ML_TE	= 0	The firm is (technical) efficient in period t, i.e. the firm is on the transformation frontier in period t
ML_TE	> 0	The firm is (technical) inefficient in period t, i.e. the firm is off the transformation frontier in period t
ML_TE	> 1	The firm is closer to the transformation frontier in period t+1 than it was in period t
ML_TE	= 1	The firm is at the same distance from the transformation frontier in period t+1 as it was in period t
ML_TE	< 1	The firm is further from the transformation frontier in period t+1 than it was in period t
ML_TC	> 1	The transformation frontier shifted in the direction of 'more goods and fewer bads'
ML_TC	= 1	The transformation frontier did not shift
ML_TC	< 1	The transformation frontier shifted in the direction of 'fewer goods and more bads'

There are many variations of the ML index, dependent on the assumptions upon which the distance functions are based.

5. RECOMMENDATIONS FOR THE FUTURE AND CONCLUSIONS

Earlier we stated that productive efficiency theory is a more general approach than DEA. It is indeed possible to define hyperbolic distance functions (and consequently hyperbolic ML indices). Hyperbolic distance functions allow for the simultaneous contraction of bad outputs and inputs on the one hand, and expansion of (good) outputs on the other hand. However, they remain radial measures of efficiency, allowing for simultaneous decreases and increases only in an equi proportionate manner. Unfortunately, the mathematical programming methods used to calculate the hyperbolic distances from the transformation frontier cannot always be easily converted to robust LP models. Non-radial measures of efficiency are even more problematic in terms of computational burden, and several methodological issues still need to be resolved.

One of the biggest practical hurdles as yet may well be the availability and quality of data, or rather lack thereof.

> The conclusions of the MEPI project as a whole will, among other, emphasise the need to improve the systematic collection of data on the most significant variables. "Collect more data on a more regular basis on less variables" is one of our key recommendations, both to the public authorities and agencies responsible for collecting information on environmental performance, and to the companies' managers. (Tyteca *et al*., 2002, p.12)

Or,

> The data shortages either at the industrial or at the branch level form a severe hindrance for the compilation of the eco-efficiency indicators. Current testing revealed that shortcomings in data at the sub-sectoral level takes place both with respect to concern both economic and environmental indicators. Accordingly, better statistics as well as development of additional tools for modelling of branch contributions to environmental pressures would be necessary in a short term. (Anite, 1999, p. 55).

Most operative indicators at present are still concerned with only one single type of environmental pressure, even though industrial activities are related to a multitude of environmental pressures. There is difficulty in constructing a relatively small number of integrated indicators at the firm level, enabling relevant comparisons of environmental performance among firms. To assess overall environmental performance, there is need for weighting different types of environmental pressures into a single indicator. Since undesirable outputs or bads do not, in general, have observable market prices, it is difficult to assign the appropriate weights to these goods. This is where the use of economic theory of productive efficiency, or its engineering equivalent, 'Data Envelopment Analysis', may be of invaluable significance.

Finally, the standardized, integrated indicators should be easily computed with data commonly available. In this respect, better data and a detailed, transparent monitoring system to collect and measure these data is essential.

Our review of the mostly theoretical literature has shown that standardized, *integrated* indicators, based on productive efficiency theory, may play an important role in promoting VAs as an indispensable complement to more traditional approaches. The quantitative assessment of the environmental performance of firms, including their energy efficiency, is crucial when evaluating environmental VAs in general or energy benchmarking covenants in particular. Also, quantitative targets are very important for the succes of VAs. If the targets are appropriately set, monitored and enforced, then VAs could prove to be a serious instrument of significant potential for future use in addressing environmental issues, such as climate change.

6. NOTES

[1] Rank Xerox pioneered benchmarking in its modern form to meet the Japanese competitive challenge of the 1970s.
[2] Spendolini (1992) examined 49 definitions of benchmarking.
[3] D=driving force, P=pressure, S=status, I=impact, R=response
[4] A potency factor (e.g. Global Warming Potential in CO_2 equivalents / kg of substance A) helps to convert one environmental performance indicator (e.g. kg of a global warming gas emission of substance A) into another environmental performance indicator with a common unit (e.g. kg Global Warming Potential). EPIs can thus, in principle, be added up after weighting with their relative environmental impact potential. VNCI (2001) e.g. lists potency factors or "Unique Weight Factors" for 250 chemicals, to construct 7 core theme EPIs.
[5] GRI is still affiliated with the United Nations through its status as a Collaborating Centre of the United Nations Environment Programme.
[6] Although GRI initially emphasized the use of the Guidelines by corporations, governmental and non-governmental organizations are now included as well.
[7] BCSD became the WBCSD in 1995 (following a merger between BCSD and WICE in Paris).
[8] It is also possible to define "eco-intensity", the ratio of an environmental indicator divided by an economic indicator. Eco-intensity measures the use of nature per unit of output. This approach is adapted by the EEA. Ellipson (2001) oddly uses the term eco-efficiency for the ratio of environmental performance divided by financial performance, e.g. MJ energy used per euro of value added.
[9] Value added = sales minus costs of goods and services purchased. Value added at factor costs describes the gross income from operating activities, after adjusting for operating subsidies and indirect taxes.
[10] Nature is either used as a 'source' or as a 'sink'.
[11] It is in fact assumed that all firms, whose efficiency is being compared, use the same kind of production technology (belong to the same sector).
[12] One possible method is to restrict the weights to lie within a specified range.
[13] The above production model is just one of many possible representations. One could just as easily impose weak disposability on bad inputs, and / or on good outputs (treating bad and good outputs symmetrically instead of asymmetrically), impose null-jointness between bads and (good) inputs, use an input instead of an output distance function, etc.
[14] In case of multiple outputs, one speaks of transformation rather than production.
[15] The econometric approach is called Stochastic Frontier Analysis (SFA). Many analysts refrain from using SFA because it depends on arbitrary assumptions about the distribution of residuals.
[16] There is a limited body of research on stochastic DEA (SDEA), maximising the efficiency of each firm, but allowing a certain probability that the normal DEA constraints will not be satisfied. SDEA has not been widely used in empirical studies.
[17] These quantity indices satisfy Fisher's important tests like homogenity, time reversal, transitivity and dimensionality.

7. REFERENCES

Adriaanse A (1993), "Environmental Policy Performance Indicators", SDV Publishers, The Hague.

Aggeri, F. (2000), "Resource development and monitoring issues Towards a contingency assessment of NA's", paper for the Neapol Closing Conference Ghent 30/11 - 01/12/2000, Ecole des Mines de Paris.

Altham W. (2001), *"The Application of Benchmarking to Encourage the Adoption of Cleaner Production in the Drycleaning Industry"*: paper presented at the 7th European Roundtable on Cleaner Production, Lund, Sweden, 2-4 May 2001, Centre of Excellence in Cleaner Production, Curtin University of Technology, Perth.

Anite (1999), "*A first set of eco-efficiency indicators for industry: Pilot study*", final report 13 December 1999, Anite Systems, Luxembourg.

ANSI/ISO 14031-1999: "*Environmental management - Environmental performance evaluation – Guidelines*", Quality Press, 1999.

Azzone G., Noci G., Manzini R., Welford R. and Young W. (1996), "Defining Environmental Performance Indicators: An Integrated Framework", *Business Strategy and The Environment*, 5 (1), pp.69-80.

Ball, V. E., Lovell, C. A. K., Nehring, R. F. and Somwaru, A. (1994). "Incorporating undesirable outputs into models of production: an application to U.S. agriculture", *Cahiers d'économie et sociologie rurales* 31, pp. 60-74.

Berkhout F. (coord.), Azzone G., Carlens J., Hertin J., Jasch C., Noci G., Olsthoorn X., Tyteca D., Van der Woerd F., Van Drunen M., Wagner M., Wehrmeyer W. and Wolf O. (2001), *"MEPI – Measuring the Environmental Performance of Industry*", Final report, EC Environment and Climate Research Programme: Research Theme 4 - Human Dimensions of Environmental Change - Contract No: ENV4-CT97-0655. February 2001. Web site: http://www.environmental-performance.org.

Bogan, C.E. and English, M.J. (1994), *"Benchmarking for Best Practices: Winning Through Innovative Adaptation"*, New York: McGraw-Hill.

Camp R. (1995), "*Business Process Benchmarking: Finding and Implementing Best Practices*", ASCQ Quality Press, Milwaukee.

Charnes A., Cooper W. and Rhodes E. (1978), "Measuring the Efficiency of Decision Making Units", *European Journal of Operational Research*, 2, pp. 429-444.

Charnes, A., Cooper, W. W., Lewin, A. Y., & Seiford, L. M. (Eds.). (1995). *Data envelopment analysis: Theory, methodology, and applications*. Boston: Kluwer

Chidiak, M. (1999), *"Voluntary Agreements for Energy Efficiency in Five EU Countries,"* in *Energy Efficiency and CO_2 Reduction: The Dimensions of Social Change*, 1999 European Council for an Energy-Efficient Economy Summer Study, May 31-June 4, Mandelieu, France.

Chung Y.H., Färe R. and Grosskopf S. (1997), "Productivity and Undesirable Outputs: A Directional Distance Function Approach", *Journal of Environmental Management*, 51, pp. 229-240.

Cooper, W.W., Seiford, L.M. and Tone, K. (2000). *Data Envelopment Analysis. A Comprehensive Text with Models, Applications, References and DEA-Solver Software*. Kluwer Academic Publishers, Boston, MA.

Couder J. and Verbruggen A. (2003), "*Technical Efficiency Measures as a Tool for Energy Benchmarking in Industry?", Energy* and Environment, 14 (5), pp. 705-724.

Croci E. and Pesaro G. (1997), "Voluntary agreements in Italy: a new approach in environmental policy", *ENER BULLETIN*, 20.97.

Croci E. and Pesaro G. (1999), "*Voluntary Agreements in the Environmental Sector – The Italian Experience*", CAVA Working Paper no 98/11/5, January 1999.

Demming W.E. (1993), "*The new economics for industry, government, and education*", Massachusetts Institute of Technology, Center for Advanced Engineering Studies, Cambridge, MA.

Diebäcker M. (2000), "Environmental and social benchmarking for industrial processes in developing countries: a pilot project for the textile industry in India, Indonesia and Zimbabwe", *Integrated Manufacturing Systems*, 11 (7), 2000, pp. 491-499.

Kraft (1997), "*The Department of the Navy Benchmarking Handbook: A Systems View*", Department of the Navy Total Quality Leadership Office (TQLO), Washington, D.C.

EEA (1997), "Environmental Agreements, Environmental Effectiveness", *Environmental Issues series*, 1 (3), European Environment Agency (EEA), Copenhagen, July 1997.

EEA (1999), "*Environmental indicators: typology and overview*", Technical report No 25, European Environment Agency. Office for Official Publications of the European Communities, Luxembourg.

Färe R., Grosskopf S. and Hernandez-Sancho F. (2000), "*Environmental Performance: An Index Number Approach*", Department of Economics Working Paper, Oregon State University, Corvallis, revised, February.

Färe R., Grosskopf S. and Pasurka C.A. (2001), "Accounting for air pollution emissions in measures of state manufacturing productivity growth", *Journal of Regional Science*, Vol. 41, No. 3, pp. 381-409.

Gallopin G.C. (1996), "Environmental and Sustainability Indicators and the Concept of Situational Indicators. A Systems Approach", In: *Environmental Modeling and Assessment*, 1, pp. 101-107.

Gallopín G.C. (1997), "Indicators and their use: information for decision making", in Moldan B. and Billharz S. (Eds), *Sustainability Indicators. Report on the project on Indicators of Sustainable Development,* John Wiley and Sons, Chichester.

Geisler E. (1999), "*The metrics of technology evaluation: where we stand and where we should go from here*", Stuart Working Paper 99-03, presented at the 24th Annual Technology Transfer Society Meeting, July 15-17, 1999. Stuart Graduate School of Business, Illinois Institute of Technology.

Gerrits R. and Oudshoff B. (2003), "*Energy efficiency through Long-term agreements; Broadening the horizon in the new LTA approach*", in: American Council for an Energy Efficient Economy. Proceedings 2003 Summer Study on Energy Efficiency in Industry. ACEEE, New York.

GRI (2002a), "*Sustainability Reporting Guidelines*", Global Reporting Initiative, Boston, 2002.

GRI (2002b), "*Energy Protocol For use with the GRI 2002 Sustainability Reporting Guidelines*", Global Reporting Initiative, Boston, December 2002.

Hammond A., Adriaanse, A., Rodenburg E. Bryant D., and Woodward, R. (1995), "*Environmental Indicators: A Systematic Approach to Measuring and Reporting on Environmental Policy Performance in the Context of Sustainable Development*" World Resources Institute, Washington, DC. 43 pp.

Hansen K., Johannsen K.S.and Larsen A. (2002): "Recommendations for Negotiated Agreements", *Government and Policy - Environment and Planning*, 20(1), pp.19-37.

Higley C.J., Convery F. and Lévêque F. (2001), "1. Voluntary Approaches: An Introduction", in *Environmental Voluntary Approaches: Research Insights for Policy-Makers* (Higley C.J. and Lévêque F. editors), presented at the International Policy Workshop on the Use of Voluntary Approaches at the Centre Borchette, Brussels, on 1 February, 2001, Centre d'économie industrielle Ecole Nationale Supérieure des Mines de Paris (CERNA), May 2001.

Keffer C., Shimp R. and Lehni M. (1999), *"Eco-Efficiency Indicators & Reporting. Report on the Status of the Project's Work in Progress and Guideline for Pilot Application*", World Business Council for Sustainable Development (WBCSD), Geneva, April 6, 1999.

Krarup S. (2001), "4. Can Voluntary Approaches be Environmentally Effective and Economically Efficient?, in Environmental *Voluntary Approaches: Research Insights for Policy-Makers* (Higley C.J. and Lévêque F. editors), presented at the International Policy Workshop on the Use of Voluntary Approaches at the Centre Borchette, Brussels, on 1 February, 2001, Centre d'économie industrielle Ecole Nationale Supérieure des Mines de Paris (CERNA), May 2001

Krarup, S. and Ramesohl S. (2000), "*Voluntary Agreements in Energy Policy – Implementation and Efficiency",* The final report from the project Voluntary Agreements Implementation and Efficiency. AKF Forlaget, Copenhagen.

Müller K. And Sturm A. (2001), "*Standardized Eco-Efficiency Indicators*", Report 1: Concept paper, Revision: 1.0.5 / January 2001, Ellipson, Basel.

OECD (1999), "*Voluntary Approaches for Environmental Policy: An Assessment",* edited by Börkey P., Glachant M. and Lévêque F., Organisation for Economic Co-operation and Development (OECD), Paris, 1999.

OECD (2003), "*OECD Environmental Indicators Development, Measurement and Use*", reference paper, Organisation for Economic Co-operation and Development (OECD), Paris, 2003.

Phylipsen G.J.M, Blok K. And Worrell E. (1998), *"Handbook of International Comparisons of Energy Efficiency in the Manufacturing Industry"*, Dept. of Science, Technology and Society, Utrecht University.

Pieri C., Dumanski J., Hamblin A. and Young A. (1995), *"Land Quality Indicators*", World Bank Discussion Papers. No. 315. The World Bank, Washington, DC.

Rigby D., Howlett D. and Woodhouse Ph. (2000), *"A Review of Indicators of Agricultural and Rural Livelihood Sustainability",* Sustainability Indicators for Natural Resource Management & Policy, Working Paper 1, February 2000. (website: http://les.man.ac.uk/jump/indicators.html)

Shephard R.W. (1970), *"Theory of Cost and Production Functions*", Princeton: Princeton University Press.

Shephard R.W. and Färe R. (1974), "Laws of Diminishing Returns", *Zeitschrift für Nationalökonomie*, 34, pp. 69-90.

Spendolini M.J. (1992), *The Benchmarking Book*, Amacom, New York, NY.

Tyteca D. (1996), "On The Measurement of Environmental Performance of Firms – A Literature Review and a Productive Efficiency Perspective", *Journal of Environmental Management*, 46, pp. 281-308.

Tyteca D. (2001), "Economic and Business Indicators", in *Measuring the Environmental Performance of Industry (MEPI)* Final Report – Appendices, Part II, Appendix 2: Environmental Performance Indicators: State-of-Art Reviews, February 2001. Web site: http://www.environmental-performance.org.

Tyteca D., Carlens J., Berkhout F., Hertin J., Wehrmeyer W. and Wagner M. (2002), "Corporate Environmental Performance Evaluation: Evidence From The Mepi Project", *Business Strategy and the Environment*, 11, pp. 1-13.

UN CSD (1996), "*Indicators of Sustainable Development Framework and Methodologies*" United Nations Commission on Sustainable Development, New York. (Website: http://www.un.org/esa/sustdev/isd.htm)

Verfaillie H.A. and Bidwell R. (2000), *"Measuring eco-efficiency, a guide to reporting company performance"*, World Business Council for Sustainable Development (WBCSD), Conches-Geneva, June 2000.

VNCI (2001), *"The 'EPI-method' (version 1.1)*", Environmenal Performance Indicators for the Chemical Industry, Guideline, Association of the Dutch Chemical Industry (VNCI), Leidschendam, October 2001.

Wagner M. and Wehrmeyer W. (2001), "Physical Indicators", in *Measuring the Environmental Performance of Industry (MEPI)* Final Report – Appendices, Part II, Appendix 2: Environmental Performance Indicators: State-of-Art Reviews, February 2001. Web site: http://www.environmental-performance.org.

Yaisawarng, S and Klein, J. D. (1994), "The effects of sulfur dioxide controls on productivity change in the U.S electric power industry", *Review of Economics and Statistics*, pp. 447-460.

CHAPTER 6

ENVIRONMENTAL VOLUNTARY AGREEMENTS IN POLICY MIXES

ENVIRONMENTAL AGREEMENTS USED IN COMBINATION WITH OTHER POLICY INSTRUMENTS

N.A. BRAATHEN*

OECD, Environment Directorate, Paris, France

Abstract. This chapter discusses the impacts on environmental effectiveness and economic efficiency of applying environmental agreements in combination with other instruments in environmental policy. The findings are relatively negative: while the administrative costs of such combinations can be high, there is a distinct possibility that the environmental effectiveness will be lower than if the voluntary approach had not been part of the policy mix.

1. INTRODUCTION

In 2003 OECD issued the book *Voluntary Approaches for Environmental Policy: Effectiveness, efficiency and usage in policy mixes,* cf. OECD (2003a). The book was based on a number of new case studies and an extensive review of recent literature. The book concluded *inter alia* that while the environmental targets of most – but not all – voluntary approaches seem to have been met, there are only a few cases where such approaches have been found to contribute to environmental improvements significantly different from what would have happened anyway. The book also indicated that the economy-wide efficiency of voluntary approaches is generally low, as they seldom incorporate mechanisms to equalise marginal abatement costs between all polluters. This is *inter alia* because environmental targets of the approaches tend to be set for individual firms or sectors, rather than at a national level. However, traditional "command and control" policies also rarely equalise abatement costs at the margin between different polluters, and voluntary approaches can offer a higher economic efficiency than such policies, by providing firms increased flexibility in how environmental improvements are to be accomplished.

Voluntary approaches are very seldom used as "stand-alone" instruments. Instead they tend to form part of policy packages involving one or several other instruments, like some type of "command-and-control" regulations, taxes, tradable permits, etc. Part II of OECD (2003a) analysed in particular voluntary approaches used in mixes with other policy instruments. The present chapter draws to a large extent on that analysis. The purpose is to discuss the *marginal* impacts of using voluntary approaches in policy mixes: Which are the additional impacts of combining a voluntary approach with one or several other instruments? How does

E. Croci (ed.), The Handbook of Environmental Voluntary Agreements, 335 - 364

the fact that the approach is used in combination with other instruments impact on environmental effectiveness, economic efficiency, administrative costs, etc?

Any instrument evaluation should be explicit on the comparisons that are made. To what is a given policy compared? Is it compared to a "no policy" or "Business-as-Usual" scenario, or to an alternative "new" instrument that could achieve similar policy objectives? Smith (2003) pointed to two relevant dimensions of any comparison: the impact on the government budget and the impact on the environment. If the impacts on the government budget of two policies differ, the comparison could be dominated by the macroeconomic impact of the change in the government deficit. Any difference in environmental impact between two policy options ought to be quantified in physical terms and be given an economic valuation. Next, this value should be brought into a comparison of other costs and benefits of the policies in question.

A number of different comparisons will be made in this chapter. On several occasions, reference will be made to "first-best" instruments like well-designed taxes and tradable permit systems, which would assure that marginal abatement costs are equalised between sources of a given problem, thus achieving a given target at the lowest overall cost to society. It should, however, be emphasised that such instruments require the existence of institutions that have the authority to adopt and enforce them. This could be a national parliament and a well-functioning national enforcement and legal system. Within the European Union one has also created the necessary institutions to adopt and enforce taxes or tradable permits at a supra-national level.

However, no institution has similar authorities at a global level. Hence, a global tax or a global emission trading system is at present not a realistic option to address an environmental problem – neither is a global "command and control" regulation.[1] This fact emphasises the potential usefulness of global voluntary instruments like the "OECD Guidelines for Multinational Enterprises" (cf. OECD, 2000 and 2003b)[2], in spite of the somewhat negative findings of OECD (2003a).

2. A FRAMEWORK FOR ANALYSIS

Figure 1 illustrates some of the impacts, links and considerations that in principle should be taken into account when discussing policy mixes. To promote sustainable development, policy makers are interested in, *inter alia*, the environmental, economic and social impacts of different policy instruments and instrument combinations. The three oval shapes in the figure represent these dimensions. The rectangles with similar shading as the ovals indicate some relevant aspects of the three dimensions. Some of the inter-dependencies between different aspects are represented by the thin arrows in the figure, marked by the letters A to N, which will be briefly commented below.

The large box to the left in Figure 1 lists some of the instruments that a voluntary approach might be combined with. Some such combinations will be discussed in greater detail below, with examples from current practises in OECD countries. For simplicity, the discussion will often look at impacts of combining a voluntary approach with *one* particular other type of instrument, but in practice it is frequent that three or more instruments are used to address a given target.

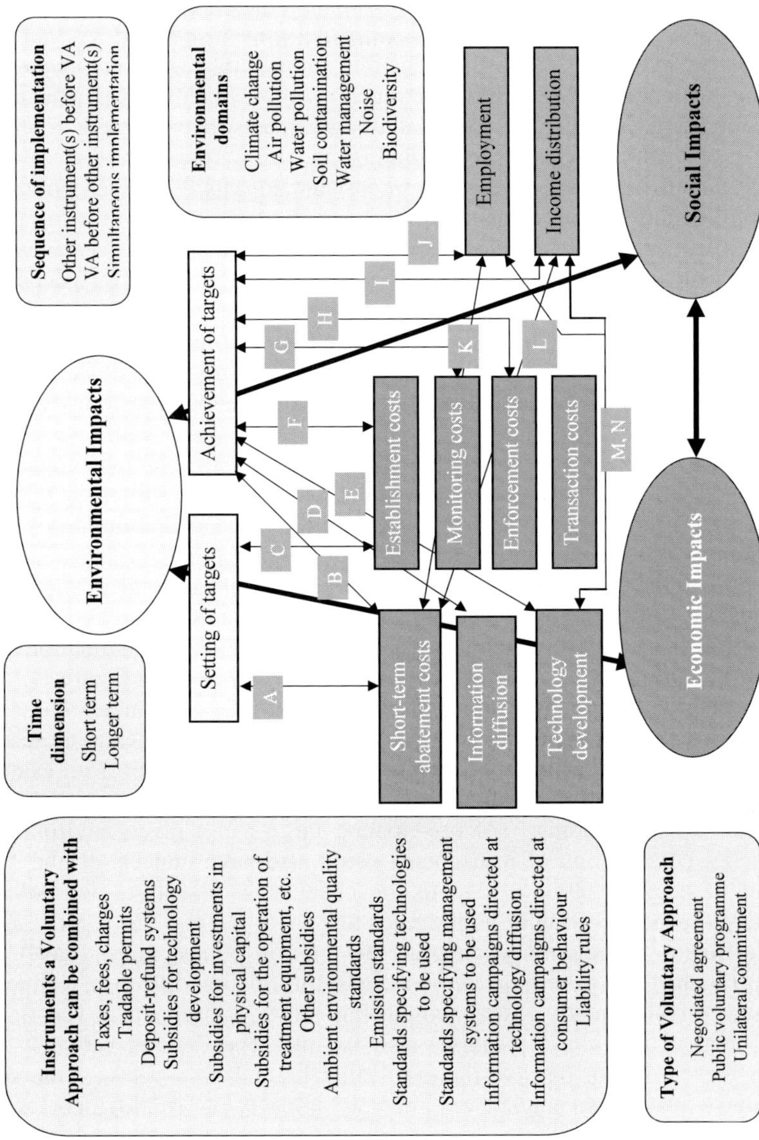

Figure 1. Analysing policy mixes involving voluntary approaches

The analysis should also distinguish between different types of voluntary approaches, as indicated in the lower left corner of the figure. Impacts of combining a negotiated agreement with a given other instrument will generally be different than combining a public voluntary programme with the same instrument. As unilateral commitments by individual firms or industries often imply little involvement of public policy, such approaches will not be much addressed below. Most attention in the discussion here is given to negotiated agreements between environmental authorities and industrial sectors or individual firms.

The time dimension of the analysis is important: impacts of a given policy mix can differ significantly between the short term and the longer term – for instance if the incentives for technology development are altered when several instruments are used in combination. Impacts of instrument combinations might also vary depending on whether the policies are meant to address climate change, local air pollution, waste handling, etc., as exemplified in the box on the right-hand side of the figure. For instance, impacts on related technology developments might be more important concerning policies implemented to address climate change than for policies used to protect biodiversity.

Finally, the sequence of implementation can matter. For example, impacts of combining a tax and a voluntary approach will be different if a voluntary option is added to, or partly replace, a pre-existing tax than if a tax is added on top of a pre-existing voluntary scheme.

I return now to the different aspects concerning the environmental, economic and social dimensions of sustainable development highlighted in the figure. Of particular relevance here is the extent to which using several instruments in combination affect the *nature of the links* that are indicated between different aspects in Figure 1.

Two aspects concerning the environmental impacts of policy mixes have been singled out: impacts of the combination on the *setting* of environmental targets and impacts of the combination on the *achievement* of a given target. For example, replacing a pre-existing tax by a negotiated agreement could come in parallel with a softening – or a strengthening – of pre-existing targets in a given environmental domain. Such a combination of instruments could also have impacts on the actual achievement of a given target – due to the changes in incentives for pollution abatement that the modification of policy leads to.

Seven different aspects of economic impacts of policy mixes have been identified in Figure 1: Short-term abatement costs, information diffusion, technology development, establishment costs, monitoring costs, enforcement costs and transaction costs. There can, of course, also be interdependencies between these aspects, but these will not be investigated further here. Two aspects of the social dimension have also been singled out, namely impacts on employment and on income distribution.

Both the setting of targets and the achievement of targets are inter-linked with the short-term abatement costs, cf. the arrows marked A and B. If, for instance, the abatement costs are believed to be very high, this could – and should – have impact on the targets set. Both the expected and actual short term abatement costs can impact on the degree of target achievement, and vice versa.

The strictness of the target set could have impact on the difficulties – and costs – of establishing a certain policy package involving a voluntary approach, cf. the arrow C. There can also be links between the resources spent on establishing a policy, and the actual achievement of the targets, cf. F. The target achievement would in this connection also be linked to the monitoring costs [cf. G] and the enforcement costs [cf. H]. If few resources are devoted to monitoring of performance and enforcement, actual target achievement is likely to be low.

A more long-term, dynamic set of links are related to (technology) information diffusion and technology development on the one side and achievement of environmental targets on the other side, cf. the links D and E. It is, for example, *possible* that combining a voluntary approach with an information campaign, or with fiscal incentives for technology development, could lead to better achievement of the targets of the voluntary approach. One should, however, ideally still undertake a cost-benefit analysis of these measures to see if they merit implementation – and also consider whether other policy combinations could provide similar results at lower costs.

Environmental and social impacts are also related. Achievement of environmental targets can, for instance, in some cases affect the *sectoral* employment situation [cf. J], for example within the fisheries, agriculture and tourism sectors. Any employment impacts could in turn affect the income distribution, cf. I. On the other hand, changes in the income distribution could influence the ability to reach a given environmental target. Possible examples relate to the cleaning of sewerage and the collection of waste among poor people in low-income countries.

Finally, most – if not all – of the economic and social aspects of sustainable development are inter-linked. There can, for instance, be links between the abatement costs in the short run and both the employment situation and the income distribution, cf. links K and L. Any impacts of a policy mix on technology developments can also affect both employment and income distribution [cf. M, N], etc.

The purpose here is not to give a comprehensive discussion of all the links between the environmental, economic and social dimensions of sustainable development. Instead, the focus is on how the use of policy mixes involving voluntary approaches might impact directly on some of the aspects of the three dimensions, and to consider if – and how – the nature of some of the links between them might be affected by the use of such mixes. The remainder of this chapter looks more in detail at a number of policy mixes involving voluntary policy approaches. Particular focus is given to mixes involving taxes and tradable permits.

3. VOLUNTARY APPROACHES USED IN COMBINATION WITH ENVIRONMENTAL PERMIT SYSTEMS

In OECD countries any major polluter needs an environmental permit to be allowed to operate. The permit might specify upper limits on the emissions to different environmental media (air, water, etc.) that can take place, often depending on characteristics of the recipient of the pollution. The permit can set upper limits on the absolute amounts of emissions over various time-spans, or on concentrations of pollutants during a short or longer period of time. A large polluter might also be given responsibility to keep pollution levels in a given recipient below certain thresholds.

Given the widespread use of environmental permit systems, very many voluntary policy approaches will in practice be part of a policy mix with such systems. In most cases, the sequence of implementation is likely to have been that a permit system already existed before a voluntary approach was launched.

It can be useful to briefly consider the impacts one *a priori* could expect from such combinations. Looking at environmental issues already addressed by the pre-existing permit system – which are the most relevant cases when discussing "policy mixes" – one would generally expect that environmental authorities would aim to tighten the existing targets with the introduction of a new voluntary approach. Even if the targets set would be stricter than before, it is, however, not given that actual environmental achievements are improved. In order for companies to come forward and participate voluntary in a more "ambitious" scheme, some combination of "carrots" or "sticks" would normally be required. Among possible "carrots" are various types of financial assistance, or promises of increased flexibility in how the targets are to be obtained. Some kind of public recognition – which can be used to position the firm as "environmentally friendly" vis-à-vis its customers – could also serve as a "carrot".[3] Possible "sticks" include *credible* threats of introducing new mandatory instruments, like stricter standards, new taxes, etc.

The environmental permitting system can in some cases seem very inflexible – and unnecessarily so. *For example*, OECD (2002b) and (2003a) refers to provisions in the Clean Air Act in United States, under which facilities must obtain approval from the permitting authorities each time they make a manufacturing change. Under the so-called "Project XL" individual firms could negotiate regulatory relief in exchange for pollution reductions in excess of status quo standards. Such relief clearly could improve the economic efficiency of the regulation for the – relatively few – firms affected. A general switch to performance standards – where *all* firms can decide themselves how a given environmental performance is to be reached – could be a better option.

As described in OECD (2002b) and (2003a), it is also unclear how much – if any – environmental improvement beyond "Business-as-Usual" the Project XL agreements provided. In addition, the administrative costs of establishing the agreements were considerable. As of January 2003 EPA is no longer accepting proposals for new XL projects.

The Environmental Protection Agency in New South Wales, Australia, has developed an interesting combination of instruments, involving an environmental permitting system, a load-based licence fee for large emitters, and a set of "Load Reduction Agreements". These agreements grant polluters a temporary reduction in licence fees if they undertake to reduce their emissions within a three-year period. The impacts of this policy mix are discussed further in the section on combinations involving taxes below, as the voluntary element in this case is more related to the licence fees than to the environmental permits as such.

Given the great diversity of combinations of environmental permit systems and voluntary approaches, it is difficult to draw generally valid conclusions of their impacts. In fact, a major part of all the (quite substantial) literature on voluntary approaches does explicitly or implicitly deal with this type of policy mix. One repeated finding is, however, that it is often very difficult to document that a given voluntary approach in practice leads to an environmental performance superior to what would in any case have taken place.[4] Some economic savings for individual firms due to *increased flexibility* have been found in several examples, and various information campaigns and forums for participants in a voluntary scheme can enhance technology diffusion compared to traditional "command-and-control" regulation.

It is not clear whether adding a voluntary approach "on top of" a pre-existing environmental permitting system would make marginal abatement costs of different polluters more equal than before. This would *inter alia* depend on which firms participate in the voluntary approach. It can seem likely that a truly voluntary public programme – incorporating e.g. some public recognition of participating firms – would attract in particular firms with low-cost abatement options. If so, there would tend to be *some* equalisation of marginal abatement costs, compared to the previous situation.

The incentives for technology development are generally rather weak under a traditional environmental permitting system, and it does not seem likely that they would be much strengthened by the introduction of a voluntary approach. Mechanisms where a positive "shadow price" on the size of permitted emissions is introduced – like with the load-based licence fees in New South Wales – could strengthen such incentives significantly.[5]

4. VOLUNTARY APPROACHES USED IN COMBINATION WITH ENVIRONMENTALLY MOTIVATED SUBSIDIES

All subsidy schemes are by definition in themselves "voluntary", but the focus here is on schemes that combine subsidies and voluntary abatement schemes. It should also be kept in mind that certain types of subsidies for pollution abatement can contradict the "polluter-pays principle", as it is formally adopted by OECD member countries.

Depending on the definition of "subsidies" being used, there are several ways voluntary approaches and environmentally motivated subsidies can be used in policy mixes. One relatively common policy package is that subsidies to stimulate the development of environmentally benign technologies are used as a "carrot" to make polluters volunteer to abate their emissions.

Referring to Figure 1, the subsidies and the voluntary approach would then normally be introduced simultaneously. Presumably the combination of the two instruments would lead to more rapid, and more environmentally focused, development of new technologies than if the voluntary approach had been introduced in isolation. It is, however, more unclear what is the marginal impact in this respect of adding the voluntary approach to a subsidy scheme – meaning that only firms that take part the voluntary approach can obtain subsidies e.g. to develop new abatement technologies, etc. It is *conceivable* that the obligation to take part in the voluntary approach to obtain the technology subsidy makes the selection of companies that apply for the subsidies biased, and leads to a slower technological progress than if the subsidies had been used in isolation – where also firms in other sectors, e.g. specialising in the production of pollution abatement equipment, could apply for support[6]

The environmental impacts of such policy combinations would, *inter alia*, depend on the extent to which the technology developments actually succeeds – and on the longer-term structural changes in the economy that the technology changes generate.[7]

In relation to the UK Emissions Trading Scheme for greenhouse gases, British authorities used a different approach to allocate environmentally motivated subsidies. Here a given amount of subsidies targeted reductions in greenhouse gas emissions directly – *i.e.* the subsidies were not given for *e.g.* technology development, etc. Companies were instead invited to participate in an auction, where they would voluntarily commit to achieve *absolute* emission reductions at progressively lower prices. The Government's objective was to obtain the maximum amount of reductions for the incentive money made available (£215 million over a five-year period). Thirty-four organisations bid successfully to join the scheme. Over the five year period of the scheme, the participating companies have pledged to reduce their annual greenhouse gas emissions by more than four million tonnes of CO_2.[8]

In this case firms voluntary take on a legally binding emission reduction obligation beyond what was imposed by other regulations, in return for the subsidies they obtain. Through the auction mechanism the scheme should help find the cheapest ways to realise a given amount of emission reductions, and thus stimulate economic efficiency.[9] The fact that the participating companies can use an emissions trading mechanism to fulfil their reduction obligations enhance economic efficiency of the scheme as a whole further, cf. a further discussion below.[10]

Lyon and Maxwell (2004) compare an emissions tax and a public voluntary program, where participation is stimulated by some sort of subsidy. They conclude:

> "The most important lesson of this chapter is that public VAs typically arise from weakness, not from strength. They should not be regarded as some new and superior policy instrument. Rather, they should be viewed as a limited tool that may be useful in settings where more powerful policy instruments are infeasible. Indeed, policy makers should approach VAs with caution, since their very availability may increase industry resistance to the use of more powerful regulatory tools. This resistance increases because the hope of obtaining a subsidy (through a public VA) strengthens industry's resolve to fight traditional regulatory tools of taxes and standards, which impose direct costs on the industry.
>
> There is a second risk associated with the increased use of public VAs by policy makers. Just as they may undermine more stringent regulatory tools, they may undermine industry's incentives to undertake environmental improvement under its own initiative. Instead, industry may prefer to wait until government offers a "carrot" before agreeing to improve its environmental performance. Industry may have incentives to preempt the imposition of a tax or standard, but it does not want to risk preempting a handout."

5. VOLUNTARY APPROACHES USED IN COMBINATION WITH TAXES OR CHARGES

5.1. Introduction

Under a tax regime, firms' compliance costs are equal to abatement costs *plus* tax payments for residual emissions. A number of countries combine certain taxes or charges with voluntary schemes, where for instance some sectors are completely exempted from a tax – or pay lower tax rates than other sectors – on the condition that they "voluntarily" undertake certain abatement measures. Such arrangements are often introduced due to a fear that the international competitiveness position of the firms concerned would be compromised if they had to pay the "full" tax rate. If this position was significantly weakened, plant closures could result, with subsequent transition costs related to capital losses and increases in unemployment, *sometimes* in regions with limited employment opportunities. Such repercussions could be found to jeopardise the social dimension of sustainable development, cf. Figure 1.

The environmental effects of a tax will generally come about through the subsequent increase in the prices of the tax-bases in question, and the price elasticities of the tax-bases in question. The price elasticities tend to differ in the short term (when available technology options are given) and in the longer term (when changes in relative prices can trigger new technological developments). Bjørner and Jensen (2002) used a large micro-panel database covering the majority of Danish industrial companies over the period of 1983–1997, and found the average price elasticity of energy to be -0.44 in Danish industry.[11]

To the extent that the sectoral competitiveness arguments often used in favour of special provisions for energy-intensive firms are valid, infra-marginal price increases due to large tax increases could, however, also trigger plant closures that (obviously) would eliminate energy use at a given plant. It is doubtful that the price elasticity estimates presented above incorporate such impacts, as energy-intensive firms in Denmark (as elsewhere) have enjoyed special tax privileges all through the estimation period. Expected demand reductions in response to significant tax increases could, hence, be higher than what the presented estimate indicate.

As mentioned, one should always consider what would *realistically* be the alternative policy when discussing the impacts of combining an environmentally related tax or charge with, *e.g.*, tax exemptions in return for negotiated agreements with some firms or sectors. If the alternative policy is a flat tax rate for all relevant polluters, at the same ('high') level as used for some sectors in the combined policy, the introduction of a voluntary option for some polluters would likely represent a weakening of the environmental target and/or a lower degree of achievement of a given target. Even if a negotiated agreement would oblige the polluters to abate emissions – and leave them increased financial resources to invest in pollution abatement, through the forgone tax revenue – it is not given that this would outweigh the emission reductions that "ordinary" price responses – and possible plant closures – under a "full tax regime" would have brought about.

Impacts on technology development could also be important: adding the voluntary option could give the affected firms more financial resources to undertake research and development, but their incentives to actually achieve technology improvements – and their profits from doing so – could be severely reduced. When the "shadow price" on emissions approaches zero, the firms have low incentives to find ways to reduce them. Over the longer term, this could have important environmental repercussions.

If the realistic alternative to a voluntary approach is a much lower tax rate for the firms included in the voluntary approach than for other firms, the significance of the points above would be reduced accordingly.

In both cases – but to a varying degree – replacing a tax by a voluntary approach will induce a revenue-loss for the government. As discussed by *e.g.* Fullerton and Metcalf (2001) and Goulder, Parry and Burtraw (1997), this revenue-loss can represent a significant efficiency cost. "Scarcity rents" created by the environmental policy are left with the private companies. Public authorities could, for instance, have used the revenues foregone to lower distorting taxes on labour, thus stimulating employment.

Various types of administrative costs could increase with the introduction of a voluntary scheme. Most environmentally related taxes are relatively simple to administer, with *e.g.* the tax-bases being measured and revenues being collected at a limited number of oil refineries for most taxes on mineral oils. Introducing a conditional tax reduction can significantly increase the administrative burden, both for public authorities and for the firms involved.

The following sub-sections discuss a few concrete examples of combinations of taxes or fees with voluntary approaches.

5.2. The Energy Efficiency Agreements in Denmark

As part of the preparation of OECD (2003a), a special case study was made of the Danish agreement scheme on industrial energy efficiency.[12]

A policy package implemented in 1996 combined the introduction of SO_2- and CO_2-taxes with an agreement scheme on improved energy efficiency in industry, and subsidies for *e.g.* energy efficiency counselling and investment. All revenues from the taxes were recycled back to industry in the form of reductions in the taxation of labour and through subsidies for energy efficiency measures. Firms that entered into an agreement with the Danish Energy Agency got a rebate on their CO_2-tax. While all firms with heavy processes had the right to enter into an agreement, firms with light processes only had the right to sign an agreement – and get a tax rebate – if the tax payment on their energy consumption amounted to at least 3% of value added. In addition, the effective tax had to exceed a certain minimum value.

The agreements could be either individual or collective, covering several firms within a sub-sector with similar production processes. The basis of individual agreements *used to be* an energy audit, usually carried out by a certified consultant. The audit should map the energy consumption, list potentials for energy efficiency improvements and suggest special investigations into ways to further reduce energy consumption. In order to ensure quality, the audit report had to be verified by an independent agency assisted by a technical expert. (As from 2000, several changes to the agreement scheme were made. For example, simpler energy surveys have replaced the former energy audits.)

The collective agreements were not based on energy audits. Instead, an analysis of energy consumption and production processes in the sector was made to identify general potentials for improving energy efficiency in the relevant firms. The results of the analysis were reported to the Danish Energy Agency and used to formulate an action programme. In addition to investment projects, special investigations and energy management measures, the action programme for the sub-sector could include inter-firm projects, such as development projects, which were of interest to all firms. Each firm covered by the agreement had to sign and was committed to the action programme.

Each firm had to implement all identified energy saving projects with a payback-period of less than 4 (heavy process) or 6 (light process) years. Firms also had to introduce improved energy management systems.

The use of a payback-period criterion implies that firms with many profitable investments would have to realise relatively large energy savings, while firms with no profitable projects were not burdened with investment projects and special investigations. This contributes to an efficient allocation of energy savings between firms.

However, there were important differences in the criteria used for different firms. Firms with light processes used to be required to undertake projects with longer payback periods than firms with heavy processes. In addition, different price assumptions were used when calculating the payback-periods: For firms with heavy processes, a (hypothetical) tax of 3.3 € per tonne CO_2 was added to the pre-tax energy price of the firm, while for firms with light processes, a (hypothetical) tax of 12 € per tonne CO_2 was added. The lower the tax being applied in the analysis, the lower is the likelihood that a given project would pass the test. Hence, some relatively low-cost energy-saving projects in firms with heavy processes could be left unrealised, which would tend to increase the overall abatement costs.

It is emphasised that the agreements provided limited *additional* tax reductions to the participating companies compared to the very large tax reductions granted to any industrial firm that employ light or heavy processes, cf. Table 1 below. However, a reduction in the tax rate for, *e.g.*, firms with heavy processes from 3.3 to 0.4€ per tonne CO_2 in 2000 was, of course, in itself substantive.

The energy efficiency agreements had two opposing effects on energy use in the respective companies. On the one hand, the companies had to carry out certain activities, like realising proposed energy-saving projects from the energy audits described above, and to increase energy management activities. The effect of these activities was estimated by Bjørner & Jensen (2002) to be a 9% *reduction* in energy use in the companies concerned. On the other hand, companies with an agreement obtained a tax reduction, which was estimated to *increase* their energy use by some 1-5%. Hence it appears that the agreement scheme resulted in a reduction in energy use overall. In other words, the agreement companies would, according to the findings of Bjørner & Jensen, have used more energy if they had not been offered the agreement, but just had paid the 'normal' tax.[13]

Table 1. *Levels of CO_2- and energy taxes in Denmark, 1996-2000. € per tonne CO_2.*

	1996	1997	1998	1999	2000
Space heating[14]	26.7	53.3	80	80	80
Light processes					
– Without agreement	6.7	8.0	9.3	10.7	12.0
– With agreement	6.7	6.7	6.7	7.7	9.1
Heavy processes					
– Without agreement	0.7	1.3	2.0	2.7	3.3
– With agreement	0.4	0.4	0.4	0.4	0.4

Source: OECD (2003a).

5.3. The Climate Change Agreements in United Kingdom

Before the introduction of the Climate Change Levy in United Kingdom from 1.4.2001, energy-intensive sectors were given the option to obtain an 80% reduction in the tax rate if they entered into Climate Change Agreements on improving energy efficiency or reducing carbon emissions.[15]

Agreements were made with 44 sector associations, covering more than 5,000 separate operators and 10,000 facilities.[16] They were negotiated with the relevant sector trade associations on behalf of the companies within the sectors concerned. Facilities identified in the agreements were eligible for the 80% tax discount until 31 March 2003. Eligibility for discount from 1 April 2003 depended on whether the first targets set in the agreements were met.[17] The agreements span the period up to 2010, with "review points" in 2004 and 2008, when the stringency of the targets will be considered again.

The agreements set target both for sectors and for each separate facility. Some sectors use a common percentage reduction target for all facilities concerned, while other sectors have internally negotiated other ways of sharing the burden. If a sector as a whole fulfils its target, each facility in that sector is deemed to be in compliance. If a sector fails to meet its overall target, those facilities that have not met their own targets loose the 80% tax discount for the next 3 years.

The fact that it is enough for the sector to meet the overall target for all the facilities to maintain their discount could – in isolation – stimulate "free-riding", where under-performing facilities would try to benefit from abatement efforts at other plants. However, facilities that do better than required have the possibility to sell the surplus reduction into the UK Emission Trading Scheme. Hence, in practise, each facility must make sure that they meet their own target.[18]

Concerning the environmental impacts of the Climate Change Agreements, ETSU (2001) states, *inter alia*, that

> "The sector targets add up to a saving of around **2.5 MtC** per year, compared to the Business As Usual scenario. (…) This is a very satisfactory result, especially given the assumption of unlimited management time and capital availability used (…). It supports the qualitative assessment, namely that there must be step change in behaviour if the negotiated targets are to be achieved. This is the change which the climate change levy is intended to deliver.

For reference, it is estimated that the price effect of the levy on its own, i.e. assuming the levy is in place with no negotiated agreements and associated discounts, would give rise to a saving of 0.25 MtC per annum."

The statement concerning the price effect of the levy on its own could be an underestimate. The estimated figure was based on an average price elasticity, with an *explicit assumption that no plants be closed down*. Application of the full rate of the climate change levy would likely have led to *some* plant closures in the most energy-intensive sectors in the period up to 2010, and to larger emission reductions than 0.25 MtC per year.

ETSU (2001) concludes:

"In summary, every sector's target represents a significant improvement beyond 'Business As Usual'. The total target saving across all sectors is a satisfactory % of the pre-set benchmark which is acknowledged to be based on certain optimistic assumptions. Review points provide the opportunity to reassess those detailed issues where agreement was not achieved. Recognising these points, noting that the process was one of negotiation, and believing that in every sector a step change in behaviour will be needed to deliver the proposed targets, ETSU is of the view that the targets represent a reasonable basis for the climate change agreements."

It should, however, be noted that ETSU was responsible for negotiating the sector agreements on behalf of the UK Government. ACE (2001) criticises the calculations concerning the impacts of the agreements, stating that the "Business-as-Usual" energy efficiency improvements were likely to be higher than ETSU assumed, based *inter alia* on efficiency improvement forecasts published by the UK Department of Trade and Industry and by the European Commission. ACE's view could seem to be supported by the relative ease *some of* the sectors seem to have had in meeting their targets, cf. DEFRA (2004b).[19]

The agreement scheme includes provisions for a revision of the targets, cf. DEFRA (2001):

"The Secretary of State shall carry out a review at the end of the year 2004 of the sector targets and sub-sector targets for the final three target periods and shall carry out a further review at the end of the year 2008 of the sector target and sub-sector targets for the final target period.

Any such review shall be to ensure that the sector targets and sub-sector targets being reviewed continue to represent the potential for cost effective energy savings taking account of any changes in technical or market circumstances."

In a discussion of revisions to the CCA targets for 2006, in the context of a discussion of the allocation of emission allowances under the European Union's CO_2 emission trading scheme that is to come into effect from 2005, DEFRA (2004c) explains that

"The purpose of the CCA target review is to ensure that the sector targets continue to represent the potential for cost effective energy savings between 2006 and 2010 taking into account any changes in technical or market circumstances. The starting point for revising the 2006 SEC [Specific Energy Consumption] targets will be that each sector's target shall be reduced by the higher of actual overachievement [of the target set for 2002] ... or 5% (representing an average of overachievement by all CCA sectors)."

This "starting point" can seem to create some perverse incentives against technological improvements. Any sector that overachieved its target for 2002 by more than 5% would, according to the above, be "punished" as from 2006 through a *particular* tightening of its target. Firms could well remember such a procedure ahead of future target revision points, and make sure that they do not overachieve their target "too much".

An average overachievement of 5% does, by the way, give credit to ACE's view that many of the original targets of the Climate Change Agreements hardly represented much environmental improvement compared to a "Business-as Usual" scenario.

However, rather than trying to correct any instance of "regulatory capture" through *ex post* adjustments of the targets of the agreements – even if the possibility of such adjustments were foreseen in the agreements from the outset – it might have been better to let the firms concerned benefit from their negotiating skills through the sale of "unneeded" emission allowances in the coming EU emission trading scheme. Even if this could give some firms a windfall profit that could be considered to be "unfair", incentives for new technological development would have been maintained.[20]

de Muizon and Glachant (2004) discuss the combination of the Climate Change Levy, the Climate Change Levy Agreements and the UK emissions trading system. They conclude that the performance of this policy mix would not have been affected by an absence of the agreements.

5.4. The Intention Agreement on SO_2 emission reductions in Norway

A tax on the sulphur content of fuels has been in place for many years in Norway, covering at present about 27% of all SO_2 emissions in the country – with a tax rate of approximately 2€ per kg SO_2. From the outset, emissions from refineries, from the use of coal and coke, the use of mineral oils in the petroleum extraction activity on the continental shelf, and from supply-ships of this activity, were completely exempted from the tax. In 1999 these emission sources were included in the tax, with a reduced tax rate of about 0.35€ per kg SO_2.

However, from 1.1.2002 emissions from refineries and from the use of coal and coke (largely in industrial processes) were once again completely exempted from the tax. In return, the Federation of Norwegian Process Industries had signed an "Intention Agreement" with the Ministry of Environment, [Cf. PIL (2001)], committing the member companies to reduce SO_2 emissions by 5,000 tonnes by 2010, and to prepare a plan on how emissions could be reduced in a cost-effective way by a further 2,000 tonnes. Together this would equal the total emission reductions Norway expected to have to make to fulfil its obligations under the Gothenburg protocol of the UN ECE convention on Long-range Transboundary Air Pollution, capping total Norwegian SO_2 emissions at 22,000 tonnes from 2010.

Studies undertaken by the Norwegian Pollution Control Authority [cf. SFT (2001)] indicate that the most cost-effective measures to reduce SO_2 emissions in Norway can be found in the process industry. In 1999, firms taking part in the agreement emitted more than 16,000 tonnes of SO_2, compared to total Norwegian emissions estimated to be about 29,000 tonnes.

Statistics Norway (2004) indicates, by the way, that Norwegian SO_2 emissions have already decreased considerably since 1999, *inter alia* due to plant closures and temporary production reductions in the process industries, cf. Figure 2 below.

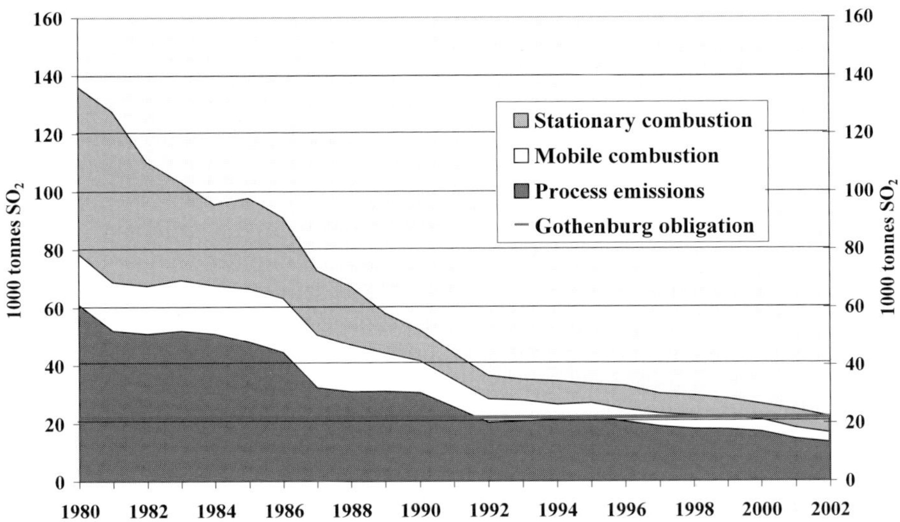

Figure 2. *SO_2 Emissions in Norway*

The Federation of Norwegian Process Industries have stated that the tax rate of about 0.35€ per kg SO_2 was not environmentally effective, as it was cheaper for firms to pay the tax than to install cleaning equipment that would be required to reduce emissions. This was underpinned by the findings of SFT (2001), where all potential abatement measures (with one exception) were found to have a marginal cost of 0.45€ or more per kg SO_2 abated. Most low-cost measures to reach a 7,000 tonnes emission reduction in total were found to have marginal costs of between 1 and 1.5€ per kg SO_2 abated.

The Federation of Norwegian Process Industries further stated that if the tax rate had been set so high that it would be profitable for the firms in question to install the cleaning equipment, the firms would not survive economically.[21]

The "Intention Agreement" is *not legally binding* for the two parties. Until the measures covered by the agreement have been implemented, by 2010 the latest, the ordinary environmental emission permit system will be the main policy instrument addressing the emissions from the sources concerned. According to PIL (2001), the pollution control authorities will seek to design future emission permits in such a way that industry can fulfil their obligation by joint measures.[22]

The process industry has – based on a *legally binding* "implementation agreement" involving all the firms that used to pay the lower tax rate – set up an "environmental fund", organised as a self-owned foundation, and financed by fee payments similar to the previous tax. An "action plan" for how the Intention Agreement is to be fulfilled was developed in 2003. It is estimated that the fund will raise about 30 million € in revenues through the fees paid by the member companies. The resources of the fund will be used to – fully or partially – finance development, implementation and operation of abatement measures and other measures suitable in the pursuit of the targets of the implementation agreement, including support to closure of activities that leads to lasting emission reductions. Measures are to be implemented where they will contribute the most to reduce emissions, until the targets of the Intention Agreement have been reached. Consideration will also be given to where emission reductions will contribute most to improve local air quality. In general, measures will be supported based on applications from the participating firms. If not enough applications should be made to reach the targets of the Intention Agreement, a site might be *instructed* to undertake a measure financed by the fund.

It seems that the most realistic alternative policy to this combination of a negotiated agreement and the sulphur tax being applied to other sectors in Norway would have been a reduced rate in the sulphur tax for industry. And as it seems less costly for the firms to pay the tax rate that was applied between 1999 and 2002 than to abate emissions, the chosen policy mix might lead to lower emissions from industry.[23] Similar emission reductions *could* have been obtained by increasing the tax rate sufficiently to make it cheaper for firms to abate than to pay the tax, but that might have entailed significant social costs – at least in the short to medium term – to the extent threats about plant closures are correct.

It is also important to emphasise that in this case, some mechanisms to promote economic efficiency under the agreement have been put in place. Instead of *e.g.* every firm reducing their emissions by equal percentage amounts, the firms involved have agreed to "pool" resources, and to undertake the emission reductions where they can be obtained at lowest cost. The decisions on where to make the emission reductions will be taken by representatives of the firms involved – who should have better knowledge of the actual costs of various abatement options than public authorities in many cases will have.[24]

5.5. The Load Reduction Agreements in New South Wales, Australia

Pollution control authorities in the Australian state of New South Wales (NSW) have developed an interesting policy mix to address a broad spectre of pollution issues. The Protection of the Environment Operations Act 1997 sets out which activities require a license by the NSW Environmental Protection Agency. Licensees pay an administrative fee at the beginning of each year. Ten percent of the largest activities licensed by the EPA with potential to cause environmental harm are also required to pay pollution load fees in accordance with the pollutant loads emitted – the lower the emissions, the lower the fee.

A number of incentives are provided to make industry improve their environmental performance. A weighted load discount is available under certain conditions, where a licensee reduces the *harmfulness* of the emissions but not necessarily the actual *load*. For example, 100% fee savings can be obtained for sustainable re-use of effluent.

Load Reduction Agreements (LRAs) are a voluntary incentive for licensees to reduce fees preparing to reduce loads in future. LRAs provide *immediate* fee reductions for licensees willing to commit to *future* reductions of assessable pollutant loads. Load fees are paid based on the future 'agreed' load rather than the current 'actual' loads during the term of the agreement. Money that would otherwise be paid in fees can then be used for investment in improving environmental performance. LRAs may be for a maximum period of four years, giving licensees up to three full years to upgrade operations and a final year to show they have permanently reduced pollutant loads to an agreed lower level.

The Load-based Licensing scheme commenced on 1 July 1999. A four-year phase-in plan gradually introduced the new licence fee structure, allowing industry time to adjust to the new arrangements. The first year of the scheme industry was required to monitor and report their pollutant loads, but no fee was payable. Consequently the first LRAs have been entered to apply from the second year of the scheme as load fees became payable.

By September 2002, 19 licensees had entered an LRA, with anticipated fee savings of more than 7 million AUD. Local councils, who operate sewage treatment systems, have entered the majority of agreements. This is the most common activity that is required to pay load-based fees. The difference between the actual load and the agreed load is the pollutant load reduction. For the 19 agreements this includes more than 1,865 tonnes of water pollutants and 1,650 tonnes of air pollutants.

To enter an agreement, the licensee completes an application form providing a description of the works, the time period for the agreement and nominates the agreed load for each pollutant to be included in the agreement along with an estimate of their current loads. The EPA will discuss the agreement with the licensee and verify the agreement details to check it is within the legal scope permitted. Agreements are signed at a senior level of the EPA and then provided to the licensee for their signature. The relevant environment protection licence is also varied at this time to link to the agreement.

By focusing on the end result rather than on close monitoring of the works to be undertaken, there is no annual reporting against milestones, resulting in saved administrative effort for both the licensee and the EPA.

It is too early to determine whether the industry will achieve greater or less emission reductions than planned. However, according to Stace (2002), no licensees have indicated that they will be unable to meet their goal. It is up to the licensee to consider the extent of emission reductions they are going to achieve, but EPA encourages the licensee to consider realistic goals.

Load fees are calculated on the lowest of the actual, weighted or agreed load. So in any year of the agreement including the final year, where the actual load is less than the agreed load, the licensee pays less. If an agreed load was not to be met, fee savings provided are to be repaid, with interest, at the end of the agreement period.

The agreements allow the licensee to enter for one or more pollutants as appropriate. So, for example, a sewage treatment plant may consider phosphorus removal technology and later enter a second agreement to reduce nitrogen or prepare for sustainable reuse and reduce loads of all pollutants emitted to water.

Load-based Licensing has, according to Stace (2002), already lead to emission reductions with many licensees having commenced environmental improvement works earlier than may have ordinarily occurred. Load Reduction Agreements have also provided a financial incentive for licensees to commit to lower loads and for industry manage the licence fees payable.

It is worth noticing that the price incentive to abate emissions is not much reduced in this case, as the reduction in the Load-based Licence Fee is only given for a three-year period, and it only concerns the difference between actual emissions and agreed emissions. During this period, the polluter must find measures to reduce emissions on a lasting basis. On the other hand, the somewhat lower tax payment might have *some* impact on polluters' financial ability to undertake emission-reducing measures – but the reductions achieved so far seem relatively modest.

5.6. Conclusions on combinations of taxes and voluntary approaches

Both the energy efficiency agreements in Denmark, the Climate Change Agreements in the United Kingdom and the "Intention Agreement" on SO_2 emission reductions in Norway have – to a significant degree – been motivated by a wish to prevent close-down of industrial companies that could have taken place if "full" tax rates had been applied. It seems unlikely that the agreements provide environmental benefits beyond what "full" tax rates would have done, but, in the case of Norway, it seems that the previous reduced rates applied to certain industrial sectors were too low to have any significant environmental impact – at least in the short term.

By combining taxes and a voluntary approach in these cases, policy makers have *tried* to avoid having to make trade-offs between the environmental, economic and social dimensions of sustainable development. It remains to be seen whether such trade-offs *can* be avoided in the longer term, as – for example – more ambitious climate policies are being put in place.

For the Load Reduction Agreements in New South Wales, the competitiveness issue seems to have been of less importance, as the most important polluters under the scheme are local councils, who operate sewage treatment systems. They hardly face competition from other actors.

As stated in the beginning of this section, under a tax regime, firms' compliance costs are equal to abatement costs plus tax payments for residual emissions. When firms can avoid paying for any residual emissions by taking part in a voluntary scheme, impacts of the policy on the production costs of these firms will be limited. A wish to limit such cost impacts – especially for firms facing stiff international competition – is exactly one of the reasons for which the tax relief is given. However, to the extent the firms could have shifted any part of the cost increases on to their customers – through increases in the prices of their products – applying a voluntary approach looses out on any direct impacts on the *demand* for products that cause pollution in their production. In many cases, such demand changes could provide an important part of the environmental benefits from using economic instruments.

6. VOLUNTARY APPROACHES USED IN COMBINATION WITH EMISSION TRADING SYSTEMS

There are two main categories of emission trading systems: baseline-and-credit systems and cap-and-trade systems. Voluntary approaches can be integrated with such systems in three ways that will be discussed in turn below:[25]

- Adherence to tradable permit systems can be voluntary;
- Tradable permits can be used as a means of allocating responsibilities within an industry-wide negotiated agreement; and,

- Emission reductions agreed to under voluntary agreements can be used as a means to allocate permits in a grandfathered tradable permit scheme.

6.1. Voluntary adherence to trading systems

To a certain extent, all baseline-and-credit trading schemes can be described as "voluntary approaches" to environmental regulation. Credits are issued to all firms which reduce emission below a set amount, such as the level of emissions that would prevail under a regulatory system. They can then sell these credits to firms that have emissions in excess of regulated emission levels. In both cases, involvement is voluntary. Low-cost abaters are not "required" to create credits, and high-cost abaters are not "required" to purchase them. In effect, the efficiency gains provided by the tradable permit system are the carrots which provide the incentive for firms to volunteer to be involved in the system.

However, in a cap-and-trade trading scheme the situation is quite different. If the permits were auctioned, no firm would be likely to volunteer to be involved in the absence of a regulatory threat (a "stick") or a financial inducement (a "carrot"). In the case where permits are grandfathered, the question is significantly more complicated. It is also more policy-relevant, since a number of countries have introduced, or are introducing, voluntary cap-and-trade schemes with grandfathered permits.

Allowing voluntary participation in cap-and-trade schemes based upon grandfathered permits can be a means to increase economic efficiency of abatement, as more low-cost options could be covered by the scheme. However, voluntary trading schemes are characterised by strategic behaviour and financial uncertainty. Unlike under a mandatory cap-and-trade scheme, the firm does not know what the ultimate "cap" will be, since this depends upon how many (and which) firms volunteer. More significantly, each firm must try to predict the ultimate permit price – which is the key factor in its evaluation of whether it makes sense to volunteer or not – without knowing how many and which firms are likely to volunteer. In order to make an informed choice, each firm must estimate:

a) probabilities of adherence for other firms;
b) other firms' emission levels; and,
c) other firms' marginal abatement costs.

Each firm faces a different benefit and cost schedule depending upon which other firms are involved. In some cases the net benefits will be positive and in some cases they will be negative relative to the case where they continued to adhere to some existing regulatory regime. It is possible that the distribution of costs and benefits is such that no firm will volunteer, even if it would be in their collective interest to do so.

In many instances, however, voluntary adherence is only an option for a sub-set of firms, with most firms being mandatory participants. This is the case with the US EPA's SO_2 Allowance Program. It is also the case with Pennsylvania's NO_x Allowance Retirement Program, which is mandatory for fossil fuel powered electric generating plants, but voluntary for others [cf. Stavins (2001)]. Similarly, under the Californian RECLAIM program, it is possible for mobile sources and small point sources to volunteer to become involved [see Nash and Revesz (2000)].

Allowing for voluntary adherence for some firms while preserving a core of firms for which the cap-and-trade programme is mandatory simplifies the decision for the firm considering whether or not to volunteer. If the number of potential "voluntary" firms is small relative to the number of "mandatory" firms, the permit price can be taken as given. In such cases, the potentially volunteering firm need not be concerned with the three factors mentioned above, but only its own calculus of costs and benefits. This also means that the regulator faces less uncertainty about the likely number of firms that are to be involved.

However, even in such cases voluntary adherence to a trading scheme can raise concerns. Most importantly, depending upon how the grandfathered permits are allocated, such a scheme might encourage "adverse selection". If the permit allocation is based upon historical emissions (or some variant), those firms that would be most likely to volunteer would be the firms that since the "base year" already have undertaken abatement – even in the absence of the programme.

The case of the SO_2 Allowance Program in United States is instructive. Between 1996 and 1999 the percentage of emissions that were attributable to "opt-ins" was between 12% and 13%. However, Montero (1999) found that this "substitution" provision of the program tended to be taken up by the power plants which, by doing so, were grandfathered emission permits far in excess of what would have been their "Business-as-Usual" emissions. These plants had already – for other reasons – reduced their emissions significantly between the base year used for permits allocations (1988) and the start of the program (1993).[26]

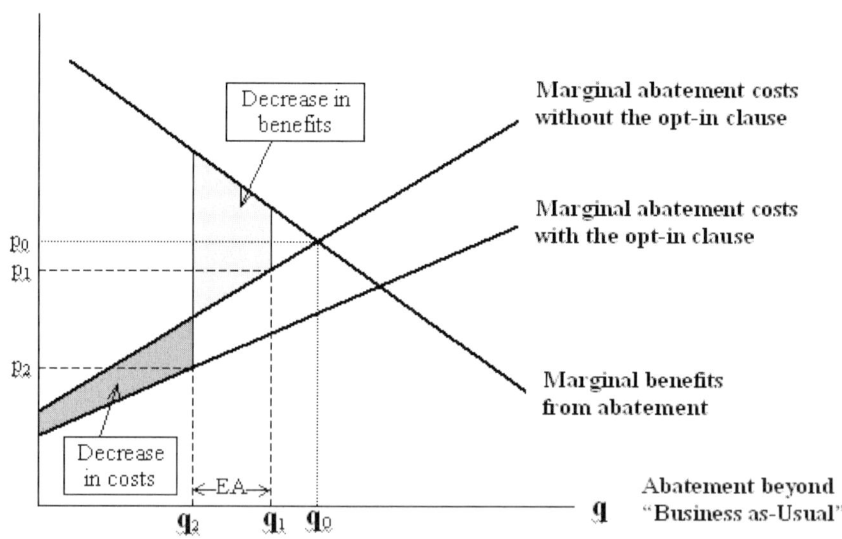

Figure 3. *Costs and benefits from voluntary compliance in a tradable permits system*

Figure 3 illustrates the impacts of including a voluntary "opt-in" possibility in a tradable permits program, and the problem related to "adverse selection" among the volunteers.[27] In the figure, which is based on Montero (1999), it is assumed that –

due to imperfect information or political constraints – the abatement effort at the outset is set at q_1, where marginal (environmental) benefits from further abatement are higher that the marginal abatement costs. Introducing the "opt-in" possibility has two separate impacts:

- A number of firms with "Business-as-Usual" emissions lower than the corresponding number of permits they would receive chose to "volunteer" to participate – while firms that had increased their emissions since the base year would tend not to participate. This is the adverse selection problem – *that reduces the environmental effectiveness of the scheme all in all,* causing the total abatement effort to *decrease* by an amount equal to the "excess allocations" (EA), from q_1 to q_2.
- Some firms will also "opt-in" because they have low abatement costs. This will shift the marginal abatement curve down.

Whereas it is given that the environmental effectiveness of the trading scheme as a whole decreases with the "opt-in" possibility, it is unclear whether total "welfare" decreases or increases. That depends – in this simplified context – on the relative size of the light and dark shaded areas in the figure, which in turn depends on *e.g.* the slope of the marginal costs and benefits curves, on the number of "excess allowances" and on how much the marginal cost curve shifts downwards. In his econometric study of the opt-in provision in the SO_2 Allowance Program in United States, Montero (1999) found that the net welfare effect was likely to have been *negative*. He also found that an increase of one standard deviation in the firm's allocation of permits relative to actual emissions increased the probability of "volunteering" from 32% to 84%.

Moreover, McLean (1997) estimated that the opt-in provision was responsible for a large share of total administrative costs of the programme:

> "… phasing in the participation of sources can complicate administration and undermine achievement of emission reduction goals and has been perhaps the most serious flaw of the SO_2 allowance program. Two types of problems can occur: a) with interconnected electric utility grids, participating sources can shift electrical load to nonparticipating sources whose emissions could increase and undermine the emission reduction goal, and b) if sources in a particular region are allowed to voluntarily participate while others in the same region can chose not to participate, there is a risk of allowances being earned by the voluntary participants and used by other participants in lieu of reducing emissions, while the nonvolunteering sources increase their emissions and cause a net increase in emissions.
>
> Administrative mechanisms to compensate for these problems can be complex and are of limited effectiveness in ensuring the environmental integrity of the program. The "substitution" and "reduced utilization" provisions employed in the SO_2 allowance program have been litigated and revised, and have become the most complicated administrative parts of the program. For example, complex allocation formulas had to be developed for substitution units (those Phase II units that volunteered to participate in Phase I) to prevent creation of large numbers of excess allowances. Further, in determining compliance of the Phase I units, it is necessary to review significant amounts of information on most of the 2000 Phase II units (to ensure that load shifting does not undermine intended emissions reductions). Approximately 75 percent of the cost of developing and implementing the permitting provisions of Title IV and at least one third of the cost of developing and operating the allowance tracking system, or about \$6.6 million, can be attributed to the complexity of Phase I. In retrospect, all affected sources should have been included from the outset in Phase I with emissions limitations tightened in Phase II to accomplish the goals of the program."

6.2. Emission trading as elements of voluntary approaches

Rather than voluntary adherence being elements of tradable permit programmes, in some cases emission trading may be the vehicle through which firms meet their commitments in voluntary approaches. These would, of course, only be relevant for approaches which are negotiated at the industry or sector level. Moreover, in most cases the trading (if it can even be labelled as such) is implicit.

For instance, in the Australian Greenhouse Challenge, aggregate agreements can be struck between the Australian Government and an industry association, on behalf of its members. The agreement describes the actions to be taken and the emission forecasts for the member companies [cf. AGO (1999)]. These actions and forecasts are the outcome of negotiations between these companies. To a great extent the agreement can be considered as a springboard towards a voluntary tradable permit system, with industry serving as a "bubble" and the firms negotiating "off-sets" internally.

In some cases industry itself has initiated discussions concerning the introduction of formal trading as a means to reduce negotiating/bargaining costs which can plague industry-level voluntary agreements. This is for example the case concerning the agreement on SO_2 and NO_x emissions reductions by the power generation industry in the Netherlands. SEP (Samenwerkende Elekticiteits-Produktiebedrijven / Co-operating electricity production companies) had, according to Brand (2000), for a long time sought to have SO_2 emissions in the sector considered as one "bubble" – instead of each plant having to undertake abatement efforts unilaterally, and this was in the end agreed to.

6.3. Accounting for voluntary approaches in permit allocations and baseline estimation

An important additional point relates to the treatment of "voluntary" commitments in the determination of permit allocations. There have been extensive discussions in different programs about the extent to which reductions achieved through voluntary approaches should be included in the allocation of permits or in the evaluation of the baseline. To a certain extent, this relates to the ambiguous legal status of different types of voluntary approaches. There is a significant difference between the case of firms that volunteer to reduce emissions through a negotiated agreement under threat of a regulatory backstop, and firms that co-operate amongst themselves without the government playing an active role.

In the latter case the inclusion of voluntary reductions in the calculation of a later permit allocation or the baseline may be controversial. One criticism of "pure" grandfathering has been that it is biased against firms that have been "early movers", investing in abatement above and beyond regulatory requirements prior to the year (or years) which are used as the basis for the allocation of permits.[28]

This could be obviated by the use of modified grandfathering, such as the allocation on the basis of the maximum level of emissions that could have been emitted by firms, while still being in compliance.[29]

In the CEC (2001) it is stated that "the target set under the [negotiated] environmental agreements can serve as a useful basis for the allocation of allowances by Member States". This would, however, be politically difficult to achieve if the scope of the permit trading scheme is broader than the scope of the pre-existing agreement since firms which were not party to the agreement would benefit. More generally, this raises the issue of "moral hazard", making it exceedingly difficult for governments to negotiate agreements with firms in future due to the possibility of this affecting future permit allocations.

The final EU Directive on greenhouse gas trading [cf. CEC (2003)] does not contain any explicit reference to voluntary approaches.[30] But *e.g.* in the UK Draft National Allocation Plan, for sectors that form part of a Climate Change Agreement, the allocation of allowances for the first phase (2005-2007) under the EU trading scheme is based on the targets of the agreement for 2006. In DEFRA (2004d) it is stated that

> "The sectoral targets set out in the relevant Climate Change Agreements have been used as part of the methodology for allocating allowances by incorporating the targets for 2006 into the activity/sectoral sub-targets as set out in paragraphs 72-74 above. It has been necessary to adjust the targets in the agreements in two ways: firstly, to make adjustments to the targets to reflect the fact that the targets cover both electricity use (indirect emission) and direct emission [while only direct emissions are relevant in the context of the EU trading scheme]; and secondly to take account of the review of targets for 2006 and beyond which is due to take place during 2004."

In the next step, when sector allocations are to be divided between individual installations, according to DEFRA (2004c),

> "The activity level allocations are divided between installations according to each installation's average share of annual emissions over the period 1998 to 2002. To calculate each installation's allocation, we have taken an average of the historic data provided, excluding the lowest year's emissions for each installation. (...)
>
> A historic measure has been chosen to allocate at the installation level due to the complexity of projecting installation level emissions consistently and accurately. The use of an historic average also provides some compensation to potentially stranded assets. The exclusion of the lowest year's emission is intended to minimise the impact of an anomalous year with unusually low emissions on an installation's allocation."

Also in Germany the allowance allocations will be based on the existing voluntary declaration of the German industry to limit greenhouse gas emissions, cf. BMU (2004).

The interactions between voluntary approaches and permit allocations are even more important in credit-and-baseline schemes – where credit creation is affected by the choice of the baseline – than for cap-and-trade schemes. If voluntary commitments are not considered part of the baseline, the firm will create more credits than in the case where voluntary commitments are considered part of the baseline. In some cases, the distinction may result in a switch from the firm being a net seller to become net buyer of permits.

6.4. Conclusions on combinations of trading schemes and voluntary approaches

The impacts of combining a tradable permit system with a voluntary approach depend crucially upon what element of the regime is voluntary. Including "voluntary" emission reductions in the permit allocation method in a grandfathered cap-and-trade system could penalise "early movers", and would make firms very reluctant to take on ambitious targets in the future. Including voluntary reductions in the baseline of a baseline-and-credit scheme would have even more significant consequences in terms of incentives for discouraging future abatement efforts by firms.

Effectively, in both cases the regulatory authority would have decided *ex post* to convert a voluntary scheme into a mandatory scheme. In one case this is reflected in initial permit allocations and in the other cases it is reflected in the possibilities for credit creation. This may increase the economic efficiency of the program, but is better understood as a reflection of the inadequacies of voluntary approaches than as a case for their use in combination with tradable permit regimes.

The use of tradable permit schemes by firms within negotiated agreements is surprisingly rare. Considering that many such agreements involve significant bargaining costs which would be avoided by a tradable permit system, the benefits of adding a trading mechanism "on top of" a voluntary approach could be important. However, the trading scheme would depend upon tight monitoring and enforcement, attributes that are rare in most voluntary agreements. A firm would not want to buy a permit from a competitor unless it could be certain that the permit represented a real property right. However, when satisfactory monitoring and enforcement are present, a case could be made for the government serving as an "honest broker" in order to help firms set up credible trading schemes.

Voluntary adherence to tradable permit systems might have a role to play. While an entirely voluntary scheme is unlikely to result in many adherents for the reasons discussed above, using voluntary provisions to expand the coverage of a trading scheme *may* increase economic efficiency by bringing in firms with very different abatement costs. Further, an increased number of potential traders can increase the liquidity of the scheme. However, the important danger of adverse selection must be confronted. This can only be done by ensuring that the permit allocation mechanism does not grant "excess permits" to firms that can then volunteer for the programme. For a given number of permits being issued to the "mandatory" firms, the inclusion of "voluntary" firms with "excess permits" would lower the market price of permits and reduce the environmental effectiveness of the whole scheme.

7. OVERALL CONCLUSIONS ON THE USE OF VOLUNTARY APPROACHES IN POLICY MIXES

From the preceding discussion it is clear that combining a voluntary approach with other policy instruments can have a number of different impacts, depending on the circumstances.

Combining "command-and-control" regulations with a voluntary scheme can provide increased flexibility in how to abate, and thus increase the economic efficiency for the firms affected. However, the negotiation of such agreements can be time-consuming and costly, especially if it is to be done at the level of individual firms. A greater reliance on performance standards – where firms are left to decide themselves how a given environmental performance is to be reached – could be a better option.

There is a risk that the use of voluntary approaches in combination with various types of subsidies in the long term may undermine industry's incentives to undertake environmental improvement under its own initiative. Industry may prefer to wait until government offers a "carrot" before agreeing to improve its environmental performance.

If participation in a voluntary scheme is set as a condition for receiving subsidies for technology development, there is a possibility that the selection of firms participating in the subsidy scheme will be biased, and that technological progress thus be slowed down.

Rather than applying various voluntary approaches, it will generally be better to give firms a direct economic incentive to improve their environmental performance, through internalisation of the externalities they cause. This can for example be done through a tax on measured or estimated emissions or on products that in particular cause emissions, through a fee on the size of the permitted emissions – or through the introduction of a emissions trading system, preferably through the auctioning of emission permits.

A prime reason for why voluntary approaches are introduced in connection with what could be "first-best" instruments, like taxes and tradable permits, is that policy makers want to avoid making a trade-off between environmental targets and social, industrial or employment targets: they want to avoid hurting the international competitiveness position of the most polluting, hence most affected, sectors and firms.

In fact a trade-off is nevertheless often made: frequently the environmental achievements are *de facto* lowered when voluntary approaches are included in the policy mix. In other words, protection of other policy targets through the use of voluntary approaches tends to come at the cost of reduced environmental protection.

The introduction of negotiated agreements to supplement or replace tax payments for certain sectors can entail a considerable administrative burden, both in terms of negotiating the targets and concerning monitoring and enforcement of the "voluntary" obligations. Similarly, combining a trading system with e.g. an "opt-in" possibility can significantly increase the administrative costs of that system.

To the extent that tax payments would cause plant closures that *in practise* would represent a welfare loss to the society as a whole, rather than introducing special tax preferences for certain firms or sectors in exchange for them signing up to some form of negotiated agreement, part of the revenues raised through the taxes could – temporarily – be recycled back to the firms in question. This could *e.g.* be done through the lowering of other taxes, through subsidies for technology development or in proportion of volumes produced.[31]

Similarly, in an emission permit trading system, rather than introducing the possibility for some firms to "opt-in", thus jeopardising the environmental effectiveness of the scheme, it would seem better to include all relevant firms in the scheme from the beginning. If some element of "grandfathering" of permits is found necessary for sectoral competitiveness reasons, it is important – from the point of view of the society as a whole – not to overcompensate the firms in question, by letting them keep the major share of the "scarcity rents" generated through the environmental policy.

Whatever measure is taken to address a perceived "competitiveness problem", it is important to make sure that firms' incentives to abate emissions are maintained, both in the short and long term. While a tax relief in return for a "voluntary commitment" can give firms increased financial resources to develop new technological solutions to improve the environment, their incentives for actually succeeding in these efforts would decrease significantly.

8. NOTES

* The opinions expressed in this chapter are those of the author and do not necessarily reflect the views of the OECD or its member countries.

1 The flexibility mechanisms under the Kyoto protocol go some way to create a global emissions trading system, but – as is well-known – it is optional for countries to adhere to the protocol.

2 The Guidelines are recommendations addressed by governments to multinational enterprises operating in or from adhering countries. They provide voluntary principles and standards for responsible business conduct in a variety of areas including employment and industrial relations, human rights, environment, information disclosure, competition, taxation, and science and technology.

3 An example could be an eco-labelling scheme.

4 See OECD (2003a) for a large number of references.

5 It is, however, emphasised that the positive "shadow-price" in the case from New South Wales stems from the load-based fee that was introduced, not from the Load Reduction Agreements.

6 Before passing a judgement on whether or not such a policy mix all in all is beneficial to society, one should also take into account the costs of the financing of any subsidies. This is especially important where the distortionary costs of raising an additional unit of tax revenue are high.

7. When promoting technology development in highly polluting sectors, one should take into account that this could improve the domestic and/or international competitiveness of these sectors, leading to increased production – and perhaps to an *increase* in overall pollution levels, even if emissions *per unit* in the sector decrease.

8. For more information on the UK Emission Trading Scheme, see DEFRA (2004a). Firms being party to the Climate Change Agreements related to the Climate Change Levy can also on certain conditions participate in the Emission Trading Scheme. Enviros (2003) indicates that 80% of emissions of firms participating in the Emission Trading Scheme were declining at the time of entry into the scheme. Over half of these were declining due to abatement activities, but a substantial proportion was due to declining business activity. They conclude *inter alia*: "Emissions trading markets are difficult to establish effectively through voluntary mechanisms. Mandatory enforcement with a wide number of participants provides a better basis for creating an efficient environmental trading market."

9. It is – as always – possible that some of the emission reduction measures being subsidised would have been undertaken anyhow, implying a "dead-weight loss" of the policy. Enviros (2003) indicates that this might indeed have been the case.

10 DEFRA (2004d) discusses modifications to the scheme that the introduction of an EU-wide CO_2 emission trading scheme will require.

11 They also found that the price elasticities depended on the level of the energy prices firms were facing at the outset. For firms at the 10% decile when ranked in increasing order according to energy prices they were facing, the estimated price elasticity of energy was about -0.4. For firms at the median, the price elasticity was found to be about -0.6, while for firms at the 90% decile, the estimated price elasticity of energy was about -0.7.

12 Cf. OECD (2002). The study focused on how the scheme was designed up to 2000.

13 The estimated increase in energy use due the tax reduction was 5% in 1993 and 1995, but only 1% in 1997, when the tax reduction offered to participating companies was much smaller. In later years, the tax reduction has once again increased, which would tend to add to the associated increases in energy use. It is underlined that their calculations only take into account the specific tax concessions granted to firms participating in energy efficiency agreements. The impacts of the – much bigger – tax reduction granted to all firms that employ light or heavy industrial processes were not estimated.

It is also interesting that Bjørner and Jensen (2002) found no statistically significant impact on energy use of subsidies that used to be granted to investments in energy-saving projects.

14 The numbers represent the total energy and CO_2 tax rate for space heating. The CO_2 tax rate alone was 13.4€ each of the years 1996-2000. The Danish Economic Council (2002) evaluates the Danish environment and energy policies in the 1990s. Based on a cost-benefit analysis, that study concludes that "the tax rate applied to the energy consumption by households and the energy consumption for room heating purposes by companies is too high. On the other hand, the tax on companies' energy consumption in manufacturing is too low."

15 Energy-intensive sectors were defined as those that are subject to existing UK PPC Regulations, largely similar to the sectors that are covered by the European Union's Integrated Pollution Prevention and Control Directive. There are ten major energy intensive sectors (aluminium, cement, ceramics, chemicals, food & drink, foundries, glass, non-ferrous metals, paper, and steel) and thirty-four smaller sectors.

16 Where an energy-intensive installation uses less than 90% of the energy within a site, the facility which is covered by a Climate Change Agreement must be sub-metered so that the energy used by the facility is known accurately. The cost of this metering has to be borne by the companies concerned. Costs may be in the region of £1000 to £5000 per meter, possibly more where the energy supply arrangements are particularly complicated. In some sectors that require more metering, such as motor vehicles, the eligible activities are already sub-metered so there is no additional cost. In other sectors, such as supermarkets and aerospace, a programme of installing sub-meters was agreed. Both the supermarkets and master bakers sectors have *over a thousand sites* with activities which are eligible to be covered by an agreement and where additional metering is required if they are to be included in an agreement. This is one reflection of the considerable administrative costs negotiated agreements can entail.

17 According to DEFRA (2004b), 10,698 facilities were recertified after the results of the first target-period had been assessed.

18 This thus provides one example of a beneficial impact of combining a voluntary approach with another instrument.

19 For example, according to DEFRA (2004b), the adjusted target in 2002 for the aluminium sector was a performance ratio of 0.714, while actual performance in that year was 0.681. The adjusted target for the master bakers was 1,548 kWh_p/£k, while actual performance was 1,494 kWh_p/£k.

20 Cf. also the discussion below on the treatment of pre-existing voluntary approaches when allocating allowances in an emission trading scheme.

21 Several emission-intensive plants have nevertheless been closed since 1999 – for reasons not related to the Intention Agreement. Rather than retrofitting plants that anyway have a modest or low profitability, it can be a low-cost option for society as a whole to reach environmental targets by closing down some of these plants.

22 Hence, the present policy mix to address sulphur emissions in Norway includes the environmental permit system, a tax on the sulphur content of fuels (with two different tax rates), the Intention Agreement with the process industry – in addition to regulations on the maximum content of sulphur in fuels and a special incentive in the tax on diesel to favour the use of diesel with a sulphur content below 0.005%.

23 The environmental target for the country as a whole is in this case given by the obligations under the Gothenburg protocol. The issue at stake is to what extent the emission reductions should be undertaken in the process industry – where abatement costs are the lowest – but where there is also a fear for the international competitiveness position of the firms concerned. It should, however, be emphasised that the protection of the sectoral competitiveness of the process industry comes at the price of increased costs – and lower competitiveness – for the remainder of the Norwegian economy.

24 In this particular case, the Norwegian Pollution Control Authority had made detailed studies of the costs of various abatement options, cf. SFT (2001). I have no foundation for questioning any of their

findings – but in general there is an asymmetry in the information available to the firms involved and to public authorities.

25 This section draws heavily on a paper prepared by Nick Johnstone, a colleague of mine in the Environment Directorate of OECD.

26 Ellerman et *al*. (2003) wrote: "For example, in the Acid Rain Program, evidence indicates that many of the voluntary participants received credits for having emissions below the pre-specified baseline even though they took no abatement actions. The simple emissions baseline had been set higher than these facilities' actual emissions, so at least some of the credits they received did not represent real emissions reductions."

27 Enviros (2003) indicates that adverse selection can have been a problem among direct participants in the UK CO_2 Emission Trading Scheme, i.e. the firms that bid for subsidies to take on emission reduction obligations.

28 A variant of the same issue could be how to treat any voluntary commitments in permit or 'allowance' allocations based on projected future emissions. It could be seen as 'unfair' *if* a firm or sector had taken on a *really* ambitious target for a future year, with an understanding that they would not be "punished" if they failed to reach the target despite 'best efforts' to do so, at a later stage finds that its permit allocation had been based on that "overly ambitious" target. However, the *practical relevance* of this point seems limited.

29 The problem could be reduced to the extent allowances in the trading system were auctioned instead of being grandfathered. If all allowances were auctioned, no special treatment would be required for firms taking part in a pre-existing voluntary approach. Any firm that had participated in the voluntary scheme "in good faith" – and reduced its emissions beyond "Business-as-Usual" – would benefit directly from this in needing to buy fewer allowances. The drawback with this option is the negative impact on the international competitiveness of the firms concerned.

30 Article 10 of CEC (2003) states that "For the three-year period beginning 1 January 2005 Member States shall allocate at least 95 % of the allowances free of charge. For the five-year period beginning 1 January 2008, Member States shall allocate at least 90 % of the allowances free of charge." This provision means that almost all the "scarcity rent" related to this limitation of CO_2 emissions is transferred to the affected firms for free.
Based *inter alia* on the findings of Bovenberg and Goulder (2003), this represents a very significant *overcompensation* of the firms in question. They found that "By enabling firms to retain only a small fraction of these potential rents – e.g., by grandfathering a small percentage of CO_2 permits, or by exempting a small fraction of emissions from the base of a carbon tax – the government can protect firms' profits and equity values."

31 Recycling of revenues in proportion to production volumes is applied for the Swedish charge on NO_x emissions at energy plants.

9. REFERENCES

ACE (Association for the Conservation of Energy) (2001), *ACE Policy Update: Climate Change Levy.* Available at www.ukace.org.

AGO (1999), *Greenhouse Challenge: Evaluation Report.* Australian Greenhouse Office, Canberra. Available at www.greenhouse.gov.au/challenge/about/report.html..

Bjørner T.B. and Jensen H.H. (2002), *Energy taxes, voluntary agreements and investment subsidies - a micro panel analysis of the effect on Danish industrial companies' energy demand*, Resource and Energy Economics, 24(3), pp. 229-249.

Bovenberg A. L and Goulder L.H. (2003), *Confronting industry-distributional concerns in U.S. climate-change policy.* Les séminaires de l'Iddri n° 6. Available at www.iddri.org/iddri/telecharge/mardis/06_goulder.pdf.

BMU (Bundesministerium für Umwelt, Naturschutz und Reaktorsicherheit) (2004*), Antwort der Bundesregierung auf die Große Anfrage der Abgeordneten Dr. Peter Paziorek, Marie-Luise Dött, Dr. Klaus W. Lippold (Offenbach), weiterer Abgeordneter und der Fraktion der CDU/CSU,Nationale Umsetzung des Emissionshandels"- Drs. 15/1282*. [Reply from the federal Government to questions from members of the German Parliament regarding emissions trading.] Available in German at www.bmu.de/files/kleine_anfrage_15_1282.pdf.

Brand E. (2000), *The Netherlands: Agreement on the Reduction of SO_2 and NO_x Emissions by the Power Generation Industry*. Paper prepared for the closing conference of the NEAPOL project. Available at http://fetew.rug.ac.be/neapol/Index.htm.

CEC (Commission of the European Communities) (2001), *Proposal for a Directive of the European Parliament and of the Council establishing a scheme for greenhouse gas emission allowance trading within the Community and amending Council Directive 96/61/EC*. Available at http://europa.eu.int/eur-lex/en/com/pdf/2001/en_501PC0581.pdf.

CEC (2003), *Directive 2003/87/EC of the European Parliament and of the Council of 13 October 2003 establishing a scheme for greenhouse gas emission allowance trading within the Community and amending Council Directive 96/61/EC*. Available at http://europa.eu.int/eur-lex/en/search/search_lif.html.

Danish Economic Council (2002), *Evaluations of Danish Environmental and Energy Policies in the nineties*. English summary of a chapter in the report "Danish Economy, Spring 2002". Available at www.dors.dk/english/index.htm.

DEFRA (Department for Environment Food and Rural Affairs) (2001), *Umbrella Climate Change Agreement for the Aluminium, Titanium and Magnesium Sector*. Available at www.defra.gov.uk/environment/ccl/agreements.htm.

DEFRA (2004a), *Climate Change Agreements and the Climate Change Levy*. Available at www.defra.gov.uk/environment/ccl/index.htm.

DEFRA (2004b), *Climate Change Agreements and the Climate Change Levy. First target-period results*. Available at www.defra.gov.uk/environment/ccl/results.htm.

DEFRA (2004c), *EU Emissions Trading Scheme. UK Draft National Allocation Plan for 2005-2007*. Consultation Draft, January 2004. Available at www.defra.gov.uk/corporate/consult/eu-etsnap/draft-nap.pdf.

DEFRA (2004d), *EU Emissions Trading Scheme. Consultation paper on the UK Draft National Allocation Plan for 2005 to 2007*. Available at www.defra.gov.uk/corporate/consult/eu-etsnap/consultation.pdf.

Ellerman A,. Denny P., Joskow L. and Harrison D.(2003), *Emissions trading in the U.S. Experience, Lessons, and Considerations for Greenhouse Gases*. Prepared for the Pew Center on Global Climate Change. Available at www.pewclimate.org/document.cfm?DocumentID=216.

Enviros (2003), *A Qualitative Study of the Direct Entry UK Emissions Trading Scheme*. Enviros Consulting Ltd. Available at www.enviros.com/index.cfm?fuseaction=13.1&id=15.

ETSU (2001), *Climate Change Agreements - Sectoral Energy Efficiency Targets*. Report prepared for the Department for Environment, Food & Rural Affairs, by ETSU, AEA Technology Environment, United Kingdom. (ETSU has since changed its name to 'Future Energy Solutions'). Available at www.defra.gov.uk/environment/ccl/analyses.htm.

Fullerton D. and Metcalf G. (2001), *Environmental Controls, Scarcity Rents, and Pre-Existing Distortions*, Journal of Public Economics, Volume 80(2). A preliminary version, with calculations, is available at http://papers.nber.org/papers/W6091.

Goulder L. H., Parry I.W.H.and Burtraw D.(1997), *Revenue-Raising versus Other Approaches to Environmental Protection: The Critical Significance of Preexisting Tax Distortions*. Rand Journal of Economics, Volume 28, No. 4, pp. 708-731. Available at www.rje.org/abstracts/abstracts/1997/Winter_1997._pp._708_731.html.

Lyon T.P. and John Maxwell W. (2004), "Mandatory and Voluntary Approaches to Mitigating Climate Change". In Baranzini A. & Thalmann P. (Eds.): *Voluntary Approaches in Climate Policy*. Cheltenham (UK): Edward Elgar.

Mclean B.J. (1997), *Evolution of Marketable Permits: The U.S. Experience with Sulfur Dioxide Allowance Trading.* International Journal of Environment and Pollution, Vol. 8, Nos. 1/2, pp. 19-36. Available at http://www.epa.gov/airmarkt/articles/mclean.

Montero J.-P. (1999), *Voluntary Compliance with Market-Based Environmental Policy: Evidence from the U.S. Acid Rain Program.* Journal of Political Economy, vol. 107, no. 5.

De Muizon G. and GlachanM. (2004), "The UK's Climate Change Levy Agreements: combining voluntary agreements with tax and emission trading." In BaranziniA. & Thalmann P. (Eds.): *Voluntary Approaches in Climate Policy.* Cheltenham (UK): Edward Elgar.

Nash J. R. and Revesz R.L. (2001), Markets and Geography: Designing Marketable Permit Schemes to Control Local and Regional Pollutants. Ecology Law Quarterly, Vol. 28, pp. 559-661. Available for purchase at www.law.berkeley.edu/journals/elq/articles.html.

OECD (Organisation for Economic Co-operation and Development) (2000), *OECD Guidelines for Multinational Enterprises: Revision 2000.* OECD, Paris. Available for purchase at http://new.sourceoecd.org.

OECD (2002), *Voluntary Approaches: Two Danish Cases*, env/epoc/wpnep(2002)13/final. OECE, Paris. Available at www.oecd.org/env/va.

OECD (2003a), *Voluntary Approaches for Environmental Policy: Effectiveness, efficiency and usage in policy mixes.* OECD, Paris. Available for purchase at http://new.sourceoecd.org.

OECD (2003b), *OECD Guidelines for Multinational Enterprises. A key corporate responsibility instrument.* Policy brief. OECD, Paris. Available at www.oecd.org/department/0,2688,en_2649_34889_1_1_1_1_1,00.html.

PIL (Prosessindustriens Landsforening) (2001), *Agreement of Intent - on reduction of the emissions of sulphur dioxide (SO_2) from the process industry.* Available at www.pil.no/NR/rdonlyres/2D50E960-18B7-45FE-97FF-3BFB90254D37/436/AgreementofIntent.doc.

Statistics Norway (2004), *Sulphur emissions reduced since 1990.* News release 13.02.04. Available at www.ssb.no/agassn_en/.

SFT (Statens Forurensningstilsyn) (2001), *Reduksjon av SO_2-utsleppa i Norge* [Reduction of SO_2-emissions in Norway], SFT-Rapport 1814/2001 from the Norwegian Pollution Control Authority. Available (in Norwegian) at www.sft.no/english/publications.

Smith S. (2003), *Environmental Taxes and Competitiveness: An Overview of Issues, Policy Options, and Research Needs.* OECD, Paris. Available at www.oecd.org/env/taxes.

Stavins R.N. (2001), "Experience with Market-Based Environmental Policy Instruments" in Mäler K.-G. and Vincent J. (eds.) *Handbook of Environmental Economics.* North-Holland/Elsevier Science, Amsterdam.

USING THE BENCHMARKING COVENANT FOR ALLOCATING EMISSION ALLOWANCES: ARE WE STILL MOVING AHEAD?

An inquiry into policy effectiveness

A.W.N. VAN DRIL

ECN Policy Studies, Petten, The Netherlands

Abstract. In this chapter, an analysis will be made of Dutch energy efficiency covenant policies and the conversion of these policies into the EU emissions trading scheme. A simple model is used to assess the effectiveness of these policies. The regulation of the eighties and voluntary agreements of the nineties concerning energy efficiency meet most of the conditions regarding policy effectiveness. The current Benchmarking Covenant suffers from lack of transparency and the absence of an emission reduction target. The EU emissions trading scheme that is currently developed, at least has a well defined cap, but whether it will enforce real emission reduction in the future remains to be seen. The conversion of the Benchmarking Covenant into a cap for emissions trading currently does not reduce CO_2 emissions but increases emissions of the participants involved. The cap that is derived exposes the lack of stringency of this policy.

1. INTRODUCTION

The Netherlands was one of the first countries to introduce voluntary agreements on energy and environmental issues between governments, business organisations and companies. These policies have been widely implemented on many subjects. Government and industries negotiate on feasibility of environmental targets and thus foster commitment. Therefore, instead of using the term voluntary agreement, the preferred wording is "negotiated agreement" or "covenant". For energy efficiency, there is almost fifteen years experience with covenants. Several evaluations have claimed that covenants are successful and effective in reaching policy targets (Das et.al., 1997; Benchmarking Committee, 2002). Other studies, including this one, are more critical.

In this chapter, an analysis will be made of Dutch energy efficiency covenant policies and the conversion of these policies into the EU emissions trading scheme. First, a simple model is introduced to establish conditions for effective policies. Then, a brief historic overview is given regarding energy policies for Dutch industry. In section 4, the current Benchmarking Covenant is evaluated using the policy model. In section 5, the European Emission trading scheme is presented. In section 6, an evaluation is made for the present conversion of the covenants into a

E. Croci (ed.), The Handbook of Environmental Voluntary Agreements, 365 - 380

national ceiling under the European scheme for greenhouse gas emissions trading. Section 7 concludes.

2. A SIMPLE ENERGY POLICY MODEL

2.1. Conceptual framework

In this section, is a general policy framework is presented for policies aimed at energy efficiency improvement and CO_2 emission reduction. With this framework, these policies are evaluated for their effectiveness to obtain desired changes in the physical environment. The actor, a person or organization executing these changes, makes a rational decision considering an investment, a change in operations, a change in purchases and consumption, or a change in behavior. A policy maker tries to induce or influence these changes in a desired way, using policy instruments.

According to the model, there are 2 times 3 conditions to be fulfilled in order to accomplish these desired changes. One set of conditions applies to the domain of the policy design, another set of conditions applies to the actor and the physical domain.

2.2. Three conditions policies have to meet

In the domain of the policy maker, it is essential to define the desired changes and the ways by which they are attained. This is part of the general policy accounting framework in Dutch government. Here we use the first two steps of this accounting framework called "From Budget to Policy Accounting" (VBTB, 2002). These steps are phrased in the questions "what do we want to accomplish" and "what are we going to do for that purpose". A policy is effective if the target is accomplished by means of the actions proposed. To evaluate the outcome of policies and policy activities, according to VBTB, monitoring systems are to be used when possible. In the field of energy efficiency and CO_2 emissions, these systems are based on quantified physical effects. We do not yet include the cost effectiveness or efficiency of the policies in this simple evaluation framework, although it is also part of VBTB.

The first condition is to identify the desired physical changes. A general identification of possible improvements has to be made, resulting e.g. in an energy savings or CO_2 reduction potential. Without a target in terms of quantified physical changes policy, quantitative evaluation is not possible. The level of detail to which these changes have to be identified and quantified is one of the main policy dilemmas. Creating a "sustainable energy system" or "reducing greenhouse gas emissions by 80% worldwide" is a general way to express the desired physical changes. However, this general approach is problematic when other conditions have to be met. A policy has to be directed to the actor, the decision maker that has to bring about the desired changes. Furthermore a reasonable time frame is needed and possibilities for monitoring and verification.

Therefore, the second condition is that decision makers and their scope and preferences have to be clearly identified. For many policies, this condition seems to be fulfilled. Subsidies or regulatory policies always relate to certain groups or persons by legal definitions. When these specific policies are designed, a clear definition of the target groups has to be made, to assess their possibilities and understand their behavior. Almost by definition, negotiated agreement policies are aimed at specified actors, being the participants in the agreement. However, for policies aimed at the general public like taxes or information campaigns, actors are often not well defined. In that case, the mechanisms by which the policy works cannot be identified and estimated accurately. In general, it is hard to monitor and verify the physical effects of taxation or information campaigns.

The third condition is to identify occurrences of decisions regarding energy efficiency. Replacement of capital goods is not a continuous process, industrial installations and buildings often remain largely unchanged for decades. And even when they are replaced, policies have to be directed to such a moment. Outside energy companies or energy intensive industry, decisions regarding energy efficiency often are not consciously made by the identified actor. When purchasing vehicles or new machines, actors are often not aware that this is also a decision on energy efficiency. Looking at strategic decisions on investments as a continuous flow of occurences that is simply caught by energy policies, therefore may disregard large time lags and underestimate policy effects. As the occurence of a decision moment is an essential condition for a policy mechanism to work, the timeframe has to be adequately long and to include these moments.

2.3. Three conditions to meet for the actors

When policies are specified well, still the actors have to be inclined act upon them. The three conditions for the actor's situation defined here are similar to what is found in management literature. The conditions still need further development and empirical foundation.

Attention is the first condition for actors. The actor has to be aware of the policy and its purpose and target. He has to know about the policy makers view on how to bring about physical changes. When the actor faces a situation where a decision has to be made on energy, the physical changes desired by policy makers has to be on his list of alternative strategies. An even stronger attention policy would be to create these situations of decision making. Notification of new regulation would be a typical example of the latter policy. For the attention condition, there is no active position of the actor required other than learning about the purpose of policies and policy instruments and how they affect him.

The second condition for decision makers is the possibility condition. First of all the actor has to become active and develop and define the physical measure. This is a consequence of considering it as a serious alternative, the first condition. However, an effort is required from the actor. His activities may include research, engineering, making specifications, calculating costs, checking regulations, making an implementation planning and setting up commissioning contracts. Other possible consequences of the measure have to be considered. Although these activities refer to an industrial setting, the same activities hold for consumers when they renovate their home. Of course, the lengthy process described is not always necessary in case of ready products, like vehicles or light bulbs. However, a selection of options and an assessment of all relevant properties is still required. Alternatives do not only include distinct technologies, but also reducing energy demand, postponing, outsourcing, relocation, etc.

The third condition for the decision maker is the expected advantage. The alternatives that embody the desired physical change have to be the most advantageous ones for the decision maker. The advantage can simply be expressed in terms of costs, e.g. high fuel costs savings. Other advantages like improved product quality or yield, environmental benefits or improved company image can be slightly harder to express in financial terms. An obvious advantage is also avoiding a sanction for not complying to environmental regulation or a voluntary agreement. This advantage is strong when sanctions are strict. Disadvantages have to be taken into account including high investments, operational uncertainties and financial uncertainties. Financial decision making is very common in energy models and a subject of energy policy literature. However, very obvious gaps still occur in fields like information costs and the assessment of ancillary costs and benefits, both of which can be substantial. E.g. when investment costs are low compared to information costs, the latter may remain a decisive disadvantage.

2.4 Model mechanism

According to the simple modeling concept, physical changes due to energy or climate policies only occur when all conditions are met. Policies can be aimed at all conditions. Therefore, successful policies are the ones that address and fulfil all the missing conditions. In other words, only when all lights are green, passage is possible. This is depicted in Figure 1.

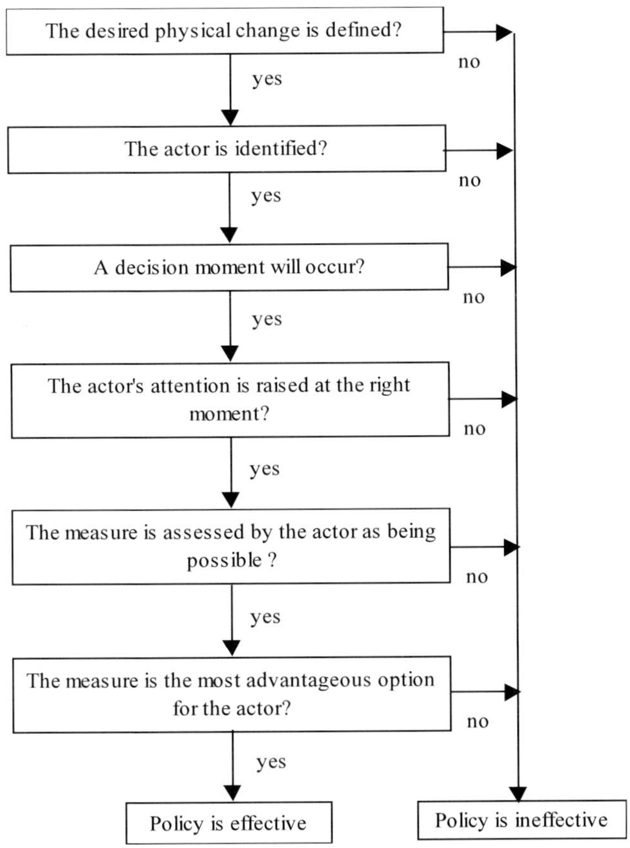

Figure 1. *Policy model flow sheet*

3. A BRIEF OVERVIEW OF ENERGY POLICIES IN THE NETHERLANDS

3.1 Motives for energy and climate policies

Policies for energy conservation largely developed around 1980 as a consequence of the two oil crises of the seventies. Oil prices quadrupled and since natural gas prices were linked to oil prices, the impact on the Dutch natural gas based economy was strong. Motives for energy conservation were security of supply and cost reduction. In the late eighties, environmental issues like acidification and climate change became additional motives for energy conservation. During the nineties, climate change and more specifically CO_2 reduction became the focus for energy policies. The Kyoto protocol in 1997 and the subsequent burden sharing agreement in the EU became a focal point for energy policies since. Only lately, with the liberalisation of energy markets, supply security is becoming a policy issue once again.

3.2. Instruments for energy and climate policies

Energy policies seem to be induced by the perceived problems in society and therefore follow a largely random path in time. For the energy policy instruments however, a trend can be observed throughout the past decades. This trend runs from prescribing detailed solutions to setting general targets. In the early stages of energy efficiency policy, governments prescribed the physical measures to be taken. Public energy agencies and energy companies adviced about insulation, efficient boilers, efficient lighting etc. In building regulations, insulation measures were compulsory for new buidings. During the oil crises, storage of large oil buffers was implemented by government, and on appointed sundays, the use of cars was forbidden. Policies to reduce acidification in the eighties were largely regulation too. Standards were set for emissions of large sources and measures like the catalyst on cars were prescribed.

Gradually, in the eighties, costs and efficiency of reduction policies became relevant factors. Pollution taxes were considered as an alternative for regulation but not yet implemented. The notion gained ground that environmental policies may be costly and inefficient since governments do not always know what the best solutions are. Actors in industry wanted to take responsibility for targets and decide for themselves about the measures to reach these targets. Around 1985, the government did not prescribe new measures anymore, but still stimulated energy efficient technology like combined heat and power generation (CHP). Regulation became focused on targets, and not on specific technologies or materials. At first, energy efficiency and emission standards were specified for building parts or installations. Subsequently, targets were specified for entire buildings and for industrial plants. During the nineties, the long term agreements (LTA) on energy efficiency emerged in the Netherland, specifying improvement targets for entire sectors. Most of these targets were specified as 20% energy efficiency improvement in 2000 compared to 1989. For the following decade up to the first budget period of the Kyoto Protocol, the Benchmarking Covenant largely replaced the LTA's. This Covenant too fits in the pattern of the goverment taking evermore distance. Under Benchmarking, companies are allowed to derive their own standards within the bounds of the Covenant. Presently, in 2004, the European Union emission trading scheme is implemented, were the goverment sets a cap for all participating sectors together. These latter two policies are dealt with in separate sections.

3.3. Evaluation: has energy policy been effective?

According to the policy model, regulation specifying measures is effective. Targets are quantified in detail, actors and decision occurences follow from legal procedures like building permits and environmental permits. Attention also is enforced by legal procedures and the technical possibilities have been verified. Sanctions appear to be strong enough for actors to comply although measures are costly. However, regulation policies ("command and control") have lost popularity by both actor and policy makers. Specified regulation policies have not lost popularity for reasons of ineffectiveness. They have been abandoned because of costs, both costs of measures and procedural costs. Nevertheless, regulation policies in the Netherlands are still effective, for instance for new buildings high energy efficiency is accomplished.

Relating to the energy policy model, the trend from specified solutions to general targets bears the risk that the actors are no longer identified. Decision makers in the field read about policy targets in newspapers but may not feel committed. Also, general targets may be in conflict with the actual possibilities. The LTA Covenants for energy efficiency have addressed these risks adequately. Effort has been put in bridging the gap between sectoral targets and industrial decision makers, by using energy efficiency agencies and monitoring systems.

4. THE BENCHMARKING COVENANT

4.1. General description and participants

In 1999 the Energy Efficiency Benchmarking Covenant was signed by the Dutch government and industry partners. The energy-intensive industry is committed to be among the world leaders in terms of energy efficiency no later than 2012. Government has agreed not to impose any extra specific national measures governing energy conservation or CO_2 reduction on the participating companies.

The covenant was signed by:

- The Ministers of Economic Affairs (EZ) and of Housing, Spatial Planning and the Environment (VROM).
- The Inter-Provincial Consultative Forum (IPO) on behalf of the provinces.
- Industry, represented by the Confederation of Netherlands Industry and Employers (VNO-NCW) and by sectoral organisations from various industrial sectors and the power sector.

Companies commit themselves to the covenant by means of a Declaration of Participation. Companies in the Netherlands can join the covenant if they operate plants with an energy consumption of at least 0.5 PJ primary energy annually. In table 1, the number of companies and installations in various sectors is shown.

Table 1. *Number of companies and installations participating in the Benchmarking Covenant as per 31 January 2003. Source: Benchmarking Committee (2003)*

Sector	*Number of companies*	*Number of installations*
Oil refineries	4	4
Iron, steel and non-ferrous	5	6
Breweries	4	8
Cement	1	3
Chemical industry	45	88
Miscellaneous	9	56
Glass	6	8
Paper mills	22	25
Sugar	2	5
Sub-total	*98*	*203*
Electricity sector	7	30
Total	*105*	*233*

The total amount of primary energy covered by the Benchmarking Covenant presently exceeds 1000 PJ. The coverage for manufacturing industry is 81% whereas for the power sector there is almost 100% coverage.

4.2 Motives for Benchmarking

The Benchmarking Covenant was proposed by the industry itself as a successor of the LTA's. In the LTA's the industry was committed to a target of mostly 20% efficiency improvement in 2000 compared to 1989. On average, this target is reached by most sectors. To be committed to another target of 20% for the next 10 years was deemed to be very difficult by industry. Physical or practical boundaries were expected to limit further efficiency improvements. The relatively cheap and easy improvements, so called "low hanging fruit", were already made. Therefore, the target for the Benchmarking Covenant was to be among the worlds best performing plants, but not to be pressed any further than competitors abroad.

By february 2002, when a first inventory was made, most of the industry in terms of energy consumption was already able to reach this benchmark (Benchmarking Committee, 2002). This has raised questions about the stringency of the Covenant (PWC, 2003). On the other hand, instead of sectoral targets, individual plants all have to comply to reach the benchmark. For older plants, this may pose a substantial effort. Also, the benchmark has to be updated every four years, so efficiency improvements abroad may oblige Dutch companies to set the target more stringent. This may pose a risk in future. However, according the the Covenant, in the third round in 2008, companies may fulfil their obligation to be world's best by improving energy efficiency outside the company, using the flexible mechanisms of the Kyoto Protocol.

Many companies decided to join the Covenant, even more than expected. The covered share is more than 84% of the eligible plants, and still rising (PWC, 2003). This proves that the vast majority expects participation to be favourable, considering the alternatives. The direct alternative is to be submitted to conditions posed by local authorities requiring all energy efficiency measures according to the ALARA principle (As Low As Reasonably Achievable).

4.3. The Benchmarking process

The Benchmarking process is confidential and the consultant that has to establish this benchmark is hired by the industrial company itself. However, an independent Verification Bureau has to approve the process of Benchmarking. The method of Benchmarking and the plant population to be compared with can be chosen to some extent. For instance, coal fired power plants are not compared with efficient gas fired power plants, integrated blast furnace steel plants are not compared with EAF, Corex or DRI plants. Old generations of plants in some cases do not have to be compared to new generations. The choices made for the Benchmarking method also depend on more subtle differences in processes and are influenced by data availability. Participating companies are allowed to customize benchmarking methods in their favour, provided that the Verification Bureau approves their choices. Both benchmarking results and methods for individual companies are confidential information. Apart from the report establishing the Benchmark, the industry also has to submit an Energy Efficiency Plan including measures to reach the Benchmark. The plan also includes information about the costs and the planning of measures.

4.4. CHP is favoured disproportionally

Combined heat and power installations have been stimulated in the Netherlands long before the Benchmarking Covenant was in place. For the LTA's up to 2000 this already was an important and widely used energy efficiency improvement option. Also in the Benchmarking Covenant, CHP is the cornerstone of energy efficiency for many plants. Large CHP installations up to 400 MW have been installed, and CHP covers up to 40% of the national electricity demand. Apart from producing chemicals, food products and paper, industry therefore is a large and efficient producer of electricity.

When calculating the efficiency improvement, all electricity deliveries to the grid are subtracted from total primary energy demand. Electricity is valued using the average efficiency of monogeneration of 40% in the base year. This is normal practise when these deliveries are small compared to total consumption. However, total deliveries to the grid from industrial CHP represents about 20% compared to total primary energy consumption of participating industry. For individual cases, more extreme subtractions are made. Efficient combined cycle installations are operated with electric efficiencies of about 45% and heat/power ratios of 0,7 and feed 90% of their produced electricity into the grid. This can distort monitoring of the actual efficiency developments for a chemical or paper production process completely. Calculated savings up to 50%-80% for the industrial heat consuming processes are not uncommon under Benchmarking. For the installation specified above, calculated savings would be even more than 100%.

One would argue that the best installations abroad that are used for comparison in the Benchmarking process would also operate large and efficient CHP. Calculation primary energy consumption in the same manner would therefore allow a fair comparison. However, according to the Covenant, Benchmarks may be developed by using best regions instead of best plants for comparison. This is applied where information on specific plants is incomplete. For CHP in "best regions", information indeed was said to be incomplete, therefore an average penetration is calculated. The Verification Bureau approved a benchmarking method whereby single plants with CHP in the Netherlands are compared with fictitious plants with partial CHP abroad. This puts companies with CHP in the Netherlands in a favourable position.

4.5 Evaluation: is the Benchmarking Covenant effective?

Following to the policy model, all conditions for successful policies are checked. The actual decision makers are very well committed by signing a declaration of participation to the covenant. Better than the LTA's 1989-2000, individual plant managers are aware of the task of proving to be the world 's best. Decision occurences are well identified in the Energy efficiency plans, and so are the technical possibilities. The advantage for participants is not directly clear. From the evidence so far, measures taken under the Benchmarking Covenant appear to be cost effective. More than 50% of the companies in energy terms has already proved to meet the benchmark, whereas presently until 2004, the covenant requires only measures with an internal rate of return exceeding 15%. Also the impending sanctions and the high participation rate may suggest that meeting the requirements of the Covenant is advantageous in the eyes of the industry.

However, the very first condition for policy makers is not met for the Benchmarking Covenant. Under the Benchmarking Covenant and the EU emissions trading scheme, there is no clear view of what has to change physically. Therefore it is unclear whether "being the world best" or "meeting Kyoto protocol obligations" means in terms the physical targets, and actually reduces the climate problem. There is no policy goal expressed in terms of emissions or other physical improvement by the government regarding the Benchmarking Covenant. Therefore, there is no target formulated for the expected contribution to the reduction of greenhouse gas emissions. In other words, we do not know whether the world's best will be good enough.

5. THE EUROPEAN UNION EMISSIONS TRADING SCHEME

5.1. General description

The European Commission has launched the European Climate Change Programme (ECCP), including a wide range of proposals and initiatives to reduce GHG emissions within the EU. The centrepiece of this programme is the Directive for the establishment of an EU Emissions Trading Scheme (EU ETS). This scheme is planned to be introduced in 2005 and may cover some 40-50 percent of EU greenhouse gas emissions in 2010 (CEC, 2003). The first phase of the proposed EU ETS is supposed to run from 2005 to 2007, followed by a second phase which overlaps with the first commitment period of the Kyoto Protocol (2008-2012). Participants in the scheme include electricity generators, oil refineries and energy intensive installations in manufacturing sectors such as iron and steel, paper, and minerals. Overall, it is estimated that initially the EU ETS will cover some 10,000-15,000 installations, accounting for approximately 45-50 per cent of total CO_2 emissions in the EU during the period 2008-2012, and of some 36-40 per cent of total GHG emissions in these years. It is envisaged, however, that the scope of activities and emissions covered by the EU ETS will be gradually extended over time.

Under emissions trading, companies have to surrender emission allowances in accordance with their actual emissions to the authorities. They can choose between reducing actual emissions or buy allowances, whatever they think is cheapest or profitable in the long run. National governments hand out the lion's share of allowances for free, according to a publically approved allocation plan. In this way average costs for industry are low, but marginal costs for allowances may be an incentive to reduce emissions, similar to a tax on energy. For the national allocation plans (NAP's) there is no additional effort required beyond the national commitments of the burden sharing agreement. Therefore, in its first phase, the trading scheme is not a policy aimed at reducing emissions but at cutting reduction costs. In later phases, the scheme is intended to achieve actual emission reductions at lowest costs for participants.

5.2. Evaluation: is emissions trading effective?

Again, according to the policy model, the EU emission trading system can be evaluated. First, conditions for the policy makers are well met in theory, since the emission is maximised to a fixed level also known as the "cap". The permit system can define actors and moments of decision adequately. For the actors themselves attention is also warranted when they have to surrender the allowances. The advantage for actors in the first place comes from avoiding sanctions when not delivering allowances according to the actual emissions. In the EU ETS, this is covered with a financial penalty. The advantage furthermore can be found in reducing emissions at a cost lower than the allowance price. This leaves one condition still unfulfilled: possibilities regarding emission reduction are not known or made available under the EU ETS. The allowance price has to become a signal strong enough to incur the supply of reduction measures. The mechanism of emissions trading will automatically raise the allowance price when supply of reduction technology is low. It is still unclear whether in future the cap can be set tight enough to put this technology generating mechanism into action. Possibly, high prices incur a policy reaction to relax the cap before meaningful physical reductions occur. Uncertainty about future price levels may not be a favourable climate for investments in R&D and breakthrough emission reduction technologies.

6. COMBINING BENCHMARKING WITH THE EU ETS

6.1. Overlaps

For the Netherlands, in terms of coverage, there is a large overlap of the Benchmarking Covenant and the EU ETS. Almost all participants in the Covenant also qualify for the emissions trading scheme, and direct emissions of CO_2 are largely coinciding with primary energy consumption. This creates a potential conflict between the two policies. BC requires improving of energy efficiency mostly on the company sites, whereas EU ETS opens up the possibility to buy allowances. According to the BC, only after 2008 and when companies proved to have taken all reasonable measures, any shortcomings to the benchmark can be settled by emissions trading. The interactions between the EU ETS and the BC are extensively described (Sijm and van Dril, 2003). According to theory, the most efficient way to reduce emissions is to abolish the Benchmarking Covenant and all other policies in favor of emissions trading. However, most participating companies want to retain the covenant, although is has to be adapted to emissions trading.

6.2. Using the Benchmarking Covenant for the National Allocation plan

Most of the attention from both policy makers and industry regarding the EU ETS is drawn towards the initial allocation of allowances. When a full market for allowances emerges, allocation is identical to distribution of wealth. According to the EU directive annex III, criteria for national allocation plans are formulated. A fair allocation can include elements of both actual emissions and benchmarks, see for example PWC/ECN (PWC/ECN 2003). Both industry and government in the Netherlands have chosen to use the Benchmarks as a base for distribution of allowances. The idea of using the available information for fair distribution was appealing and seemed an excellent opportunity to convert one policy instrument into the other. Fair allocation was to be based on the Benchmarks. The restrictions in the Benchmarking Covenant described above were to be solved by adapting it. However, a large number of issues emerged that complicated the allocation process substantially. Apart from several legal issues, the following four issues illustrate the principal differences between EU ETS and BC.

6.3. Fixed amounts versus efficiency targets

An important difference between the Covenants and the EU ETS is the definition of targets. Under the LTA and under Benchmarking, the actual target is a ratio and not an absolute amount of energy or emission. Energy efficiency together with the amount of output produced determine the emission. A fair allocation therefore has to include a fair estimate of the output growth. In the Dutch allocation process this has caused considerable discussions, lasting almost a whole year. The growth issue has so far (march 2004) been settled by using an adapted version (Boonekamp, 2003) of the 2001 National Reference Outlook (ECN/RIVM, 2001).

6.4. Process emissions enter the stage

Since producing energy by combustion of fossil fuels is not the only source of CO_2, other sources are also included in the EU ETS. These other sources are the release of CO_2 in chemical reactions, using fuels or other raw materials. Some examples are the reduction of iron oxides to produce iron in blast furnaces, and the production of lime from limestone in cement industry. Since these chemical reactions are physically proportional to the production output, emissions per unit of product cannot be reduced. Process emissions and emissions from non-energy use of fuels are not included in the Benchmarking Covenant. In the allocation for emission allowances they had to be added, and the general reduction factor was not applied to these emissions.

6.5. Heat and electricity leave the stage

Under the Benchmarking Covenant, all energy carriers were included in determining the energy efficiency. For purchased heat and electricity, consumption was converted with a fixed calculating procedure into the primary fuel consumption used to generate these secundary energy carriers. Under the EU ETS, however, independent heat and power producers would get the allowances. Large heat consumers would forfeit the allowances they would have received if they would generate heat or electricity themselves. This would make them reluctant to retain heat contracts or venture new CHP projects. Therefore the position of large joint venture CHP was feared (COGEN, 2003). Basically, these issues were solved by after ample negotiations.

6.6. Double counting of emission reductions from CHP

Under the Benchmarking Covenant, all savings due to CHP were allowed to be subtracted from primary energy consumption, as was described previously. According to this favourable method for benchmarking, industry operating CHP would get far more allowances than needed to cover actual emissions. This would put the national emission balance at a deficit. Moreover, power producers operating joint venture CHP with industry also booked CHP energy savings for their benchmark. Converting benchmarks to allocation here produced a large double counting of CO_2-reductions. This issue took also ample negotiations and could not be solved satisfactory. The Benchmarking method for CHP had to be adapted, and in the end a misty compromise has been devised (Commissie Benchmarking, 2004).

6.7. The draft allocation plan is still controversial

The results from the draft allocation plan are that the amount of emissions allowed for industry and power sector is 115 Mton annually for 2005-2007. This amount exceeds the projected emissions of 112 Mton calculated to meet the Kyoto target for 2008-2012. The amounts mentioned here include emissions of non-participants from manufacturing industry. Allocation in 2005 also exceeds projected emissions in 2005 with more than 5%, which is considerable. The risk of the European Commission or other stakeholders rejecting the allocation plan may still jeopardize the merger of EU ETS and BC.

Figure 2. *Projected emissions and emission cap for the Netherlands.*

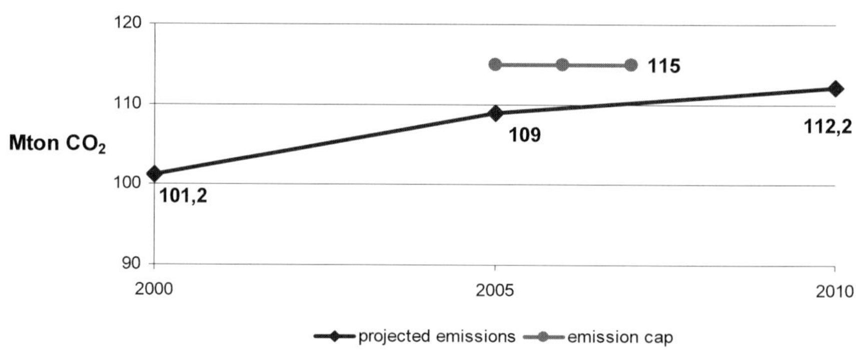

6.8. Is the combination of EU ETS and BC an effective policy?

When joining EU ETS and BC, the conditions derived from the policy model are all met. The "possibility condition" now is solidly founded in Benchmarks and Energy Efficiency plans of the Covenant. The operation therefore may prove very well to be successful. However, the patient is worse off than before. The policy result which is the actual physical change that is also well identified, brings the Netherlands further from its goal than the Benchmarking Covenant did in isolation.

7. CONCLUSIONS

To conclude, some remarks are made about the policy evaluation model and about the evaluated policies. First, the policy model is a simple tool for the first evaluation of policies and identification of weak spots in policy design. The model does not include cost effectiveness of policies, but cost effectiveness can only be assessed when a policy has passed the test described in this chapter.

The policies evaluated in this chapter all are intended to achieve meaningful energy savings or emission reductions. Over time, policies have become less prescriptive and more focused on overall targets. This trend addressed the need for more responsibility for actors and cost effectiveness of measures. For government, the drawback of this trend is that vital conditions for effective policies may not be fulfilled.

In classical regulation policies, all conditions for policy effectiveness are fulfilled, since targets, actors, decision moments and measures are legally defined. Sanctions for not taking the prescribed measures warrant the advantage of taking them.

The long term agreements on energy efficiency of 1989-2000 (LTA's) still adressed most conditions for effective policies. They had the relative, but meaningful target of 20% efficiency improvement. The possibility of reaching the required target was assessed in advance and awareness was raised by policy agencies. Advantage for the actor stems from the sanction of regulation rather than taxes. Since measures often had positive revenues from energy savings, the advantage condition was often already fulfilled.

The Benchmarking Covenant has no target that can be used by policy makers to demonstrate actual emission reduction. This is the weak spot of this policy which makes policy effectiveness hard to prove.

The European emissions trading scheme, by nature, does not specify the required physical measures. Participants are left in uncertainty about the long term reductions required and therefore aim there efforts at widening the emission cap. So far, the trading scheme is not expected to generate a price signal high enough to incur physical measures additional to current policies.

An interesting example of a policy that increases emissions is the use of the Benchmarking Covenant for the National Allocation Plan of the Netherlands. This combination of policies pushes widens CO_2 reduction targets for participants in the period 2005-2007.

8. REFERENCES

Benchmarking Committee (2002), *Convenant Benchmarking Energie-Efficiency* – Tussenstand Februari, The Hague, www.benchmarking-energie.nl

Das M.C., Driessen P.P.J., Glasbergen P., Habermehl N., Vermeulen W.J.V., Blok K., Farla J.C.M., Korevaar E.M (1997), *Evaluatie Meerjarenafspraken over Energie-Efficiency* - Report commisioned by the Ministry of Economic Affairs, Utrecht University, Vakgroep Milieukunde, Vakgroep Natuurwetenschap en Samenleving, (No.97063)

PriceWaterhouseCoopers (2003), Evaluate *Benchmark Convenant Energie-Efficiency*, Utrecht, august

PWC, ECN (2003), *Allowance allocation within the Community-wide emissions allowance trading scheme*, Utrecht, may, http://europa.eu.int/comm/environment/climat/pdf/allowance_allocation.pdf

VBTB (2002), *Van Beleidsbegroting tot Beleidsverantwoording: Regeling Prestatiegegevens en Evaluatieonderzoek Rijksoverheid.* January, http://www.minfin.nl/vbtb

van Dril: T. (1997), Government-industry agreements on greenhouse gas reductions, and pg. 102, Long term agreements on energy efficiency: the Dutch experience. European network for energy economics research, *ENER Bulletin* 20.97, pg. 89

Sijm J.P.M. and Dril A.W.N. van (2003), *The Interaction between the EU Emissions Trading Scheme and Energy Policy Instruments in the Netherlands*, ECN-C--03-060, september 2003

PWC/ECN (2003), *Allowance allocation within the Community-wide emissions allowance trading scheme*, Utrecht, may

Boonekamp P.G.M., Daniels B.W., Dril A.W.N. van, Kroon P., Ybema J.R., Wijngaart R.A. van den (RIVM) (2003) *Sectorale CO2-Emissies tot 2010: Update Referentieraming ten behoeve van besluitvorming over Streefwaarden*, ECN-C--03-095 Petten, december

ECN/RIVM (2002),*Reference Projection for Energy and CO_2 2001-2010*, ECN Petten

CEC (2003): *Directive of the European Parliament and of the Council establishing a scheme for greenhouse gas emissions allowance trading within the Community and amending Council Directive 96/61/EC,* 2001/0245(COD), 16 September, Brussels.

Boonekamp et.al. (2003), *Sectorale CO2-emissies tot 2010, Update Referentieraming ten behoeve van besluitvorming over Streefwaarden*, ECN-C--03-095, Petten, december

COGEN (2003), *Will Emissions trading put Combined Heat and Power at Risk?*, http://www.cogen.org/Downloadables/Publications/Leaflet_CHP_and_ET_June2003.pdf

Commissie Benchmarking: Allocatieplan (2004), *CO_2-emissierechten 2005 t/m/ 2007*, inspraakversie, februari

CONTRIBUTORS

Nicholas A. Ashford

Nicholas A. Ashford is Professor of Technology and Policy at the Massachusetts Institute of Technology, where he teaches courses in *Environmental Law and Policy*; *Law, Technology, Law and Public Policy*; and *Sustainability, Trade and Environment*. Dr. Ashford is a Faculty Associate of the Center for Technology, Policy and Industrial Development in the School of Engineering; the Institute for Work and Employment Research in the Sloan School of Management; and the Environmental Policy Group in the Urban Studies Department. He holds both a Ph.D. in Chemistry and a Law Degree from the University of Chicago, where he also received graduate education in Economics. Dr. Ashford also holds adjunct faculty positions at the Harvard and Boston University Schools of Public Health. He was a public member and chairman of the National Advisory Committee on Occupational Safety & Health, served on the EPA Science Advisory Board, and was chairman of the Committee on Technology Innovation & Economics of the EPA National Advisory Council for Environmental Policy and Technology. Dr. Ashford serves as an advisor to the United Nations Environment Programme and is also legislation, regulation, and policy editor of the *Journal of Cleaner Production*. He currently serves as co-chair of the US-Greece Council for the Initiative on Technology Cooperation with the Balkans. Dr. Ashford's research interests and publications include regulatory law and economics; the design of government policies for encouraging both technological innovation, and improvements in health, safety and environmental quality; pollution prevention and cleaner/inherently safer production; the effects of liability in improving product and process safety; the consequences of low-level exposure to chemicals; sustainability, trade and environment; labor's participation in technological change; and environmental justice. Dr. Ashford's research activities include work for the United Nations Environment Programme, the OECD, and the European Union, as well as for U.S. regulatory agencies and the U.S. Office of Technology Assessment.

Massachusetts Institute of Technology
77 Mass Ave., MIT E40-239
Cambridge, MA, 02139
USA
nashford@MIT.EDU
http://web.mit.edu/ctpid/www/tl/

Nils Axel Braathen

Nils Axel Braathen (1956) holds a Masters degree in Economics from the University of Oslo, Norway. He has previously worked in the Ministry of Industry and as deputy director general in the Department for long-term policy analysis and planning in the Ministry of Finance in Norway. Since 1996 he has been working in the Environment Directorate of OECD in Paris, inter alia on environmentally related taxes, macroeconomic modelling, voluntary approaches, the economics of waste and on instrument mixes for environmental policy.

OECD, Environment Directorate
National Policies Division
2 rue André-Pascal
F-75775 Paris Cedex 16
France
Nils-Axel.Braathen@oecd.org
www.oecd.org/env

Roeland Bracke
After finishing his study in Economics at the Ghent University, he continued his career as a staff member at the Faculty of Economics and Business Administration. As a research assistant associated with the Centre for Environmental Economics and Environmental Management (CEEM), his research focuses on the use voluntary instruments in environmental policy. He participated in the NEAPOL research project (Negotiated Environmental Agreements: Policy Lessons to be Learned from a Comparative Case Study Analysis) funded by the European Commission. Based on this experience, He took part in a policy advisory study on negotiated agreements for the Belgian Federal Government and a policy evaluation study on the use of environmental policy agreements in Flemish waste management for the first Report on the Environment and Nature: Policy Evaluation of the Flemish Environmental Agency.

University of Ghent
Sint-Pietersnieuwstraat 25
B-9000 GENT
Belgium
roeland.bracke@ugent.be

Hans Th.A. Bressers
Prof. Dr. Hans Th.A. Bressers (1953) is professor of Policy Studies and Environmental Policy at the University of Twente in the Netherlands and scientific director of the Center for Clean Technology and Environmental Policy of that university. Furthermore he has been chair of government advisory committees on environmental policy and the Dutch social science association for environmental and energy research, SWOME. Currently he is an independent scientific member of the Commission on Sustainable development of the Dutch Social-Economic Council (SER). In over two hundred articles, chapters, reports, papers and books (both in Dutch and in English) he published on policy-instruments, -implementation, -evaluation and policy networks, mostly applied on environmental policies. He has been researcher and project leader of numerous externally funded projects, including several projects funded by EU research frameworks, Dutch national science foundation, Dutch ministries, etc.

Center for Clean Technology and Environmental Policy
School of Business, Public Administration and Technology

University of Twente
PO Box 217
7500 AE Enschede
The Netherlands
j.t.a.bressers@utwente.nl

Keith Brouhle
Keith Brouhle graduated with a B.A. from Grinnell College and a Ph.D. in Economics from the University of Illinois at Urbana-Champaign. His research interests include Public Economics and Environmental and Resource Economics. He has taught at the University of Illinois at Urbana-Champaign and Indiana University Purdue University at Indianapolis and has worked at the U.S. Environmental Protection Agency. He is currently an Assistant Professor at the University of Alberta.

Department of Marketing, Business Economics and Law
University of Alberta
114 St – 89
keith.brouhle@ualberta.ca

Roger Burritt
Dr Roger Burritt is a Reader in the School of Business and Information Management and International Coordinator of the Asia Pacific Centre for Environmental Accountability (APCEA) at The Australian National University, Canberra, Australia. Environmental accounting has been the focus of his research attention for the last 15 years. Roger is coauthor of books on this subject. He also has published widely in environmental accounting and reporting issues in the private and public sectors. As a member of the United Nations international experts group on Environmental Management Accounting Dr Burritt contributes to theoretical and practical developments related to the promotion of voluntary engagement of organisations with environmental issues.

School of Business and Information Management
Faculty of Economics and Commerce
Building 21
The Australian National University
Canberra ACT 0200
Australia
roger.burritt@anu.edu.au
http://ecocomm.anu.edu.au/people

Manuel F. M. Cabugueira
Manuel F. M. Cabugueira is Assistant Professor and Researcher at the Centre for Applied Economics Studies at the Portucalense University, Porto.

Universidade Portucalense - Departamento de Economia
Centro de Investigação - Gab.: 213
Rua Dr. António Bernardino de Almeida, 541/619

4200-072 Porto
Portugal
mmc@upt.pt

Charles C. Caldart

Charles C. Caldart, J.D., M.P.H., is a Lecturer in law and policy at the Massachusetts Institute of Technology, and is the Litigation Director of the nonprofit National Environmental Law Center (NELC) in Boston, Massachusetts and Seattle, Washington. At NELC, Mr. Caldart brings environmental enforcement actions in federal courts across the United States on behalf of state and local citizen groups. He is coauthor of books and has written a number of articles on environmental and occupational health law and policy in legal, medical, and scientific journals.

Civil and Environmental Engineering
Technology and Policy
MIT E40-266
USA
cccnelc@aol.com

Johan Couder

John Couder is researcher at the University of Antwerp. Main topics of research are building simulation models to estimate the costs of emissions abatement in different industrial sectors, industrial pollution prevention, finding ways on how to improve the quality of environmental statistics, the construction of energy and environmental indicators, benchmarking and analysis of the energy sector in Flanders.

University of Antwerp
Prinsstraat 13
B-2000 Antwerp
Belgium
johan.couder@ua.ac.be

Edoardo Croci

Edoardo Croci (1961) is Vicedirector of IEFE, the research center of Bocconi University in Milan devoted to economics and policy of energy and the environment. He graduated with honours in Economic and Social Sciences at Bocconi University and has been Visiting Scholar at New York University (1991-1992). Since 1987 He works at IEFE. His research focus is on environmental policy and in particular on market and information-based instruments. He is author of publications about environmental voluntary agreements, public information and participation, emission trading. He is member of the Coordination Committee of the Master in economics and management of the environment at Bocconi University and Coordinator of the economic program of the Master in management of environmental resources at Bicocca University in Milan. He has served as President and Administrator of the Environmental Protection Agency of Lombardy Region, Member of the Board of Directors of the Gran Paradiso National Park and Vicepresident of FEDARENE (European Federation of Regional Energy and

Environment Agency). He is Member of the Italian Governmental Committee for the implementation of the national research policy in relation with the sixth European program for scientific research and innovation. He is Member of the Editing Committee of the review *Economia delle fonti di energia e dell'ambiente*.

IEFE – Università Bocconi
Viale Filippetti 9
20122 Milano
Italia
edoardo.croci@unibocconi.it
www.iefe.unibocconi.it

Theo J.N.M. de Bruijn
Dr. Theo J.N.M. de Bruijn (1964) is a senior researcher at the Center for Clean Technology and Environmental Policy at the University of Twente and an associate professor of Sustainable Development at Saxion University of professional education, both in the Netherlands. In his work he focuses on issues of environmental governance, industrial transformation, liveability and sustainability. He has been the project leader of several evaluation studies on the effectiveness and efficiency of voluntary environmental agreements. He has published an edited volume with the MIT Press on voluntary approaches both in the USA and in Europe. De Bruijn is a member of the editorial boards of the journal 'Business Strategy and the Environment' (John Wiley & Sons) and of the book series 'Eco-efficiency in Science and Industry' (Kluwer Academic Publishers). He has acted as European Coordinator of the Greening of Industry Network since 1996.

Center for Clean Technology and Environmental Policy
School of Business, Public Administration and Technology
University of Twente
PO Box 217
7500 AE Enschede
The Netherlands
theo.debruijn@utwente.nl

Marc De Clercq
Marc De Clercq is full professor of general economics at Ghent University (Department of General Economics), where he directs also the Centre for Environmental Economics and Environmental Management (CEEM). He teaches classes in general economics, environmental economics and European Economic Integration. His main research theme is the use of economic instruments in environmental policy. In this field he worked also as a consultant of several Belgian and international organizations, a. o. the EU-Commission, the OECD, the King Baudouin Foundation. Between 1993-1999 he was chairman of the Belgian Ecotax Commission. Since 2001 he is the vice-rector of Ghent University.

University of Ghent
Sint-Pietersnieuwstraat 25
B-9000 GENT

Belgium
Marc.DeClercq@UGent.be

Matthieu Glachant
Research fellow at CERNA, Matthieu Glachant is working on environmental policy issues. He got a PhD in Economics at the Ecole des mines de Paris in 1996 and was a Jean Monnet Fellow at the European University Institute in 2000-2001. He also gratuated in engineering from Agro. He regularly works for the OECD, the French Commissariat Général au Plan, the French Ministry of the Environment and ADEME on water, waste or climate change policies and on voluntary approaches.

CERNA, Ecole des Mines de Paris
60, boulevard St Michel
F-75006 Paris
France
matthieu.glachant@ensmp.fr

Charles Griffiths
Charles Griffiths is an Economist in the U.S. Environmental Protection Agency's National Center for Environmental Economics. He earned his PhD in Economics from the University of Maryland and a Masters in Economics from the University of Zimbabwe. His current areas of research are valuing ecological benefits, estimating morbidity effects of air pollution, and improving the use of risk assessment for benefits assessment in policy making. Dr. Griffiths has taught courses on the Guidelines for Preparing Economic Analyses used by the EPA. Prior to joining the EPA, Dr. Griffiths has worked at a macroeconomic forecasting group at the University of Maryland (INFORUM) and for the World Bank's Development Economic Research Group (DECRG) and taught at Gettysburg College.

National Center for Environmental Economics
Room 4316f, EPA West, Mail Code 1809T
U.S. EPA
1200 Pennsylvania Avenue, N.W.
Washington, D.C. 20460
USA
griffiths.charles@epamail.epa.gov

Lars Gårn Hansen
Lars Gårn Hansen (1959) is M.Sc./Ph.D from the Institute of Economics at Copenhagen University 1986/1997. He was Head of Section at the Ministry of Economics in 1987-1990. From 1990 until now assistant professor and then associate professor at AKF (Institute of Local Government Studies - Denmark). His main research interests are environmental economics and regulation.

AKF (Institute of Local Government Studies)
Nyropsgade 37
DK - 1602 Copenhagen V
Denmark

lgh@akf.dk
www.akf.dk

Karli James
Karli James manages the Sustainable Products program at the Centre for Design at RMIT University. Prior to this she worked as the LCA Researcher for six years at the Centre for Packaging, Transportation and Storage at Victoria University on behalf of the CRC for International Food Manufacture and Packaging Science. In the past seven years she has worked on life cycle assessment projects and eco-design projects with companies from the packaging industry and furniture sector. She has also undertaken work for state and federal government departments. She is the founding Secretary of the Australian Life Cycle Assessment Society (ALCAS), and co-writer of ALCAS columns in the International Journal of Life Cycle Assessment. Dr James completed her PhD thesis in 2003.

Centre for Design
RMIT University
GPO Box 2467V
Melbourne VIC 3001
Australia
Karli.James@rmit.edu.au

Helen Lewis
Helen Lewis is Director of the Centre for Design at RMIT University. The Centre undertakes research, consulting and training activities in environmental design of products and buildings. Prior to this she spent over 10 years working in Government and industry, managing product stewardship programs. She has a Bachelor of Economics and a Masters of Environmental Science.

Centre for Design
RMIT University
GPO Box 2476V
Melbourne VIC 3001
Australia
helen.lewis@rmit.edu.au

Piotr A. Mazurkiewicz
Dr. Mazurkiewicz is an economist specialized in corporate social and environmental responsibility (CSR), and sustainable development with experience in both development community (UNDP and World Bank) and the private sector. His expertise includes strategic studies, building consensus with various stakeholders on their development priorities, and facilitating project implementation. He has direct work experience in several professional sectors including private sector development, environmental protection, human development, and rural development. He holds MA and Ph.D. in international economics from the Warsaw University. He is a member of the International Association of Impact Assessment.

The World Bank
Development Communications Division
1818 H Street, NW
Washington, DC 20433
MS U 11-1102
USA
pmazurkiewicz@worldbank.org

Philippe Menanteau

Philippe Menanteau, CNRS Researcher at LEPII-EPE, University of Grenoble, is an engineer, graduate from the "Ecole Spéciale des Travaux Publics" and also doctor in Energy Economics of the "Institut National des Sciences et Techniques Nucléaires". He has an experience in the management of technical co-operation between French and European local authorities on urban energy and environment policies and is currently member of the scientific council of the "Fonds Français pour l'Environnement Mondial. His research is focused on the analysis of technological change in the energy sector with a specific interest in the comparison of policy instruments aimed at stimulating the creation and the diffusion of new or more efficient energy technologies. He has been working for several years on the relation between energy labelling and performance standards policies and technical change in the sector of electrical appliance. His current research work consists on analysing the efficiency of incentive instruments used for the development of renewable energy sources in a dynamic perspective that includes technological progress.

LEPII-EPE
CNRS / University of Grenoble
BP 47
38040 Grenoble Cedex 9
France
Philippe.Menanteau@upmf-grenoble.fr

Gernot Schnabl

Gernot Schnabl was born in Innsbruck (Austria) in 1953. He studied law at Innsbruck University. He experienced a civil service career with the Austrian Federal Ministry for Economic Affairs. Since 1996 He works at the European Commission, Directorate-General Environment.

European Commission
B 1049 Brussels
Belgium
gernot.schnabl@cec.eu.int

Lynn Price

Lynn Price is a Scientist and Deputy Group Leader of the International Energy Studies Group in the Energy Analysis Department, Environmental Energy Technologies Division of the Lawrence Berkeley National Laboratory. Ms. Price has been an author for a number of Intergovernmental Panel on Climate Change reports including the Second Assessment Report, the report on Policies and Measures for

Mitigating Climate Change, the Special Report on Emissions Scenarios, and the Third Assessment Report. She will also be a lead author for Assessment Report 4. Current research areas include mitigation of greenhouse gas emissions in the industrial sector, development of global scenarios of greenhouse gas emissions, international benchmarking of energy use and greenhouse gas emissions, and effectiveness of voluntary agreements for use in China's industrial sector.

Energy Analysis Department
Environmental Energy Technologies Division
Lawrence Berkeley National Laboratory
1 Cyclotron Road, MS 90-4000
Berkeley, CA 94720
USA
LKPrice@LBL.gov

Jonathan Sinton
Since 1990, Jonathan Sinton has been employed mainly at the Energy and Environment Division, Lawrence Berkeley National Laboratory. Dr. Sinton compiles the *China Energy Databook*, in collaboration with the China's Energy Research Institute, which remains the most complete English-language sourcebook on China's energy system. Dr. Sinton has published in *Science*, *Energy Policy*, *Annual Review of Energy and Environment* and other journals, and has also prepared reports for and made many presentations to U.S. government departments and multilateral agencies (Asian Development Bank, World Bank, International Energy Agency, World Health Organization) on the technology, economics, and policy of energy supply and demand in China, evaluation of specific energy-efficiency investment projects, and environmental-protection issues in China.

Energy Analysis Department
Lawrence Berkeley National Laboratory
1 Cyclotron Road, MS 90-4000
Berkeley, CA 94720
USA
jesinton@lbl.gov
http://china.lbl.gov

A.W.N. van Dril
Ton van Dril is senior scientist at ECN Policy Studies and presently group leader energy use and emission reduction, graduate engineer and economist, specialist on energy consumption and energy efficiency in manufacturing industry. At ECN has been responsible for the national energy outlooks for industry for 10 years, specialist on covenants and long term agreements on energy efficiency. He has made various sector studies on the position of the energy intensive industry, integrating market, technological and environmental aspects. He is an expert on the market position of CHP in the Netherlands. Before joining ECN, He had eleven years experience as a consulting engineer at Royal Haskoning on environmental issues for industry.

ECN - Energy Research Foundation
P.O. box 1
1755 ZG Petten
The Netherlands
vandril@ecn.nl

Aviel Verbruggen

Trained in engineering and economics at Louvain, Antwerp and Stanford University, his energy research covers electricity economics (cogeneration, planning, costing and pricing in power systems, distributed generation and grid access) and energy efficiency. Professor Verbruggen is co-founder of research and consultant units STEM, CENERGIE and FINES. He conceived, supervised and edited the State of the Environment Reports in Flanders (1993-98) and was the first president of the Environmental Advisory Council (1991-95) and principal advisor to the Minister of the Environment (1999-01). He contributes to the IPCC Third Assessment Report WGIII, Ch. 5 (Barriers and Opportunities) as co-author and as review editor of a chapter in the fourth report.

University of Antwerp
Prinsstraat 13
B-2000 Antwerp
Belgium
aviel.verbruggen@ua.ac.be

Ann Wolverton

Ann Wolverton is an economist at the National Center for Environmental Economics at the U.S. Environmental Protection Agency. Her research to-date has focused on the geographic and socioeconomic distribution of pollution, the use of market-based incentives, and the effectiveness of voluntary programs for environmental policy. Prior to joining the U.S. EPA, she worked at ICF Consulting. She holds a Ph.D in economics from the University of Texas at Austin.

National Center for Environmental Economics
U.S. EPA
1200 Pennsylvania Ave., NW, MC 1809T
Washington, DC 20460
USA
Wolverton.Ann@epamail.epa.gov

Ernst Worrell

Ernst Worrell (Ph.D.) joined the Environmental Energy Technologies Division of Lawrence Berkeley National Laboratory, USA, in 1998. His work includes research and evaluation projects in industrial energy efficiency improvement, energy policy, energy and materials, and waste processing. He has experience in multi-national research teams and has lead studies for various governments, European Commission, World Energy Council and the United Nations. He was a coordinating lead author of the Special Report on Technology Transfer, a lead-author of the IPCC special report on Emission Scenarios and the Third Assessment Report and will be an author of the Assessment Report 4. He is Editor-in-Chief of the peer-reviewed journal *Resources, Conservation and Recycling* and associate editor of *Energy, the International Journal* and editorial board member of *Waste Management*.

Energy Analysis Department
Lawrence Berkeley National Laboratory
1 Cyclotron Road, MS 90-4000
Berkeley, CA 94720
USA
Eworrell@lbl.gov

ENVIRONMENT & POLICY

1. Dutch Committee for Long-Term Environmental Policy: *The Environment: Towards a Sustainable Future*. 1994 ISBN 0-7923-2655-5; Pb 0-7923-2656-3
2. O. Kuik, P. Peters and N. Schrijver (eds.): *Joint Implementation to Curb Climate Change.* Legal and Economic Aspects. 1994 ISBN 0-7923-2825-6
3. C.J. Jepma (ed.): *The Feasibility of Joint Implementation*. 1995 ISBN 0-7923-3426-4
4. F.J. Dietz, H.R.J. Vollebergh and J.L. de Vries (eds.): *Environment, Incentives and the Common Market.* 1995 ISBN 0-7923-3602-X
5. J.F.Th. Schoute, P.A. Finke, F.R. Veeneklaas and H.P. Wolfert (eds.): *Scenario Studies for the Rural Environment*. 1995 ISBN 0-7923-3748-4
6. R.E. Munn, J.W.M. la Rivière and N. van Lookeren Campagne: *Policy Making in an Era of Global Environmental Change.* 1996 ISBN 0-7923-3872-3
7. F. Oosterhuis, F. Rubik and G. Scholl: *Product Policy in Europe: New Environmental Perspectives*. 1996 ISBN 0-7923-4078-7
8. J. Gupta: *The Climate Change Convention and Developing Countries: From Conflict to Consensus?* 1997 ISBN 0-7923-4577-0
9. M. Rolén, H. Sjöberg and U. Svedin (eds.): *International Governance on Environmental Issues*. 1997 ISBN 0-7923-4701-3
10. M.A. Ridley: *Lowering the Cost of Emission Reduction: Joint Implementation in the Framework Convention on Climate Change*. 1998 ISBN 0-7923-4914-8
11. G.J.I. Schrama (ed.): *Drinking Water Supply and Agricultural Pollution.* Preventive Action by the Water Supply Sector in the European Union and the United States. 1998 ISBN 0-7923-5104-5
12. P. Glasbergen: *Co-operative Environmental Governance: Public-Private Agreements as a Policy Strategy*. 1998 ISBN 0-7923-5148-7; Pb 0-7923-5149-5
13. P. Vellinga, F. Berkhout and J. Gupta (eds.): *Managing a Material World.* Perspectives in Industrial Ecology. 1998 ISBN 0-7923-5153-3; Pb 0-7923-5206-8
14. F.H.J.M. Coenen, D. Huitema and L.J. O'Toole, Jr. (eds.): *Participation and the Quality of Environmental Decision Making.* 1998 ISBN 0-7923-5264-5
15. D.M. Pugh and J.V. Tarazona (eds.): *Regulation for Chemical Safety in Europe: Analysis, Comment and Criticism.* 1998 ISBN 0-7923-5269-6
16. W. Østreng (ed.): *National Security and International Environmental Cooperation in the Arctic – the Case of the Northern Sea Route*. 1999 ISBN 0-7923-5528-8
17. S.V. Meijerink: *Conflict and Cooperation on the Scheldt River Basin*. A Case Study of Decision Making on International Scheldt Issues between 1967 and 1997. 1999 ISBN 0-7923-5650-0
18. M.A. Mohamed Salih: *Environmental Politics and Liberation in Contemporary Africa*. 1999 ISBN 0-7923-5650-0
19. C.J. Jepma and W. van der Gaast (eds.): *On the Compatibility of Flexible Instruments.* 1999 ISBN 0-7923-5728-0
20. M. Andersson: *Change and Continuity in Poland's Environmental Policy.* 1999 ISBN 0-7923-6051-6

21. W. Kägi: *Economics of Climate Change: The Contribution of Forestry Projects*. 2000 ISBN 0-7923-6103-2
22. E. van der Voet, J.B. Guinée and H.A.U. de Haes (eds.): *Heavy Metals: A Problem Solved?* Methods and Models to Evaluate Policy Strategies for Heavy Metals. 2000 ISBN 0-7923-6192-X
23. G. Hønneland: *Coercive and Discursive Compliance Mechanisms in the Management of Natural Resourses*. A Case Study from the Barents Sea Fisheries. 2000 ISBN 0-7923-6243-8
24. J. van Tatenhove, B. Arts and P. Leroy (eds.): *Political Modernisation and the Environments*. The Renewal of Environmental Policy Arrangements. 2000 ISBN 0-7923-6312-4
25. G.K. Rosendal: *The Convention on Biological Diversity and Developing Countries*. 2000 ISBN 0-7923-6375-2
26. G.H. Vonkeman (ed.): *Sustainable Development of European Cities and Regions*. 2000 ISBN 0-7923-6423-6
27. J. Gupta and M. Grubb (eds.): *Climate Change and European Leadership*. A Sustainable Role for Europe? 2000 ISBN 0-7923-6466-X
28. D. Vidas (ed.): *Implementing the Environmental Protection Regime for the Antarctic*. 2000 ISBN 0-7923-6609-3; Pb 0-7923-6610-7
29. K. Eder and M. Kousis (eds.): *Environmental Politics in Southern Europe: Actors, Institutions and Discourses in a Europeanizing Society*. 2000 ISBN 0-7923-6753-7
30. R. Schwarze: *Law and Economics of International Climate Change Policy*. 2001 ISBN 0-7923-6800-2
31. M.J. Scoullos, G.H. Vonkeman, I. Thornton, and Z. Makuch: *Mercury - Cadmium-Lead: Handbook for Sustainable Heavy Metals Policy and Regulation*. 2001 ISBN 1-4020-0224-6
32. G. Sundqvist: *The Bedrock of Opinion*. Science, Technology and Society in the Siting of High-Level Nuclear Waste. 2002 ISBN 1-4020-0477-X
33. P.P.J. Driessen and P. Glasbergen (eds.): *Greening Society*. The Paradigm Shift in Dutch Environmental Politics. 2002 ISBN 1-4020-0652-7
34. D. Huitema: *Hazardous Decisions*. Hazardous Waste Siting in the UK, The Netherlands and Canada. Institutions and Discourses. 2002 ISBN 1-4020-0969-0
35. D. A. Fuchs: *An Institutional Basis for Environmental Stewardship: The Structure and Quality of Property Rights*. 2003 ISBN 1-4020-1002-8
36. B. Chaytor and K.R. Gray (eds.): *International Environmental Law and Policy in Africa*. 2003 ISBN 1-4020-1287-X
37. F.M. Brouwer, I. Heinz and T. Zabel (eds.): *Governance of Water-Related Conflicts in Agriculture*. New Directions in Agri-Environmental and Water Policies in the EU. 2003 ISBN 1-4020-1553-4

ENVIRONMENT & POLICY

38. G.J.I. Schrama and S. Sedlacek (eds.): *Environmental and Technology Policy in Europe. Technological Innovation and Policy Integration.* 2003 ISBN 1-4020-1583-6
39. A.J. Dietz, R. Ruben and A. Verhagen (eds.): *The Impact of Climate Change on Drylands.* With a focus on West Africa. 2004 ISBN 1-4020-1952-1
40. I. Kissling-Näf and S. Kuks (eds.): *The Evolution of National Water Regimes in Europe.* Transitions in Water Rights and Water Policies. 2004 ISBN 1-4020-2483-5
41. H. Bressers and S. Kuks (eds.): *Integrated Governance and Water Basin Management.* Conditions for Regime Change and Sustainability. 2004 ISBN 1-4020-2481-9
42. *To be published.*
43. E. Croci (ed.): *The Handbook of Environmental Voluntary Agreements.* Design, Implementation and Evaluation Issues. 2005 ISBN 1-4020-3355-9

For further information about the series and how to order, please visit our Website http://www.wkap.nl/series.htm/ENPO